Agricultural Proteomics Volume 1

Ghasem Hosseini Salekdeh
Editor

Agricultural Proteomics Volume 1

Crops, Horticulture, Farm Animals, Food, Insect and Microorganisms

 Springer

Editor
Ghasem Hosseini Salekdeh
Department of Systems Biology
Agricultural Biotechnology Research
  Institute of Iran
Karaj
Iran

ISBN 978-3-319-43273-1   ISBN 978-3-319-43275-5 (eBook)
DOI 10.1007/978-3-319-43275-5

Library of Congress Control Number: 2016946323

© Springer International Publishing Switzerland 2016
This work is subject to copyright. All rights are reserved by the Publisher, whether the whole or part of the material is concerned, specifically the rights of translation, reprinting, reuse of illustrations, recitation, broadcasting, reproduction on microfilms or in any other physical way, and transmission or information storage and retrieval, electronic adaptation, computer software, or by similar or dissimilar methodology now known or hereafter developed.
The use of general descriptive names, registered names, trademarks, service marks, etc. in this publication does not imply, even in the absence of a specific statement, that such names are exempt from the relevant protective laws and regulations and therefore free for general use.
The publisher, the authors and the editors are safe to assume that the advice and information in this book are believed to be true and accurate at the date of publication. Neither the publisher nor the authors or the editors give a warranty, express or implied, with respect to the material contained herein or for any errors or omissions that may have been made.

Printed on acid-free paper

This Springer imprint is published by Springer Nature
The registered company is Springer International Publishing AG Switzerland

Preface

Although the global food production has increased in recent decades, the global food demand increases more rapidly than production. It has been reported by FAO that demand for cereals will increase by 70 % by 2050 as an outcome of both larger populations and higher per-capita consumption among communities with growing incomes. To meet higher demand, growing more food at affordable prices becomes even more important.

Agricultural proteomics can play a role in addressing the growing demand for food. The application of proteome science in agriculture has allowed researchers to identify a broad spectrum of proteins in living systems and associates them to many major traits. It may give clues not only about nutritional value, but also about yield production and food quality and how environments affect these factors. In recent years, technical improvements in the mass spectrometry, bioinformatics, protein extraction, and separation have made the high-throughput analysis of agricultural products feasible and the reproducibility of the technology has reduced errors in assaying protein levels. Meanwhile, the application of mass spectrometry-based quantification methods has become mainstream in recent year. The rapid advances of genome-sequencing tools also paved the way to sequence the full genome of many crops, animals, insects, and microorganisms. This provided Proteomics Scientist with a huge number of reference genome and genes for genome-wide proteome analysis.

An emerging field of the proteomics aimed to integrate knowledge from basic sciences to translate it into agricultural applications to solve issues related to economic values of farm animals, crops, food security, health, and energy sustainability. Given the wealth of information generated and to some extent applied in agriculture, there is a need for more efficient and broader channels to freely disseminate the information to the scientific community.

This book will cover several topics to elaborate how proteomics may enhance agricultural productivity. These include crop and food proteomics, farm animal proteomics, aquaculture, microorganisms, and insect proteomics. It will also cover several technical advances, which may address the current need for comprehensive proteome analysis.

Karaj, Iran   Ghasem Hosseini Salekdeh

Contents

1 **Applications of Quantitative Proteomics in Plant Research**...... 1
Mehdi Mirzaei, Yunqi Wu, David Handler, Tim Maher,
Dana Pascovici, Prathiba Ravishankar, Masoud Zabet Moghaddam,
Paul A. Haynes, Ghasem Hosseini Salekdeh, Joel M. Chick
and Robert D. Willows

2 **Seed Proteomics: An Overview**............................ 31
Kanika Narula, Arunima Sinha, Toshiba Haider,
Niranjan Chakraborty and Subhra Chakraborty

3 **Fruit Development and Ripening: Proteomic as an Approach to Study *Olea europaea* and Other Non-model Organisms**....... 53
Linda Bianco and Gaetano Perrotta

4 **Proteomics in Detection of Contaminations and Adulterations in Agricultural Foodstuffs**................................ 67
Javad Gharechahi, Mehrshad Zeinolabedini
and Ghasem Hosseini Salekdeh

5 **Holistic Sequencing: Moving Forward from Plant Microbial Proteomics to Metaproteomics**............................ 87
Behnam Khatabi, Neda Maleki Tabrizi
and Ghasem Hosseini Salekdeh

6 **Proteomics in Energy Crops**.............................. 105
Shiva Bakhtiari, Meisam Tabatabaei and Yusuf Chisti

7 **The Proteome of Orchids**................................ 127
Chiew Foan Chin

8 **Proteomic Tools for the Investigation of Nodule Organogenesis**... 137
Nagib Ahsan and Arthur R. Salomon

9 **Proteomic Applications for Farm Animal Management**.......... 157
Ehsan Oskoueian, William Mullen and Amaya Albalat

10	Applications of Proteomics in Aquaculture 175
	Pedro M. Rodrigues, Denise Schrama, Alexandre Campos, Hugo Osório and Marisa Freitas
11	Wool Proteomics 211
	Jeffrey E. Plowman and Santanu Deb-Choudhury
12	Proteomic Research on Honeybee 225
	Yue Hao and Jianke Li

Erratum to: Proteomics in Energy Crops E1
Shiva Bakhtiari, Meisam Tabatabaei and Yusuf Chisti

Index 253

Contributors

Nagib Ahsan Division of Biology and Medicine, Brown University, Providence, RI, USA; Center for Cancer Research and Development, Proteomics Core Facility, Rhode Island Hospital, Providence, RI, USA

Amaya Albalat School of Natural Sciences, University of Stirling, Stirling, UK

Shiva Bakhtiari Biology Department, Concordia University, Montreal, Canada

Linda Bianco ENEA, Trisaia Research Center, Rotondella, MT, Italy

Alexandre Campos Interdisciplinary Centre of Marine and Environmental Research (CIIMAR/CIMAR), University of Porto, Porto, Portugal

Niranjan Chakraborty National Institute of Plant Genome Research, New Delhi, India

Subhra Chakraborty National Institute of Plant Genome Research, New Delhi, India

Joel M. Chick Department of Cell Biology, Harvard Medical School, Boston, MA, USA

Chiew Foan Chin Faculty of Science, School of Biosciences, University of Nottingham Malaysia Campus, Semenyih, Selangor Darul Ehsan, Malaysia

Yusuf Chisti School of Engineering, Massey University, Palmerston North, New Zealand

Santanu Deb-Choudhury Food & Bio-Based Products, AgResearch Lincoln Research Centre, Lincoln, New Zealand

Marisa Freitas Department of Environmental Health, Escola Superior de Tecnologia da Saúde do Porto, CISA/Research Center in Environment and Health, Polytechnic Institute of Porto, Gaia, Portugal

Javad Gharechahi Chemical Injuries Research Center, Baqiyatallah University of Medical Sciences, Tehran, Iran

Toshiba Haider National Institute of Plant Genome Research, New Delhi, India

David Handler Department of Chemistry and Biomolecular Sciences, Macquarie University, Sydney, NSW, Australia

Yue Hao Institute of Apicultural Research, Chinese Academy of Agricultural Sciences, Beijing, China

Paul A. Haynes Department of Chemistry and Biomolecular Sciences, Macquarie University, Sydney, NSW, Australia

Behnam Khatabi Department of Biological Sciences, Delaware State University, Dover, Delaware, USA

Jianke Li Institute of Apicultural Research, Chinese Academy of Agricultural Sciences, Beijing, China

Tim Maher Department of Biological Sciences, Macquarie University, Sydney, NSW, Australia

Mehdi Mirzaei Department of Chemistry and Biomolecular Sciences, Macquarie University, Sydney, NSW, Australia

Masoud Zabet Moghaddam Center for Biotechnology and Genomics, Texas Tech University, Lubbock, TX, USA

William Mullen Institute of Cardiovascular and Medical Sciences, University of Glasgow, Glasgow, UK

Kanika Narula National Institute of Plant Genome Research, New Delhi, India

Ehsan Oskoueian Agricultural Biotechnology Research Institute of Iran (ABRII), East and North-East Branch, Agricultural Research, Education, and Extension Organization, Mashhad, Iran

Hugo Osório Instituto de Investigação e Inovação em Saúde - i3S (Institute for Research and Innovation in Health), University of Porto, Porto, Portugal; Institute of Molecular Pathology and Immunology of the University of Porto (IPATIMUP), Porto, Portugal; Faculty of Medicine of the University of Porto, Porto, Portugal

Dana Pascovici Australian Proteome Analysis Facility (APAF), Macquarie University, Sydney, NSW, Australia

Gaetano Perrotta ENEA, Trisaia Research Center, Rotondella, MT, Italy

Jeffrey E. Plowman Food & Bio-Based Products, AgResearch Lincoln Research Centre, Lincoln, New Zealand

Prathiba Ravishankar Department of Chemistry and Biomolecular Sciences, Macquarie University, Sydney, NSW, Australia

Contributors xi

Pedro M. Rodrigues Departamento de Química e Farmácia, Universidade do Algarve, CCMar, Faro, Portugal

Ghasem Hosseini Salekdeh Department of Systems Biology, Agricultural Biotechnology Research Institute of Iran, Agricultural Research, Education, and Extension Organization, Karaj, Iran

Arthur R. Salomon Center for Cancer Research and Development, Proteomics Core Facility, Rhode Island Hospital, Providence, RI, USA; Department of Molecular Biology, Cell Biology, and Biochemistry, Brown University, Providence, RI, USA

Denise Schrama Departamento de Química e Farmácia, Universidade do Algarve, CCMar, Faro, Portugal

Arunima Sinha National Institute of Plant Genome Research, New Delhi, India

Meisam Tabatabaei Microbial Biotechnology Department, Agricultural Biotechnology Research Institute of Iran (ABRII), Agricultural Research Education and Extension Organization (AREEO), Karaj, Iran; Biofuel Research Team (BRTeam), Karaj, Iran

Neda Maleki Tabrizi Department of Agronomy and Plant Breeding, College of Agriculture and Natural Resources, University of Tehran, Karaj, Iran; Department of Systems Biology, Agricultural Biotechnology Research Institute of Iran, Agricultural Research, Education, and Extension Organization, Karaj, Iran

Robert D. Willows Department of Chemistry and Biomolecular Sciences, Macquarie University, Sydney, NSW, Australia

Yunqi Wu Department of Chemistry and Biomolecular Sciences, Macquarie University, Sydney, NSW, Australia

Mehrshad Zeinolabedini Department of Systems Biology, Agricultural Biotechnology Research Institute of Iran, Agricultural Research, Education, and Extension Organization, Karaj, Iran

Chapter 1
Applications of Quantitative Proteomics in Plant Research

Mehdi Mirzaei, Yunqi Wu, David Handler, Tim Maher,
Dana Pascovici, Prathiba Ravishankar, Masoud Zabet Moghaddam,
Paul A. Haynes, Ghasem Hosseini Salekdeh, Joel M. Chick
and Robert D. Willows

Abstract Over the past two decades, we witnessed significant technological advances in Proteomics. Methodology and instrumentation have developed remarkably and proteomics has become a priority field of research in biology. Furthermore, analysis of the entire proteome of many organisms became possible due to complete genome sequencing. The advances in mass spectrometry instrumentation and bioinformatics tools have advanced quantitative proteomics techniques, resulting in important contributions to the biological knowledge of plants. In this chapter, we highlight the recent applications of proteomics in plants, in both model and non-model species. We then discuss the pros and cons of the major

---

Mehdi Mirzaei and Yunqi Wu have equally contributed in this work.

M. Mirzaei · Y. Wu · D. Handler · P. Ravishankar · P.A. Haynes · R.D. Willows (✉)
Department of Chemistry and Biomolecular Sciences, Macquarie University,
Sydney, NSW, Australia
e-mail: robert.willows@mq.edu.au

T. Maher
Department of Biological Sciences, Macquarie University,
Sydney, NSW, Australia

D. Pascovici
Australian Proteome Analysis Facility (APAF), Macquarie University,
Sydney, NSW, Australia

M.Z. Moghaddam
Center for Biotechnology and Genomics, Texas Tech University,
Lubbock, TX, USA

G.H. Salekdeh
Department of Systems Biology, Agricultural Biotechnology Research Institute of Iran,
Agricultural Research, Education, and Extension Organization, Karaj, Iran

J.M. Chick
Department of Cell Biology, Harvard Medical School, Boston, MA, USA

© Springer International Publishing Switzerland 2016
G.H. Salekdeh (ed.), *Agricultural Proteomics Volume 1*,
DOI 10.1007/978-3-319-43275-5_1

quantitative approaches implemented in plant studies. Next, we describe the most studied post-translational modifications (PTMs) in plant research, and lastly, we review the challenges of bioinformatics data analysis in the plant proteomics field.

Keywords Mass spectrometry · Quantitative proteomics · Post-translational modification

1.1 Introduction

Over the past 15 years, significant technological advances in methodology and instrumentation have developed proteomics into a powerful tool. For this reason proteomics is a priority field of research in biology. These advances combined with breakthrough technologies in genomics have allowed complete sequencing of the genome of an organism. Analysis of the entire proteome of an organism became possible for the first time [1] due to complete sequencing. These advances allowed the study of not only entire proteomes but also the changes which occur under various perturbations. These studies provide the means to understand protein regulation, function and protein interactions [2]. In particular, proteomics can now, for a specific complement of proteins present in a biological system at any one time, set out to answer the questions of "what?", "how?", "where?", and "when?" for those proteins in that system [3]. Plant sciences in particular have benefited from proteomics technology by studying and identifying metabolic pathways and protein functions, and identifying protein-protein interactions within model and crop plant systems.

A PubMed search covering the period between 2001 and 2016 using the search terms "plant and proteomics" revealed over 5500 research and review articles. One of the first proteomics plant studies was published by Pfannschmidt et al. in 2000. In this study the authors used two-dimensional gel electrophoresis for the analysis of the chloroplast polymerase A from mustard (*Sinapis alba* L.) [4].

Since then, the evolution of mass spectrometry instrumentation (MS), sample preparation protocols, and bioinformatics data analysis platforms have advanced quantitative proteomics techniques, resulting in important contributions to the biological knowledge of plants. Most of the plant proteomics papers published to date on model plants apply to the subgroups of descriptive, subcellular, and comparative proteomics [5]. Of these sub-areas, comparative proteomics is the most concerned with quantitative analysis, as it aims to highlight quantitative differences in proteins expressed between different genotypes, organelles, cell types, developmental stages and external conditions of an organism. Such studies have provided valuable information on plant biological processes, although without further functional analysis the proteomics data remain mostly descriptive [6]. A typical plant proteomics experimental pipeline from recent literature starts with a comparative proteomics analysis, followed by bioinformatic analyses of protein expression data.

1.2 Proteomics of Model and Non-model Plant Species

Quantitative proteomics approaches have largely been dedicated to model plant species [3]. In proteomics, these are species with a completely sequenced and well annotated genome. This information is typically made available in comprehensive databases such as that of the National Centre for Biotechnology Information (NCBI), which greatly assists in correct protein identification. At present, a total of 165 land plant species meet the model plant criteria. However, compared to the 3002 eukaryotic and 67,762 prokaryotic genomes sequenced to date (NCBI-April 2016), plants are highly under-represented. This relative shortfall is largely the result of plant genomes being greater in size and complexity, which makes them more difficult to sequence and annotate compared to other organisms [7]. To date, the plant species chosen as models are either species with relatively small genomes, or crop species of great economic value. The main application of plant proteomics being to better understand how crop species regulate development and respond to environmental changes [3, 6]. The first plant species to be sequenced was the dicot *Arabidopsis thaliana* in 2000 [8], which was soon followed by the complete sequencing of the economically important monocot rice (*Oryza Sativa*) in 2002 [9].

To date, proteomics studies concerned with model plants have provided key insights into a variety of protein families in plant systems, and how they are regulated and modified [6]. However, model plants alone do not possess all the features and processes of interest to plant biology [7]. Thus, 'orphan species', those without complete sequencing or adequate annotation, need to be investigated. Compared to model species the lack of genomic resources has seen orphan species largely neglected by quantitative proteomic studies [10]. Whether a species becomes a model is a trade-off between economic importance and the size and difficulty of sequencing its genome. Therefore, orphan plant species tend to be either not of great economic significance, or possess large, complicated genomes or both [11]. However, crop species often possess large genomes which is largely the reason that many remain as orphan species [11]. Most current MS-based proteomics methods rely on comprehensive sequence databases for identification and quantitation, thus the analysis of these "orphan species" poses specific challenges. These challenges can be at least partially overcome through utilising the sequence databases available from closely related species, or using databases of expressed sequence tags (EST) instead [12].

ESTs constitute a random snapshot of the expressed portion of the genome of a biological system at a moment in time in response to certain conditions. EST sequencing is seen as the most economical method of obtaining gene sequence information for orphan species [13]. Once collected, these ESTs are then added to databases for future protein identification in these species. However, the quality of EST sequences can vary greatly, so they need to be used very carefully in making confident protein identifications [14]. For those orphan species that do not have a sufficient number of ESTs available, protein identification may instead be achieved by relying on the established sequence databases of closely related model species

[15]. This method relies on the fact that many of the same genes are conserved between closely related species [16]. However, this is a less accurate method, especially for quantitative approaches, as it is not possible to know how many peptide assignments are taken into account, this due to sequence variation between species. Moreover, homologous proteins in different species often possess different functions and as the species become more distantly related the number of sequence mismatches increases [17]. Nevertheless, relying on this principle of shared protein sequence identity between related species has enabled quantitative proteomics approaches to be applied to many orphan species.

In the event that no ESTs and no closely related model species exist for a particular orphan species, the method of de novo peptide sequencing may be used [18]. This method derives the sequence of the peptide solely from the tandem mass spectra output with no need for an existing sequence database [19]. A common error in using this approach is wrongly identifying amino acids from mass spectra that share very similar masses. However, the recent development of very high resolution mass spectrometry instrumentation suitable for proteomics analysis, for example Orbitrap MS, has greatly reduced this sequence ambiguity, increasing the accuracy of peptide sequencing and providing more confident protein identifications [3]. This is still not really applicable when it comes to large-scale or shotgun proteomics studies, but it has been applied successfully in more targeted approaches.

1.3 Quantitative Proteomics Approaches

Numerous quantitative proteomics techniques have been used in proteomic research; however, not all of these are suitable for plant samples. Quantitative plant proteomics studies pose a number of unique challenges compared to animal and bacterial samples, such as: complications in protein extraction due to interfering secondary metabolites; separation of low abundance proteins; presence of incredibly high abundance proteins such as Rubisco in green plant tissues; genome multiploidy; and, most importantly, the absence of well-annotated and completed genome sequences [14]. Quantitative proteomic methodologies have undergone a number of advances which assist in overcoming these experimental challenges associated with plants. Evidently, 2-DE has been the most popular technique in plant research until approximately 2010; however, there has been a linear decrease in recent years. The second most used technique (label-free) reached its peak in 2012, however it seems the technique has lost its popularity and levelling off in the past couple of years. On the other hand, an emerging technology (chemical isobaric labelling approaches) is following an upward trend and it is estimated to keep on the same trend at least for the next few years to come (Fig. 1.1). Advances in high mass accuracy mass spectrometers, as well as multiplexing power of chemical isobaric labelling application in analysing the proteome and PTMs, are the main

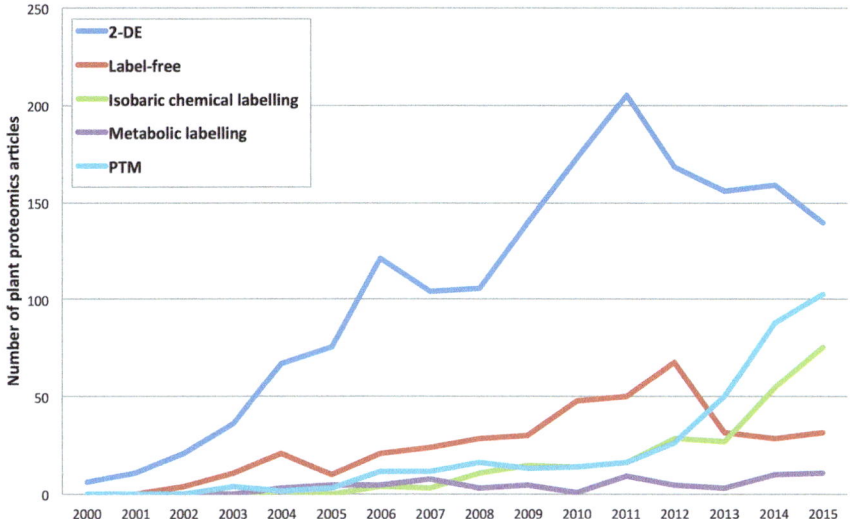

Fig. 1.1 Over all distribution of quantitative proteomics techniques used in plant proteomics studies over the period of 2000–2015

reasons behind the increasing attractiveness of chemical isobaric labeling approaches (Fig. 1.1).

In this section we briefly discuss the major quantitative proteomics approaches used in plant research.

1.3.1 Multi-dimensional Protein and Peptide Separation Methods

The first quantitative proteomics studies were performed by separating proteins via two dimensional gel electrophoresis (2-DE). In this approach, proteins are separated based on their isoelectric point (pI) in the first dimension, then separated based on molecular weight in the second dimension. Identification of proteins from 2-DE gels is obtained by excising protein spots, digesting with trypsin, extracting peptides, and then analysing the resultant peptides by mass spectrometry. One of the major advantages of 2-DE-based proteomic methods [20, 21] is that they provide a visual output for protein profiling and comparative mapping of expressed proteins between biological samples. Applications of 2-DE have been reported extensively in various plant species, such as Arabidopsis [22, 23], soybean [24, 25], rice [26–28], wheat [29, 30], and many others [31–35]. However, there are considerable drawbacks to this technique, including the fact that certain groups of proteins are poorly represented on 2D gels. Such poorly represented proteins include those with large molecular weights, highly basic proteins with high pI's, and hydrophobic proteins such as membrane spanning proteins which suffer due to poor solubility in 2-DE sample buffer. In addition, the detection of low abundance proteins is limited by total protein loading. A problem particular to green plant vegetative tissue samples is the presence of Rubisco, which can represent up to 60 % of the soluble protein in green plant tissue. Rubisco quantities interfere with the detection of many lower abundance proteins. For these reasons, 2-DE techniques are estimated to be only capable of detecting up to about 30 % of all cellular proteins. Furthermore, membrane proteins are poorly represented which is an issue for plant cells which are packed by various specialized membranous structures [36]. One particular technical drawback of 2DE technology, the detection of lower abundance proteins, has been addressed by recent advancements of fluorescent dyes like SYPRO Ruby. The main advantages of SYPRO Ruby protein gel stains are their linear quantitation range, which spans almost over three orders of magnitude [37].

Despite these problems, 2-DE still remains a suitable approach for identification and visualization of intact proteins [38]. In addition, 2-DE allows de novo sequencing of individual protein spots, which facilitates identification of proteins from plants with unsequenced or incomplete genomes. As an example, 2D-DIGE was used in characterization of the strawberry proteome during ripening and developing stages, and correlation between different genotypes [39]. Another study employed 2D-DIGE to examine proteome expression changes of bark tissues of peach (*Prunus persica* L. Batsch) in exposure to low temperature and short photoperiod [40]. Furthermore, the technique is still considered a useful tool to detect protein isoforms and modified proteins for which there are no effective and efficient enrichment strategies available [41].

To address some of the issues with 2DE, the higher throughput Mudpit technique was introduced. Multidimensional protein identification technique (MuDPIT) is a HPLC peptide separation method coupled with mass spectrometry which is more capable than 2-DE of identifying less abundant, basic, hydrophobic and membrane-spanning proteins [42]. Mudpit analysis requires that all proteins in a sample be digested into peptides before the separation steps. Differential comparative quantitation studies by Mudpit have been reported using isotope labeling in vitro [43] or in vivo [44]. One of the earliest reports in this area used Mudpit in parallel with 2-DE to characterize the proteome of rice leaf, root and seed tissues, which included the identification of more than 2500 unique proteins [45].

1.3.2 Stable Isotope Labelling (Chemical–Metabolic)

Stable isotope labelling is a powerful quantitative proteomics approach, which has been used in a large number of plant studies. These techniques are divided into two major groups; chemical and metabolic labelling. In chemical labelling techniques (ICAT, iTRAQ, TMT and dimethyl labelling), differential isotopic labels are incorporated into the samples after protein extraction and preparation, whereas in metabolic labelling (SILAC and ^{15}N), labels are added to the growth media to be metabolised by the cell and ultimately label the whole organism.

1.3.3 Chemical Labeling

1.3.3.1 ICAT

One of the first quantitative proteomics methods using chemical labeling reagents was isotope coded affinity tagging (ICAT) [46]. This is a thiol specific proteomic technique in which protein samples are labeled with light or heavy versions of ICAT reagents on cysteine thiol groups. Samples are mixed and digested by an endoprotease such as trypsin. The relative abundance changes in the proteome can be obtained by comparing the intensities of labeled protein peaks with light and heavy mass tags. The technique is capable of comparing the protein expression changes between two biological samples. This approach has several drawbacks, including the fact that it is limited to proteins containing cysteine residues, and is less efficient at labeling acidic proteins. Applications of ICAT in plant proteomics have been reported only in a few studies, mainly focused on organelle membrane protein distribution such as characterizing the endoplasmic reticulum and Golgi apparatus in *Arabidopsis* [47], mitochondrial membrane proteins in *Arabidopsis* [48] and chloroplast soluble stromal proteins in maize leaves [49].

1.3.3.2 iTRAQ and TMT

Isobaric tags for relative and absolute quantitation (iTRAQ) [50] and tandem mass tag (TMT) [51] methods are designed based on isobaric labelling reagents. The multiplexing capabilities of these techniques provide an opportunity to compare the proteome of up to 8 samples in iTRAQ, and up to 10 samples in TMT, simultaneously in a single MS analysis. The isobaric tags with the same mono-isotopic masses chemically label the N-terminus and the amine groups of the lysine side chain of tryptic peptides. Labelled peptides from different samples/conditions are combined, desalted, fractionated and analyzed using high mass accuracy mass spectrometers. Upon fragmentation of the tag attached to the peptides, reporter ion intensities are generated which are unique to the tags used in labelling of the peptides from different biological samples. One of the major advantages of multiplexing is that MS1 complexity stays at a single proteome, whereas labelling strategies that use MS1 based quantitation (except label-free) become more complex in the MS1 as more labels are added. For example, SILAC or dimethylation labelling with light and heavy reagents are twice as complex. In multiplexing, all of the samples contribute to one peak. This contributes to depth of coverage. Hence, there is no further penalty for increased multiplexing capabilities.

The reporter ion intensities represent the relative abundance of the peptides and proteins from which they are originated. Both iTRAQ and TMT techniques benefit from their multiplexing power, however they have their own unique drawbacks such as the cost of the reagents, and the need for very high mass accuracy spectrometers. Furthermore, specialised instrumentation is needed to handle the ability to purify co-isolated peptides using MS3 strategies [52].

The use of iTRAQ and TMT has been reported in a large number of plant quantitative studies in recent years (Fig. 1.1). For instance, iTRAQ was used to investigate the response mechanism of various plants to a wide range of stresses including cold [53], heat [54] drought [55] salinity [56], and heavy metals [57]. Similarly, TMT was used in comparative expression studies for a number of plant species such as *Arabidopsis* [58, 59], rice [60], and barley [61].

The number of plant study reports using iTRAQ is significantly higher than TMT, mainly because iTRAQ was introduced a few years ahead of TMT. However, the higher multiplexing capability of TMT (TMT10plex compared to 8plex iTRAQ), has already attracted the attention of plant researchers and it is expected that the number of studies using TMT will rise over the next few years (Fig. 1.1).

1.3.4 Metabolic Labeling

The recent technological advances in proteomics enable the incorporation of stable isotope labels into samples to accurately detect changes in protein abundance [62]. Metabolic labelling refers to the methods in which stable isotopes are incorporated in vivo during the translation stage of the protein synthesis in cells [63]. The cells are grown in the media supplemented with the isotopes, so that they are incorporated into the proteome metabolically. Mass spectrometric analysis is then performed on digested lysates, and quantitation is performed by distinguishing the mass shifts between light and heavy isotopes [64, 65]. Metabolic labelling is in vivo labelling, and the different techniques of in vivo stable isotope labelling include ^{15}N labeling, Stable Isotope Labeling of Amino acids in Cell Culture (SILAC), ^{13}C labelling. In plant proteomics, ^{15}N and SILAC have been used to perform comparative studies on different metabolic states of the plant cells [65–68].

1.3.4.1 SILAC

Stable Isotope Labelling by Amino acids in Cell culture (SILAC) is the in vivo incorporation of stable isotope-containing amino acids, such as arginine and lysine, that label proteins during their cellular synthesis [69]. The technique is highly efficient and reproducible and is considered a powerful tool for studying PTMs such as phosphorylation. SILAC has been successful in numerous yeast and mammalian studies [70–72]. However, this technique has not been as successful in plants.

This might be due to poor labelling efficiency of autotrophic plant cells which are able to synthesize all amino acids [65, 73]. To date, there are only three SILAC based quantitation studies reported in plants. Two of these were carried on *Arabidopsis* cell culture [65, 66] and the labelling efficiency did not exceed more than 90 %. However, a recent study by Lewandowska et al. [74] demonstrated a technique to label the whole *Arabidopsis* seedlings using stable isotope-containing arginine and lysine with more than 95 % labelling efficiency [74]. The reliability and efficiency of this method remains to be confirmed in future studies. In general, drawbacks of any MS1 based quantitation strategy stem from higher false positive rates in the regulated set of proteins that are quantified from single peptide-based protein identifications. This is because random assignments are more likely to have only one isotope identified. The chance of identifying both isotopes simultaneously for the same random peptide is much less likely.

1.3.4.2 ^{15}N Labelling

Metabolic labelling with ^{15}N appears to be more suitable for plant studies due to the fact that nitrogen is usually a limiting factor in plant growth, and ^{15}N-containing salt as the sole nitrogen source can be added to plant cell cultures or the whole plant systems and be efficiently incorporated into amino-acids and proteins. Metabolic labelling with the heavy isotope of nitrogen could be full or partial depending upon the amount of the heavy ^{15}N isotopes used. In full labelling, up to 98 % of the proteinaceous nitrogen is labelled with ^{15}N isotopes, while in partial labelling a lower percentages of the heavy nitrogen is incorporated.

1.3.4.3 Different Experimental Approaches of ^{15}N Labeling in Plants

Labeling with $K^{15}NO_3$ has been carried out in various systems such as *Arabidopsis* cell culture [75] and whole *Arabidopsis* plants grown in liquid nutrient media using the hydroponic isotope labeling of entire plants (HILEP) [68]. In labeling the cells with ^{15}N, two populations of cells (control cells and treatment cells or normal cells and stress affected cells) are grown in two distinct media containing ^{15}N and ^{14}N. The two samples are mixed, processed and subjected to mass spectrometry analysis. The peptides from both the control and the treatment samples have the same chemical properties but exhibit mass difference because of the heavy isotope labeling that is detectable by the mass spectrometer [5]. On comparing the peak intensities between the two samples the difference in the expression of the proteins is analysed [1]. The different experimental strategies that have been used to analyse plants range from full to partial labelling techniques, reciprocal labeling and pulse labeling techniques. Several experiments were performed using 15 N-labeling in plants systems with the aim of developing the method on plants including, *Glycine max* (Soyabean) [76], *Arabidopsis thaliana* [77, 78], *Solanum lycopersicum* (Tomato) [79], *Hordeum vulgare*) (Barley) [80].

1.3.4.4 Full ^{15}N-Labeling

Complete or full metabolic labeling refers to labeling all of the nitrogen in the cells with heavy nitrogen. The main challenge present in full metabolic labeling is achieving the efficient growth of an organism supplemented with N^{15} entirely [78]. Successful full metabolic labeling for proteomic investigation was demonstrated with complete efficiency of labeling intact plants in *Arabidposis thaliana* [81]. The plants to be labeled were grown in media containing 98 % of $^{15}NH_4^{15}NO_3$ and $K^{15}NO_3$ in place of natural abundance salts. Complete incorporation of about 98 % of the heavy nitrogen was achieved in the Arabidopsis seedlings. The evaluation of the performance of MASCOT, a standard MS/MS search engine was also evaluated for the combined analysis of the ^{14}N and ^{15}N-labeled peptide samples of Arabidopsis seedlings which proves the application of ^{15}N-metabolic labeling for quantitative proteomics analysis with excellent incorporation [81].

1.3.4.5 Partial ^{15}N-Labeling

An alternative strategy to use isotope labeling was proposed by Whitelegge et al. [82] which was to decrease the amount of heavy nitrogen used for labeling resulting in partial labeling. It was reported that when both the natural and the partial labeled forms were combined, the shape of the resulting isotopic envelope is used to determine the relative amount of each peptide present in the sample mixture which enables information to be extracted from partial metabolic labeling rather than full metabolic labeling [82]. A comparison of full and partial metabolic labeling in *Arabidopsis thaliana* was made by Huttlin et al. [78]. In this experiment, labeled and unlabeled mixtures of *Arabidposis* peptides were first analysed using both full and partial labeling, each technique was assessed for consistency, dynamic range and reproducibility under controlled conditions. Further analysis of light versus dark grown *Arabidopsis* using both the techniques was carried out to analyse the performance. It was found that with partial metabolic labeling allowed the more complete biological comparison of the protein expression that exhibited significant changes under conditions, where full metabolic labeling failed to give reliable quantitative information. Partial metabolic labeling also serves as a much more economic way of metabolic labeling [78] by decreasing the cost of the labelled nutrients, and allows the quantification of more peptides across the whole dynamic range. But, from the thorough comparison of the full and partial labeling techniques, partial metabolic labeling is more challenging in the automated identification of labeled and unlabeled peptide pairs and in the quantification of the change in the isotope cluster distribution [83].

1.3.5 Reciprocal Labeling

In a reciprocal labeling experimental strategy, two pools of samples for e.g. control and treatment samples or mutant and wild type samples, are inversely labeled with heavy nitrogen in such a way that the label is associated once with the treatment and once with sample and vice versa. Pair wise comparisons made from the mass spectrometric analysis would be used to evaluate the changes in the protein abundance [83]. In plant proteomics, reciprocal heavy metal labeling has been used to study stress physiology including the study of the effect on elicitors on protein phosphorylation [84], and microdomain composition of Arabidopsis and tobacco using cell cultures [85, 86]. Whole-plant studies have been carried out to characterize the effect of abscisic acid treatment on phosphorylation [86], and protein abundance changes have been monitored during heat shock responses [68] and leaf senescence [87].

1.4 Label-Free Quantitation

Label free quantitation has become popular over the years mainly owing to its ease of use and application in a wide range of biological studies [26, 88]. It is subdivided into two separate strategies; spectral counting (SC) [89] and area under the curve (AUC) [90]. In label free quantitative shotgun proteomics, both control and sample are subjected to separate LC-MS/MS analysis, and protein quantitation is performed on either peak intensity of the same peptide or the number of spectral counts for the same proteins [91]. In the SC approach the most abundant peptides will be selected for further fragmentation, hence the abundance of generated MS/MS spectra is proportional to the protein amount in data-dependent acquisition whereas in the AUC approach the protein abundances are estimated from the measurement of the changes in ion intensity of chromatographic peak areas or heights.

All quantitative MS approaches have their own advantages and drawbacks. Some issues are common among all relative quantitative MS and MS/MS approaches, such as: accounting for peptides that are shared between different gene products or protein isoforms, and issues of missing peptide ions between biological replicates. Missing peptide ions are especially problematic for low abundance proteins that are close to the detection limit of the mass spectrometer.

Label-free quantitation, whether based on the number of peptides for the spectral counts approaches, or peak intensity for the area under the curve (AUC) approaches, also has specific limitations. The specific drawbacks to spectral counting are that the actual relative fold changes in protein abundance are not easily calculated and the detection of differential protein abundance is difficult for proteins that generate low spectral counts (<5). In the case of AUC relative quantitation, there are certain concerns regarding variation in MS1 signal intensity and chemical and biological interferences between replicates due to changes in chromatography performance from run-to-run. Despite these caveats, label free quantitative proteomics remains a versatile and practical approach for global proteome profiling studies. For low mass accuracy instruments, including ion traps, spectral counting is the method of choice over AUC. However, label-free AUC quantitative workflows have been increasing in use, largely due to improved computational platforms and wider availability of high accuracy mass spectrometers.

Label-free quantitative proteomics has been the method of choice over other techniques for the majority of the plant discovery experiments, mainly due to its affordability and ease of use features. In our lab we used label-free quantitation based on spectral counting in more than 20 different plant studies; for instance Gammulla et al. [92, 93] investigated the changes in the proteome of rice leaves and cell cultures in response to high and low temperatures using a label-free based spectral counting approach in two separate studies. In both studies, SDS-PAGE was employed to fractionate proteins. Each lane was cut into 16 pieces, followed by in-gel digestion and MS analysis by nanoflow liquid chromatography-tandem mass spectrometry (nanoLC-MS/MS) using a linear ion-trap mass spectrometer. In another study, the effect of thermal stress on cabernet sauvignon grape cells was evaluated by a combination of Filter-aided sample preparation (FASP) digests and spectral counting [94]. The same technique was applied to understand the molecular mechanisms of water deficit stress in rice shoots [95, 96] and roots [26], and mid-mature peanut seeds [97].

1.5 Post-translational Modifications (PTMs)—Plant Proteomics

PTMs are known to play a key role in controlling the activity and function of a wide range of proteins within a cell. Across both animal and plant kingdoms, we find a grand diversity of post-translational modifications (PTMs). There are about 400 discrete types of PTMs which can occur in a cell [98]. Each of the modifications has the potential to significantly alter the conformational space of the protein and ultimately its function.

Mass spectrometry is considered the most suitable tool for identifying novel post translational modifications, as no prior information about the modification sites and type is essential [99].

Although the body of research into PTM function is growing, researchers have only just scratched the surface on uncovering the importance of these modifications. The most commonly studied PTM is phosphorylation—the ubiquitous on/off switch responsible for developmental pathways associated with cell growth and in response to stress factors. In addition, protein acetylation, methylation, ubiquitination, glycosylation and a whole host of niche modifications have been thoroughly studied and catalogued [100–103]. Recent large-scale plant quantitative proteomics experiments have successfully linked many PTMs to a wide range of metabolic functions and pathways operated during unfavourable conditions [104].

It is therefore important to have knowledge of what these modifications are, how they affect protein structure and function, and how they can contribute to some important downstream biological phenotypes. The aim of this chapter is not to list in detail every example of where a certain type of modification exists, but to provide an overview of the most common types of modifications occurring in plant proteins, and highlight their importance in plant development and survival.

1.5.1 Phosphorylation

Phosphorylation remains the most widely recognised and studied PTM. Simply put, phosphorylation is the addition of a phosphate group (PO_3^{2-}), usually on the hydroxyl group of threonine, tyrosine, histidine or serine residues, which increases the molecular weight by 79.9663 Da [105]. An overwhelming number of large-scale phospho-proteomics studies have been reported in rice, *Arabidopsis*, Medicago, maize, and several other plant species, which were carefully discussed in recent reviews [106–108].

Phosphorylation can be thought of as an on/off signal mechanism; proteins that are phosphorylated have phosphate groups added to specific amino acids sites to become 'activated'. In reality, adding a phosphate group may have many different consequences, and isn't simply confined to turning proteins 'on or off'. For example, the addition of a phosphate group to a protein may provide it with a new tertiary structure, conferring a new active site and thus a new function. In providing new secondary or tertiary conformations for the protein, other important functions such as protein-protein interactions may be promoted, or be interfered with or shut down [109]. It is important to remember that phosphorylation also includes the counter-process: the removal of phosphate groups. Phosphatases (removal) play just as vital a role in protein function as protein kinases (addition), in situations where the phosphate group regulates signalling mechanisms between proteins. Larger scale biological processes are also dependant on kinase activity and phosphorylation and these include differentiation, cell maintenance, development and intracellular regulation [110].

For the purpose of this overview, it is important to consider the impact of phosphorylation in the plant research field. As we have established above, changes to the protein at the phosphate level can impact cell growth and development due to the switching on of various chemical pathways. In addition, phosphorylation also plays a critical role in host-pathogen responses and can influence plant survival [111]. Some examples of current phosphorylation-related research include the role of phosphorylation in controlling mitochondrial related processes, especially related to an enzyme responsible for acetyl-CoA processing within the TCA cycle [112];

for cell wall elasticity and cell cycle regulation within *Phaseolus vulgaris* (common bean) [113]; and for elucidating the relationship between classes of proteins such as aquaporins in the water-moving capabilities of roots [114].

1.5.2 Acetylation

The post-translational acetylation of lysine residues on specific proteins is a functional regulatory mechanism, which was only discovered relatively recently [115]. It is a topic of great research interest, because it seems to be universal in nature; it is present across all the kingdoms of life [116, 117]. Plant proteins are known to be subjected to acetylation and deacetylation by acetyltransferases and deacetylases, which have been identified in the genome of various plant species. At least twelve histone acetyltransferases (HAT) and eighteen histone deacetylases (HDAC) are present in the *Arabidopsis* genome [118], while the rice genome contains 19 HDAC genes and seven HAT genes [119]. Histone acetylation plays a significant role in the regulation of cell cycle, development, flowering time, and hormone signal transduction [120]. Despite the extensive studies on dynamic and reversible changes in histone acetylation in histones, the extent of lysine acetylation in non-histone proteins in plant cells is largely unknown.

A number of published studies in different species across the animal kingdoms have shown that acetylation is an important regulatory mechanism [121]. Some of the earliest relatively large scale studies of lysine acetylation were performed in *Arabidopsis*: Konig et al. [122] reported the identification of 120 lysine acetylated mitochondrial proteins, containing 243 distinct sites of acetylation, mainly on proteins involved in protein metabolism and the tricarboxylic acid cycle; Finkemeier et al. [100] published the identification of 91 acetylated sites on 74 proteins representing diverse functional classes; and Wu et al. [123] identified 64 lysine acetylation sites on 57 proteins and showed that lysine acetylation is an important factor in the regulation of key metabolic enzymes. In the latter study, the authors identified acetylated proteins including photosystem II (PSII) subunits, light-harvesting chlorophyll a/b-binding proteins, RuBisCO large and small subunits, and chloroplastic ATP synthase (β-subunit) [123].

Similar studies have subsequently been reported in pea [124], soya bean [125], grapevine [126], and rice [127]. All of these studies have shown that acetylated proteins in plant cells are not limited to histones and seem to be involved in a diverse range of metabolic control and energy-related functions, and found in a diverse range of cellular compartments. However, the studies published so far have analysed plant cells under ideal growing conditions; none of them have addressed the question of how the acetylation status of the peptides and proteins changes in response to the imposition of external stresses.

Complex regulatory networks control protein expression in all cells. The chloroplast and mitochondria are known to be the main cellular hub for the conversion of energy and redox homeostasis in plant cells. Hence, control of protein functions in these important organelles is increasingly being found to be underpinned by reversible post-translational modifications like acetylation. It is known that environmental stresses, as well as changes in cellular nutrient availability or energy status, have a direct impact on global changes in mitochondrial protein acetylation [128]. It has been reported that over one third of proteins in mitochondria are acetylated, of which the majority are associated with pathways such as signal transduction, histone/chromatin gene expression, or protein turnover [129]. Therefore, investigation of protein acetylation has emerged as an important research topic.

1.5.3 Methylation

Most scientists would be familiar with the notion of DNA methylation—that is, the silencing of specific genes by the addition of methyl groups on CpG sites, thus preventing transcription by structural interference with the major groove of DNA. With the discovery of the arginine methyltransferase family of proteins, protein methylation is now considered another form of PTM that warrants examination. Methylation of key arginines within proteins, such as ribosome binding proteins, has been shown to interfere with the intermolecular forces present within the binding site, thus reducing protein-protein affinity [130]. Other key areas where methylation of arginines has been shown to impact on protein function include proteins responsible for transcriptional regulation, signal transduction and even DNA repair [130, 131]. However, it is also important to note that methylation is not limited to arginine residues and has been found on many other amino acid sites; but again, the overall effect of these modifications change the electrostatic potential of specific sites within the protein structure, which has downstream consequences for pathway function.

Recent examples of methylation research include the localization of a methylation enzyme responsible for regulation of a ribosomal protein to chloroplasts and mitochondria [101]. There is also a recent push to characterize the processes by which N-terminal modifications occur, as there is mounting evidence that even rare N-terminals that end with Arg residues can be subject to protein methylation [132]. In fact, N-terminal modification is an area of research that could warrant an entire review, given the functional importance and diversity of associated PTMs.

1.5.4 Ubiquitination

Ubiquitination is the process by which proteins are tagged for degradation. Protein ubiquitination involves of the covalent and reversible binding, of a single ubiquitin molecule or a number of ubiquitin molecules to Lys residues of the target protein [133]. Ubiquitination is carried out by addition of ubiquitin protein by a set of three enzymes the ubiquitin-activating enzyme—E1, the ubiquitin-conjugating enzyme—E2, and ubiquitin protein ligase—E3.

Ubiquitination is a well-observed process that spans both animal and plant kingdom. Effectively, ubiquitin transferases mark misfolded or excess proteins with an ubiquitin tag; ubiquinated structures are then read by the proteosome (a complex multimer that acts a bit like a molecular grinder) and the tagged protein is degraded into oligopeptide chains. New research, however, suggests that ubiquitination is not solely for protein degradation and actually has a role in modulating biochemical pathways [134]. In terms of the importance of ubiquitination, research groups have linked this particular PTM with pathogen defense and development structures [135, 136]. This underscores an important point—that the major function of a PTM is not the sole function, and ubiquitin is no exception, as highlighted by research showing the implications of ubiquitin patterning for the life cycle and regulation of plant cells [137]. As such, ubiquitination has been found to control subcellular processes such as localization and cross talk [138]. Hence, this important PTM has a crucial role in plant metabolism, growth and development, hormone signalling and stress response to wide range of abiotic and biotic factors [139, 140].

Recent research on *Arabidopsis thaliana* has uncovered over three thousand individual ubiquitination sites over sixteen hundred proteins [102]. Another proteomics study showed that 950 ubiquitination substrates in whole *Arabidopsis thaliana* seedlings were identified, using stringent two-step affinity methods for purifying Ub-protein conjugates [141]. Ubiquitination is also being investigated for its role in the regulation of complex biomolecular pathways, as a push from the latter half of last decade saw a newfound interest in investigating alternative ubiquitin roles [138, 142, 143].

1.5.5 Glycosylation

Protein glycosylation, however, is a different PTM class altogether. This is an enormous field of study that is encompassed by the term glycomics, which is complementary to metabolomics, proteomics and genomics. At its core, glycomics refers to the study of sugar modification that exist on proteins, namely O- and N-linked glycans, so named because of the different type of bond and hence amino acid to which the glycan is attached. In order to study glycans or glycopeptide structures, one must invest a considerable amount of time and effort as there exists no easy automated approach to understanding the structures of sugar additions based on mass spectrometry profiles. This is due to the nature of sugar branching from amino acids—even if the constituent sugar modules can be singularly identified, there is as yet no method of automatic determination to reassemble these monomers into accurate morphologies [144].

Sugar modifications often change the function of proteins in subtle ways; for instance, surface membrane proteins coated in sugars act as intracellular communication devices that can recognise self versus antigen molecules. It may be pertinent to mention that for intracellular protein modifications that use O-linked glycans, the overall effect on protein function may be similar to that of phosphorylation [103, 145]. However, studying glycan patterns requires specific expertise, in no small part due to the inherent difficulties in acquiring glycan samples from peptides, and the manual annotation analysis that needs to take place to make sense of any glycosylation results. A recent overview of cell wall glycoproteins outlines the sparse discoveries for glycomics in *Arabidopsis thaliana*; current knowledge is limited to an O-linked class of proteins in the cell wall, and a few N-linked glycans responsible for regulation of peroxidases, a mannosidase and a polygalacturonase inhibiting protein [146]. It is clear that glycomics lags behind other PTMs in plant research, hence more research needs to done to reveal the importance of glycomics in plant development and stress response studies.

1.5.6 PTMs Cross-Talk

The reversible nature of most PTMs makes some specific residues subject to modification with alternate PTMs. In fact, a first PTM can initiate the occurrence of the next PTMs. As a result the cross talk between PTMs occur on the same protein. Broadly speaking, PTM cross talk can be broken down into two easy-to-understand categories. First, positive cross talking occurs when the initial PTM make a constructive modification, such as the addition of another nearby PTM, or if the initial PTM itself forms a new binding partner with another protein. Second, negative cross talk can also occur; in this instance, the presence of a single PTM can overcrowd the target site, change the conformation of the protein to stop other PTMs from forming, or even disable or inconvenience the functioning of a pre-established PTM [147]. Cross talk analysis is effectively the coalface of functional analysis, as often the functional result of PTMs on these proteins or signals come down to the interplay between different PTM types, which is influenced by their location and numbers.

Although preparation of tissue samples for PTM analysis can be tedious and costly, especially in the case of the phosphorylation or glycan studies, any solid understanding of intracellular regulation in response to the environment or stress factors must take PTMs into account. Having gone to the effort of preparing protein samples for PTM analysis it also makes sense to conduct PTM studies in addition to shotgun or bottom-up proteomics experiment to quantify the global protein changes within the cell. Having these two pieces of information hand in hand provides a rigorous and holistic picture of the cell's response to the factor in question—the large-scale changes of protein expression, or presence and absence of proteins, informs us of the fundamental changes, while PTMs provide insight on how the cell responds in a more subtle and nuanced sense. Sometimes, these two pieces of information can overlap—in addition to fold changes, we see PTM to the proteins of interest—but often the changes on the PTM level may come from proteins that are otherwise stable expression-wise but that alternate between two states. In this sense, an alteration in expression level cannot tell us anything about PTMs and vice versa.

1.6 Bioinformatics, Data Analysis Challenges and Platforms in the Plant Proteomics Field

The quantitative protein identification strategies described above rely on matching spectra to sequences from a database in order to generate the protein identification and quantitation; as such it is crucial to have a good quality database. If the plant in question is sequenced and well-characterized the choice is easy. For non-sequenced plants, the options range from using a limited species-specific database, a related sequenced and well annotated species, available EST databases, a comprehensive database such as NCBI plants, or smaller composite databases containing sequences from several related plants; the choices are well described in the context of plants in a recent review [148]. The majority of the papers surveyed there used the comprehensive NCBI plant database; from the remaining options, from amongst the several papers carrying out comparative experiments, no clear 'winner' option has emerged, with some finding the well-annotated related species more advantageous, others the species-specific [149] or EST option [150]. From work in our own group on bioinformatics approaches available for wheat proteomic analyses [151], we found a smaller size database from a closely related species to be more manageable for multi-run iTRAQ experiments. We likewise emphasized the importance of considering issues such as database redundancy and the implication on protein grouping, as well as the availability of down-stream analysis options when choosing the database.

Viewed from the plant biology standpoint the quantitative proteomics approaches are not a goal in themselves but a means to arrive at a useful biological or agricultural outcome. Therefore, a quantitative dataset generated by any of the proteomic technologies described above is just the beginning. The aim is increasingly not only placing the proteomic results in the context of biological information, but also gaining understanding of the mechanism of a particular process, be it drought response, signalling, or pathogen resistance. Here, we give an overview of commonly used tools and strategies for bioinformatics analysis of plant proteomics datasets. For selective reviews with a focus on bioinformatics resources available for plants across several 'Omics, platforms, see [152] and [153].

The broad outline of possible analysis steps is summarized in (Fig. 1.2), where for clarity we structure the workflow in three layers: data, tools and annotation, and results and interpretation. The particular details depend crucially of the type of plant, availability of annotations, and thus resources at ones disposal. Following experiment-specific statistical analysis, all the quantitative results have to be placed in the context of available biological knowledge, such as pathways or biological

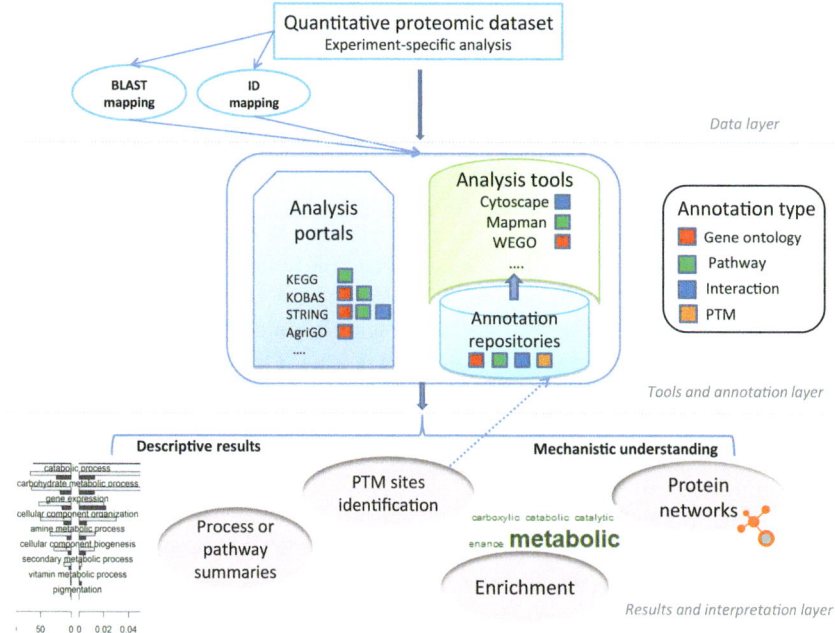

Fig. 1.2 Schematic representation illustrating the plant proteomics data analysis workflow

functions. If the organism is a well-characterised plant such as *Arabidopsis* or rice, then no further mapping of the protein identifiers is needed. If an organism has little or no annotation available then mapping the identifiers to commonly used orthologs such as Uniprot (http://www.uniprot.org/mapping/), or BLAST mapping to a well-annotated species, may be a useful first step.

In the tools layer, the most convenient option is to use an analysis portal that integrates functional or pathway data with some options for analysis, provided that the plant species used is amongst those supported. KEGG remains a popular portal [154] containing pathway information for approximately 50 plant species, with the full list of supported organisms available at http://www.genome.jp/kegg/catalog/org_list4.html. The KEGG Mapper gives the option to bulk map identifiers onto pathway images, for a short list of available types of protein ID's (KEGG, NCBI and Uniprot); other identifiers could be first mapped to Uniprot as described above. Alternatively, bioinformatics portals such as KOBAS [155] (http://kobas.cbi.pku.edu.cn/home.do) integrate the KEGG database with other resources such as additional databases like PANTHER, and gene ontology (GO) information, from a total of 1327 species. The AgriGO analysis platform [156] (http://bioinfo.cau.edu.cn/agriGO/analysis.php) is a plant-specific resource that integrates gene ontology data available at the moment for 45 plant species with six analyses options available, of which parametric gene set analysis (PAGE) can be used to integrate the abundance data from an experiment alongside the gene ontology information. The STRING [157] database and analysis portal has increased in popularity; while primarily used for visualisation of protein interactions, it also contains gene ontology and pathway information that can be overlayed on the networks.

If the plant of interest is not amongst those directly supported by one of the available analysis portals, or if a different type of analysis is required, then either software or annotation can be installed or downloaded separately. Tools such as Mapman for the analysis of pathway information [158], or Cytoscape (http://www.cytoscape.org/) for the analysis of protein networks, have matured, and their installation and use is now much more user-friendly. Mapman can integrate and visualise experimental data, for instance across a time course, but coming from the proteomics perspective it still requires downloading the right mappings from the MapMan Store, and possibly blasting the protein identifiers to the available mapping identifiers; this can be done for instance via the Plant Expression Database Blast resource (http://www.plexdb.org/). Cytoscape plugins such as BinGO [159] and ClueGO [160] offer the option to view the gene ontology data in network form. For organisms with very limited information available an option might be a gene ontology information download, which can be explored using tools such as the WEGO portal (http://wego.genomics.org.cn/cgi-bin/wego/index.pl) [161]. From our own group, we use PloGO [162] for integrating GO data with protein abundance data in the context of complex experiments. For commonly used protein identifiers such as Ensembl or Uniprot, or when mappings to these identifiers can be generated as previously described, tools such as the Ensembl Biomart or the Uniprot retrieve/mapping service can be readily used to bulk-download gene ontology information. When other identifiers are used, organism-specific resources may provide other options. The TAIR *Arabidopsis* portal maintains a list of plant model organism databases and resources, available at https://www.arabidopsis.org/portals/genAnnotation/other_genomes/index.jsp.

The established tools described above are geared towards the analysis of lists of proteins arising from proteomics experiments, but they are not directly applicable to the analysis of post translational modifications. PTM analysis in plants, whilst growing in popularity, is not yet mature [163], with data repositories emerging for a limited number of well-studied plants, and limited options for analysis. Large studies lead to identifications of PTM sites on peptides, which can be mapped onto protein sequences that can then be categorized as above in terms of gene ontology or pathway. One of the main roles of PTM experiments at this point is to accumulate information to be fed back into data repositories [164]. In the realm of plant studies, phosphorylation resources were established initially for several model plants (PhosPhAt, Medicago PhosphoProtein database), and more recently P3DB [165], currently containing phospho-site information for eight plants, but predominantly *Arabidopsis*, Medicago and rice.

While the variety of available software and resources is immense, the output of the tools and annotation layer is usually a dataset with appended biological information from various biological categories of interest. Making use of this information can range from descriptive to mechanistic. At the descriptive end, summaries of categories such as biological process or pathway and their visualisation can be provided for a set of proteins, or compared between various sets.

Enrichment analysis is described in detail in a recent review [166], and is embedded into many of the tools and portals described above; broadly it compares annotation from a protein set of interest to a baseline and suggests interesting processes or pathways to further focus on. PTM analyses in plants are in part still descriptive at this point, as information is being accumulated and stored, but can be coupled with enrichment analysis to find processes that are, for example, enriched in phosphorylation sites, or to interrogate the conservation of PTM sites in orthologs [108], across plant species. Also, targeted questions can be asked by focusing on PTM sites on protein families of interest such as families conferring resistance against pathogens [108]. Finally, analyses of protein networks and interactions are geared towards exploring relationships between proteins with the aim of delivering knowledge and a more mechanistic understanding.

1.7 Conclusions and Future Directions

Plant proteomics methodology has progressed rapidly, with numerous studies now available characterising plant proteomes in great detail. A key success of such studies has been the use of quantitative protein abundance information to shape understanding of plant biological processes, particularly environmental stress responses. With these successes and the advent of new high throughput and more sensitive technologies, we expect to see an explosion in the near future in the use of quantitative proteomic techniques addressing plant biology questions (Fig. 1.1).

Also apparent from this review of the literature is the fact that PTM analysis of plant proteomes is still in its infancy. This is particularly true for glycoproteome analysis which is at present a poorly understood and characterised PTM in plants, especially compared with PTM analysis of proteomes from other branches of the tree of life. Better methods to understand and characterise PTMs in plants are needed, and the development of these methods should be a priority.

Finally, bioinformatic methods and databases specific to plants need to be better described and utilised. It is clear that many databases and tools are adapted from mammalian and bacterial analysis systems, and these may not be ideal for analysis of plant proteomic data. In addition, there seems to be a gap in the literature between what bioinformatics predicts from the data analysis of proteomic data, and experimental verification of these predictions.

References

1. Cho WCS (2007) Proteomics technologies and challenges. Genomics Proteomics Bioinform 5:77–85
2. Patterson SD, Aebersold RH (2003) Proteomics: the first decade and beyond. Nat Genet 33:311–323

3. Jorrín-Novo JV, Pascual J, Sánchez-Lucas R, Romero-Rodríguez MC, Rodríguez-Ortega MJ, Lenz C et al (2015) Fourteen years of plant proteomics reflected in proteomics: moving from model species and 2DE-based approaches to orphan species and gel-free platforms. Proteomics 15:1089–1112
4. Pfannschmidt T, Ogrzewalla K, Baginsky S, Sickmann A, Meyer HE, Link G (2000) The multisubunit chloroplast RNA polymerase A from mustard (*Sinapis alba* L.). Eur J Biochem 267:253–261
5. Chen S, Harmon AC (2006) Advances in plant proteomics. Proteomics 6:5504–5516
6. Vanderschuren H, Lentz E, Zainuddin I, Gruissem W (2013) Proteomics of model and crop plant species: Status, current limitations and strategic advances for crop improvement. J Proteomics 93:5–19
7. Carpentier SC, Panis B, Vertommen A, Swennen R, Sergeant K, Renaut J et al (2008) Proteome analysis of non-model plants: a challenging but powerful approach. Mass Spectrom Rev 27:354–377
8. Kaul S, Koo HL, Jenkins J, Rizzo M, Rooney T, Tallon LJ et al (2000) Analysis of the genome sequence of the flowering plant *Arabidopsis thaliana*. Nature 408:796–815
9. Goff SA, Ricke D, Lan T-H, Presting G, Wang R, Dunn M et al (2002) A draft sequence of the rice genome (Oryza sativa L. ssp. japonica). Science 296:92–100
10. Armengaud J, Trapp J, Pible O, Geffard O, Chaumot A, Hartmann EM (2014) Non-model organisms, a species endangered by proteogenomics. J Proteomics 105:5–18
11. Canovas FM, Dumas-Gaudot E, Recorbet G, Jorrin J, Mock HP, Rossignol M (2004) Plant proteome analysis. Proteomics 4:285–298
12. Jorrín JV, Maldonado AM, Castillejo MA (2007) Plant proteome analysis: a 2006 update. Proteomics 7:2947–2962
13. Abril N, Gion J-M, Kerner R, Müller-Starck G, Cerrillo RMN, Plomion C et al (2011) Proteomics research on forest trees, the most recalcitrant and orphan plant species. Phytochemistry 72:1219–1242
14. Bindschedler LV, Cramer R (2011) Quantitative plant proteomics. Proteomics 11:756–775
15. Junqueira M, Spirin V, Balbuena TS, Thomas H, Adzhubei I, Sunyaev S et al (2008) Protein identification pipeline for the homology-driven proteomics. J Proteomics 71:346–356
16. Waridel P, Frank A, Thomas H, Surendranath V, Sunyaev S, Pevzner P et al (2007) Sequence similarity-driven proteomics in organisms with unknown genomes by LC-MS/MS and automated de novo sequencing. Proteomics 7:2318–2329
17. Primmer CR, Papakostas S, Leder EH, Davis MJ, Ragan MA (2013) Annotated genes and nonannotated genomes: cross-species use of gene ontology in ecology and evolution research. Mol Ecol 22:3216–3241
18. Jorrin-Novo JV (2014) Plant proteomics methods and protocols. Springer, New York
19. Ma B, Johnson R (2012) De novo sequencing and homology searching. Mol Cell Proteomics 11(O111):014902
20. Klose J (1975) Protein mapping by combined isoelectric focusing and electrophoresis of mouse tissues. Hum Genet 26:231–243
21. O'farrell PH (1975) High resolution two-dimensional electrophoresis of proteins. J Biol Chem 250:4007–4021
22. Bae MS, Cho EJ, Choi EY, Park OK (2003) Analysis of the *Arabidopsis* nuclear proteome and its response to cold stress. Plant J 36:652–663
23. Jiang Y, Yang B, Harris NS, Deyholos MK (2007) Comparative proteomic analysis of NaCl stress-responsive proteins in *Arabidopsis* roots. J Exp Bot 58:3591–3607
24. Aghaei K, Ehsanpour AA, Shah AH, Komatsu S (2009) Proteome analysis of soybean hypocotyl and root under salt stress. Amino Acids 36:91–98
25. Qiu QS, Huber JL, Booker FL, Jain V, Leakey AD, Fiscus EL et al (2008) Increased protein carbonylation in leaves of *Arabidopsis* and soybean in response to elevated [CO_2]. Photosynth Res 97:155–166

26. Mirzaei M, Soltani N, Sarhadi E, Pascovici D, Keighley T, Salekdeh GH et al (2011) Shotgun proteomic analysis of long-distance drought signaling in rice roots. J Proteome Res 11:348–358
27. Sarhadi E, Bazargani MM, Sajise AG, Abdolahi S, Vispo NA, Arceta M et al (2012) Proteomic analysis of rice anthers under salt stress. Plant Physiol Biochem 58:280–287
28. Han C, Yang P, Sakata K, Komatsu S (2014) Quantitative proteomics reveals the role of protein phosphorylation in rice embryos during early stages of germination. J Proteome Res 13:1766–1782
29. Majoul T, Bancel E, Triboi E, Ben Hamida J, Branlard G (2004) Proteomic analysis of the effect of heat stress on hexaploid wheat grain: characterization of heat-responsive proteins from non-prolamins fraction. Proteomics 4:505–513
30. Faghani E, Gharechahi J, Komatsu S, Mirzaei M, Khavarinejad RA, Najafi F et al (2015) Comparative physiology and proteomic analysis of two wheat genotypes contrasting in drought tolerance. J Proteomics 114:1–15
31. An Nguyen TT, Michaud D, Cloutier C (2007) Proteomic profiling of aphid *Macrosiphum euphorbiae* responses to host-plant-mediated stress induced by defoliation and water deficit. J Insect Physiol 53:601–611
32. Aghaei K, Ehsanpour AA, Komatsu S (2008) Proteome analysis of potato under salt stress. J Proteome Res 7:4858–4868
33. Yumiko I, Hiroshi H (2000) Effect of heat stress on tomato fruit protein expression. Electrophoresis 21:1766–1771
34. Bandehagh A, Salekdeh GH, Toorchi M, Mohammadi A, Komatsu S (2011) Comparative proteomic analysis of canola leaves under salinity stress. Proteomics 11:1965–1975
35. Taheri F, Nematzadeh G, Zamharir MG, Nekouei MK, Naghavi M, Mardi M et al (2011) Proteomic analysis of the Mexican lime tree response to "*Candidatus Phytoplasma aurantifolia*" infection. Mol BioSyst 7:3028–3035
36. Douce R, Joyard J (1990) Biochemistry and function of the plastid envelope. Annu Rev Cell Biol 6:173–216
37. Yan JX, Harry RA, Spibey C, Dunn MJ (2000) Postelectrophoretic staining of proteins separated by two-dimensional gel electrophoresis using SYPRO dyes. Electrophoresis 21:3657–3665
38. Rabilloud T, Lelong C (2011) Two-dimensional gel electrophoresis in proteomics: a tutorial. J Proteomics 74:1829–1841
39. Bianco L, Lopez L, Scalone AG, Di Carli M, Desiderio A, Benvenuto E et al (2009) Strawberry proteome characterization and its regulation during fruit ripening and in different genotypes. J Proteomics 72:586–607
40. Renaut J, Hausman JF, Bassett C, Artlip T, Cauchie HM, Witters E et al (2008) Quantitative proteomic analysis of short photoperiod and low-temperature responses in bark tissues of peach (*Prunus persica* L. Batsch). Tree Genet Genomes 4:589–600
41. Rogowska-Wrzesinska A, Le Bihan M-C, Thaysen-Andersen M, Roepstorff P (2013) 2D gels still have a niche in proteomics. J Proteomics 88:4–13
42. Washburn MP, Wolters D, Yates JR III (2001) Large-scale analysis of the yeast proteome by multidimensional protein identification technology. Nat Biotechnol 19:242–247
43. Chen R, Pan S, Yi EC, Donohoe S, Bronner MP, Potter JD et al (2006) Quantitative proteomic profiling of pancreatic cancer juice. Proteomics 6:3871–3879
44. De Godoy LM, Olsen JV, De Souza GA, Li G, Mortensen P, Mann M (2006) Status of complete proteome analysis by mass spectrometry: SILAC labeled yeast as a model system. Genome Biol 7:R50
45. Koller A, Washburn MP, Lange BM, Andon NL, Deciu C, Haynes PA et al (2002) Proteomic survey of metabolic pathways in rice. Proc Natl Acad Sci USA 99:11969–11974
46. Gygi SP, Rist B, Gerber SA, Turecek F, Gelb MH, Aebersold R (1999) Quantitative analysis of complex protein mixtures using isotope-coded affinity tags. Nat Biotechnol 17:994–999
47. Dunkley TP, Watson R, Griffin JL, Dupree P, Lilley KS (2004) Localization of organelle proteins by isotope tagging (LOPIT). Mol Cell Proteomics 3:1128–1134

48. Hartman NT, Sicilia F, Lilley KS, Dupree P (2007) Proteomic complex detection using sedimentation. Anal Chem 79:2078–2083
49. Majeran W, Cai Y, Sun Q, Van Wijk KJ (2005) Functional differentiation of bundle sheath and mesophyll maize chloroplasts determined by comparative proteomics. Plant Cell 17:3111–3140
50. Ross PL, Huang YN, Marchese JN, Williamson B, Parker K, Hattan S et al (2004) Multiplexed protein quantitation in Saccharomyces cerevisiae using amine-reactive isobaric tagging reagents. Mol Cell Proteomics 3:1154–1169
51. Thompson A, Schäfer J, Kuhn K, Kienle S, Schwarz J, Schmidt G et al (2003) Tandem mass tags: a novel quantification strategy for comparative analysis of complex protein mixtures by MS/MS. Anal Chem 75:1895–1904
52. Erickson BK, Jedrychowski MP, Mcalister GC, Everley RA, Kunz R, Gygi SP (2015) Evaluating multiplexed quantitative phosphopeptide analysis on a hybrid quadrupole mass filter/linear ion trap/orbitrap mass spectrometer. Anal Chem 87:1241–1249
53. Neilson KA, Mariani M, Haynes PA (2011) Quantitative proteomic analysis of cold-responsive proteins in rice. Proteomics 11:1696–1706
54. Liu G-T, Ma L, Duan W, Wang B-C, Li J-H, Xu H-G et al (2014) Differential proteomic analysis of grapevine leaves by iTRAQ reveals responses to heat stress and subsequent recovery. BMC Plant Biol 14:1
55. Hu X, Li N, Wu L, Li C, Li C, Zhang L et al (2015) Quantitative iTRAQ-based proteomic analysis of phosphoproteins and ABA-regulated phosphoproteins in maize leaves under osmotic stress. Sci Rep 5
56. Xu J, Lan H, Fang H, Huang X, Zhang H, Huang J (2015) Quantitative proteomic analysis of the rice (*Oryza sativa* L.) salt response. PLoS ONE 10:e0120978
57. Fukao Y, Ferjani A, Tomioka R, Nagasaki N, Kurata R, Nishimori Y et al (2011) iTRAQ analysis reveals mechanisms of growth defects due to excess zinc in *Arabidopsis*. Plant Physiol 155:1893–1907
58. Parker J, Zhu N, Zhu M, Chen S (2012) Profiling thiol redox proteome using isotope tagging mass spectrometry. JoVE (J Visualized Exp) 61:e3766–e3766
59. Turek I, Wheeler JI, Gehring C, Irving HR, Marondedze C (2015) Quantitative proteome changes in *Arabidopsis thaliana* suspension-cultured cells in response to plant natriuretic peptides. Data Brief 4:336–343
60. Neilson KA, Scafaro AP, Chick JM, George IS, Van Sluyter SC, Gygi SP et al (2013) The influence of signals from chilled roots on the proteome of shoot tissues in rice seedlings. Proteomics 13:1922–1933
61. Zeng J, He X, Quan X, Cai S, Han Y, Nadira UA et al (2015) Identification of the proteins associated with low potassium tolerance in cultivated and Tibetan wild barley. J Proteomics 126:1–11
62. Gouw JW, Tops BB, Mortensen P, Heck AJ, Krijgsveld J (2008) Optimizing identification and quantitation of 15N-labeled proteins in comparative proteomics. Anal Chem 80:7796–7803
63. Wang Y, Li H, Chen S (2010) Advances in quantitative proteomics. Front Biol 5:195–203
64. Aebersold R, Mann M (2003) Mass spectrometry-based proteomics. Nature 422:198–207
65. Gruhler A, Schulze WX, Matthiesen R, Mann M, Jensen ON (2005) Stable isotope labeling of *Arabidopsis thaliana* cells and quantitative proteomics by mass spectrometry. Mol Cell Proteomics 4:1697–1709
66. Schütz W, Hausmann N, Krug K, Hampp R, Macek B (2011) Extending SILAC to proteomics of plant cell lines. Plant Cell 23:1701–1705
67. Engelsberger WR, Erban A, Kopka J, Schulze WX (2006) Metabolic labeling of plant cell cultures with $K^{15}NO_3$ as a tool for quantitative analysis of proteins and metabolites. Plant Methods 2:1
68. Bindschedler LV, Palmblad M, Cramer R (2008) Hydroponic isotope labelling of entire plants (HILEP) for quantitative plant proteomics; an oxidative stress case study. Phytochemistry 69:1962–1972

69. Oda Y, Huang K, Cross FR, Cowburn D, Chait BT (1999) Accurate quantitation of protein expression and site-specific phosphorylation. Proc Natl Acad Sci 96:6591–6596
70. Ong S-E, Mann M (2006) A practical recipe for stable isotope labeling by amino acids in cell culture (SILAC). Nat Protoc 1:2650–2660
71. Hulce JJ, Cognetta AB, Niphakis MJ, Tully SE, Cravatt BF (2013) Proteome-wide mapping of cholesterol-interacting proteins in mammalian cells. Nat Methods 10:259–264
72. Nagaraj N, Kulak NA, Cox J, Neuhauser N, Mayr K, Hoerning O et al (2012) System-wide perturbation analysis with nearly complete coverage of the yeast proteome by single-shot ultra HPLC runs on a bench top Orbitrap. Mol Cell Proteomics 11(M111):013722
73. Hirner A, Ladwig F, Stransky H, Okumoto S, Keinath M, Harms A et al (2006) *Arabidopsis* LHT1 is a high-affinity transporter for cellular amino acid uptake in both root epidermis and leaf mesophyll. Plant Cell 18:1931–1946
74. Lewandowska D, Ten Have S, Hodge K, Tillemans V, Lamond AI, Brown JWS (2013) Plant SILAC: stable-isotope labelling with amino acids of *Arabidopsis* seedlings for quantitative proteomics. PloS ONE 8:e72207
75. Engelsberger WR, Erban A, Kopka J, Schulze WX (2006) Metabolic labeling of plant cell cultures with $K^{15}NO_3$ as a tool for quantitative analysis of proteins and metabolites. Plant Methods 2:14
76. Allen DK, Evans BS, Libourel IGL (2014) Analysis of isotopic labeling in peptide fragments by tandem mass spectrometry. PLoS ONE 9:e91537
77. Lanquar V, Kuhn L, Lelièvre F, Khafif M, Espagne C, Bruley C et al (2007) 15N-Metabolic labeling for comparative plasma membrane proteomics in *Arabidopsis* cells. Proteomics 7:750–754
78. Huttlin EL, Hegeman AD, Harms AC, Sussman MR (2007) Comparison of full versus partial metabolic labeling for quantitative proteomics analysis in *Arabidopsis thaliana*. Mol Cell Proteomics 6:860–881
79. Schaff JE, Mbeunkui F, Blackburn K, Bird DM, Goshe MB (2008) SILIP: a novel stable isotope labeling method for in planta quantitative proteomic analysis. Plant J 56:840–854
80. Nelson CJ, Alexova R, Jacoby RP, Millar AH (2014) Proteins with high turnover rate in barley leaves estimated by proteome analysis combined with in planta isotope labeling. Plant Physiol 166:91–108
81. Nelson CJ, Huttlin EL, Hegeman AD, Harms AC, Sussman MR (2007) Implications of 15N-metabolic labeling for automated peptide identification in Arabidopsis thaliana. Proteomics 7:1279–1292
82. Whitelegge JP, Katz JE, Pihakari KA, Hale R, Aguilera R, Gómez SM et al (2004) Subtle modification of isotope ratio proteomics; an integrated strategy for expression proteomics. Phytochemistry 65:1507–1515
83. Arsova B, Kierszniowska S, Schulze WX (2012) The use of heavy nitrogen in quantitative proteomics experiments in plants. Trends Plant Sci 17:102–112
84. Benschop JJ, Mohammed S, O'flaherty M, Heck AJ, Slijper M, Menke FL (2007) Quantitative phosphoproteomics of early elicitor signaling in *Arabidopsis*. Mol Cell Proteomics 6:1198–1214
85. Keinath NF, Kierszniowska S, Lorek J, Bourdais G, Kessler SA, Asano H et al (2010) PAMP-induced changes in plasma membrane compartmentalization reveal novel components of plant immunity. J Biol Chem JBC M110:160531
86. Stanislas T, Bouyssie D, Rossignol M, Vesa S, Fromentin J, Morel J et al (2009) Quantitative proteomics reveals a dynamic association of proteins to detergent-resistant membranes upon elicitor signaling in tobacco. Mol Cell Proteomics 8:2186–2198
87. Hebeler R, Oeljeklaus S, Reidegeld KA, Eisenacher M, Stephan C, Sitek B et al (2008) Study of early leaf senescence in *Arabidopsis thaliana* by quantitative proteomics using reciprocal 14N/15N labeling and difference gel electrophoresis. Mol Cell Proteomics 7:108–120
88. Monavarfeshani A, Mirzaei M, Sarhadi E, Amirkhani A, Khayam Nekouei M, Haynes PA et al (2013) Shotgun proteomic analysis of the Mexican lime tree infected with "*Candidatus* Phytoplasma aurantifolia". J Proteome Res 12:785–795

89. Liu H, Sadygov RG, Yates JR (2004) A model for random sampling and estimation of relative protein abundance in shotgun proteomics. Anal Chem 76:4193–4201
90. Podwojski K, Eisenacher M, Kohl M, Turewicz M, Meyer HE, Rahnenführer J et al (2010) Peek a peak: a glance at statistics for quantitative label-free proteomics. Expert Rev Proteomics 7:249–261
91. Zhu W, Smith JW, Huang C-M (2009) Mass spectrometry-based label-free quantitative proteomics. BioMed Res Int 2010:840518
92. Gammulla CG, Pascovici D, Atwell BJ, Haynes PA (2011) Differential proteomic response of rice (*Oryza sativa*) leaves exposed to high- and low-temperature stress. Proteomics 11:2839
93. Gammulla CG, Pascovici D, Atwell BJ, Haynes PA (2010) Differential metabolic response of cultured rice (*Oryza sativa*) cells exposed to high-and low-temperature stress. Proteomics 10:3001–3019
94. George IS, Pascovici D, Mirzaei M, Haynes PA (2015) Quantitative proteomic analysis of cabernet sauvignon grape cells exposed to thermal stresses reveals alterations in sugar and phenylpropanoid metabolism. Proteomics 15:3048–3060
95. Mirzaei M, Pascovici D, Atwell BJ, Haynes PA (2012) Differential regulation of aquaporins, small GTPases and V-ATPases proteins in rice leaves subjected to drought stress and recovery. Proteomics 12:864–877
96. Mirzaei M, Soltani N, Sarhadi E, George IS, Neilson KA, Pascovici D et al (2013) Manipulating root water supply elicits major shifts in the shoot proteome. J Proteome Res 13:517–526
97. Kottapalli KR, Zabet-Moghaddam M, Rowland D, Faircloth W, Mirzaei M, Haynes PA et al (2013) Shotgun label-free quantitative proteomics of water-deficit-stressed midmature peanut (*Arachis hypogaea* L.) seed. J Proteome Res 12:5048–5057
98. Uniprot C (2010) The universal protein resource (UniProt) in 2010. Nucleic Acids Res 38: D142–D148
99. Arnaudo AM, Garcia BA (2013) Proteomic characterization of novel histone post-translational modifications. Epigenetics Chromatin 6:24
100. Finkemeier I, Laxa M, Miguet L, Howden AJM, Sweetlove LJ (2011) Proteins of diverse function and subcellular location are lysine acetylated in *Arabidopsis*. Plant Physiol 155:1779–1790
101. Mazzoleni M, Figuet S, Martin-Laffon J, Mininno M, Gilgen A, Leroux M et al (2015) Dual targeting of the protein methyltransferase PrmA contributes to both chloroplastic and mitochondrial ribosomal protein L11 methylation in *Arabidopsis*. Plant and Cell Physiol: pcv098
102. Walton A, Stes E, Cybulski N, Van Bel M, Inigo S, Durand AN et al (2016) It's time for some "site"-seeing: novel tools to monitor the ubiquitin landscape in Arabidopsis thaliana. Plant Cell: TPC2015-00878-REV
103. Dam S, Thaysen-Andersen M, Stenkjær E, Lorentzen A, Roepstorff P, Packer NH et al (2013) Combined N-glycome and N-glycoproteome analysis of the *Lotus japonicus* seed globulin fraction shows conservation of protein structure and glycosylation in legumes. J Proteome Res 12:3383–3392
104. Friso G, Van Wijk KJ (2015) Posttranslational Protein Modifications in Plant Metabolism. Plant Physiol 169:1469–1487
105. Parker CE, Mocanu V, Mocanu M, Dicheva N, Warren MR (2010) Mass spectrometry for post-translational modifications. In: Alzate O (ed) Neuroproteomics: CRC Press, Boca Raton (FL) Chapter 6
106. Li J, Silva-Sanchez C, Zhang T, Chen S, Li H (2015) Phosphoproteomics technologies and applications in plant biology research. Front Plant Sci: 6
107. Silva-Sanchez C, Li H, Chen S (2015) Recent advances and challenges in plant phosphoproteomics. Proteomics 15:1127–1141

108. Nakagami H, Sugiyama N, Mochida K, Daudi A, Yoshida Y, Toyoda T et al (2010) Large-scale comparative phosphoproteomics identifies conserved phosphorylation sites in plants. Plant Physiol 153:1161–1174
109. Hunter T (2007) The age of crosstalk: phosphorylation, ubiquitination, and beyond. Mol Cell 28:730–738
110. Vandamme J, Castermans D, Thevelein JM (2012) Molecular mechanisms of feedback inhibition of protein kinase A on intracellular cAMP accumulation. Cell Signal 24:1610–1618
111. Xing T, Ouellet T, Miki BL (2002) Towards genomic and proteomic studies of protein phosphorylation in plant–pathogen interactions. Trends Plant Sci 7:224–230
112. Havelund JF, Thelen JJ, Møller IM (2015) Biochemistry, proteomics, and phosphoproteomics of plant mitochondria from non-photosynthetic cells. Sub-Cell Proteomics: 98
113. Zargar SM, Nazir M, Rai V, Hajduch M, Agrawal GK, Rakwal R (2015) Towards a common bean proteome atlas: looking at the current state of research and the need for a comprehensive proteome. Front Plant Sci: 6
114. Li G, Boudsocq M, Hem S, Vialaret J, Rossignol M, Maurel C et al (2015) The calcium-dependent protein kinase CPK7 acts on root hydraulic conductivity. Plant Cell Environ 38:1312–1320
115. Choudhary C, Kumar C, Gnad F, Nielsen ML, Rehman M, Walther TC et al (2009) Lysine acetylation targets protein complexes and co-regulates major cellular functions. Science 325:834–840
116. Choudhary C, Weinert BT, Nishida Y, Verdin E, Mann M (2014) The growing landscape of lysine acetylation links metabolism and cell signalling. Nat Rev Mol Cell Biol 15:536–550
117. Jeffers V, Sullivan WJ (2012) Lysine acetylation is widespread on proteins of diverse function and localization in the protozoan parasite *Toxoplasma gondii*. Eukaryot Cell 11:735–742
118. Pandey R, Muller A, Napoli CA, Selinger DA, Pikaard CS, Richards EJ et al (2002) Analysis of histone acetyltransferase and histone deacetylase families of *Arabidopsis thaliana* suggests functional diversification of chromatin modification among multicellular eukaryotes. Nucleic Acids Res 30:5036–5055
119. Hu Y, Qin F, Huang L, Sun Q, Li C, Zhao Y et al (2009) Rice histone deacetylase genes display specific expression patterns and developmental functions. Biochem Biophys Res Commun 388:266–271
120. Servet C, Conde E Silva N, Zhou DX (2010) Histone acetyltransferase AtGCN5/HAG1 is a versatile regulator of developmental and inducible gene expression in *Arabidopsis*. Mol Plant 3:670–677
121. Wang Q, Zhang Y, Yang C, Xiong H, Lin Y, Yao J et al (2010) Acetylation of metabolic enzymes coordinates carbon source utilization and metabolic flux. Science 327:1004–1007
122. Konig AC, Hartl M, Boersema PJ, Mann M, Finkemeier I (2014) The mitochondrial lysine acetylome of *Arabidopsis*. Mitochondrion 19 Pt B:252–260
123. Wu X, Oh MH, Schwarz EM, Larue CT, Sivaguru M, Imai BS et al (2011) Lysine acetylation is a widespread protein modification for diverse proteins in *Arabidopsis*. Plant Physiol 155:1769–1778
124. Smith-Hammond CL, Hoyos E, Miernyk JA (2014) The pea seedling mitochondrial N (epsilon)-lysine acetylome. Mitochondrion 19 Pt B:154–165
125. Smith-Hammond CL, Swatek KN, Johnston ML, Thelen JJ, Miernyk JA (2014) Initial description of the developing soybean seed protein Lys-N(epsilon)-acetylome. J Proteomics 96:56–66
126. Melo-Braga MN, Verano-Braga T, Leon IR, Antonacci D, Nogueira FC, Thelen JJ et al (2012) Modulation of protein phosphorylation, N-glycosylation and Lys-acetylation in grape (*Vitis vinifera*) mesocarp and exocarp owing to *Lobesia botrana* infection. Mol Cell Proteomics 11:945–956

127. Nallamilli BR, Edelmann MJ, Zhong X, Tan F, Mujahid H, Zhang J et al (2014) Global analysis of lysine acetylation suggests the involvement of protein acetylation in diverse biological processes in rice (*Oryza sativa*). PLoS ONE 9:e89283
128. Rardin MJ, Newman JC, Held JM, Cusack MP, Sorensen DJ, Li B et al (2013) Label-free quantitative proteomics of the lysine acetylome in mitochondria identifies substrates of SIRT3 in metabolic pathways. Proc Natl Acad Sci 110:6601–6606
129. Anderson KA, Hirschey MD (2012) Mitochondrial protein acetylation regulates metabolism. Essays Biochem 52:23–35
130. Bedford MT, Richard S (2005) Arginine methylation: an emerging regulatorof protein function. Mol Cell 18:263–272
131. Afjehi-Sadat L, Garcia BA (2013) Comprehending dynamic protein methylation with mass spectrometry. Curr Opin Chem Biol 17:12–19
132. Rowland E, Kim J, Bhuiyan NH, Van Wijk KJ (2015) The *Arabidopsis* Chloroplast stromal N-terminome; complexities of N-terminal protein maturation and stability. Plant Physiol, pp: 01214–02015
133. Vierstra RD (2012) The expanding universe of ubiquitin and ubiquitin-like modifiers. Plant Physiol 160:2–14
134. Ehrnhoefer DE, Sutton L, Hayden MR (2011) Small changes, big impact: posttranslational modifications and function of huntingtin in Huntington disease. Neuroscientist: 1073858410390378
135. Devoto A, Muskett PR, Shirasu K (2003) Role of ubiquitination in the regulation of plant defence against pathogens. Curr Opin Plant Biol 6:307–311
136. Moon J, Parry G, Estelle M (2004) The ubiquitin-proteasome pathway and plant development. Plant Cell 16:3181–3195
137. Sullivan JA, Shirasu K, Deng XW (2003) The diverse roles of ubiquitin and the 26S proteasome in the life of plants. Nat Rev Genet 4:948–958
138. Wu X, Gong F, Cao D, Hu X, Wang W (2015) Advances in crop proteomics: PTMs of proteins under abiotic stress. Proteomics 16(5):847–865
139. Vierstra RD (2009) The ubiquitin–26S proteasome system at the nexus of plant biology. Nat Rev Mol Cell Biol 10:385–397
140. Guerra DD, Callis J (2012) Ubiquitin on the move: the ubiquitin modification system plays diverse roles in the regulation of endoplasmic reticulum-and plasma membrane-localized proteins. Plant Physiol 160:56–64
141. Kim D-Y, Scalf M, Smith LM, Vierstra RD (2013) Advanced proteomic analyses yield a deep catalog of ubiquitylation targets in *Arabidopsis*. Plant Cell 25:1523–1540
142. Yu F, Wu Y, Xie Q (2015) Precise protein post-translational modifications modulate ABI5 activity. Trends Plant Sci 20:569–575
143. Mukhopadhyay D, Riezman H (2007) Proteasome-independent functions of ubiquitin in endocytosis and signaling. Science 315:201–205
144. Dell A, Morris HR (2001) Glycoprotein structure determination by mass spectrometry. Science 291:2351–2356
145. Khidekel N, Ficarro SB, Peters EC, Hsieh-Wilson LC (2004) Exploring the O-GlcNAc proteome: direct identification of O-GlcNAc-modified proteins from the brain. Proc Natl Acad Sci USA 101:13132–13137
146. Albenne C, Canut H, Jamet E (2015) Plant cell wall proteomics: the leadership of Arabidopsis thaliana. Sub-Cell Proteomics: 7
147. Venne AS, Kollipara L, Zahedi RP (2014) The next level of complexity: crosstalk of posttranslational modifications. Proteomics 14:513–524
148. Champagne A, Boutry M (2013) Proteomics of nonmodel plant species. Proteomics 13:663–673
149. Bräutigam A, Shrestha RP, Whitten D, Wilkerson CG, Carr KM, Froehlich JE et al (2008) Low-coverage massively parallel pyrosequencing of cDNAs enables proteomics in non-model species: comparison of a species-specific database generated by pyrosequencing with databases from related species for proteome analysis of pea chloroplast envelopes. J Biotechnol 136:44–53

150. Huang M, Chen T, Chan Z (2006) An evaluation for cross-species proteomics research by publicly available expressed sequence tag database search using tandem mass spectral data. Rapid Commun Mass Spectrom 20:2635–2640
151. Pascovici D, Gardiner DM, Song X, Breen E, Solomon PS, Keighley T et al (2013) Coverage and consistency: bioinformatics aspects of the analysis of multirun iTRAQ experiments with wheat leaves. J Proteome Res 12:4870–4881
152. Mochida K, Shinozaki K (2010) Genomics and bioinformatics resources for crop improvement. Plant Cell Physiol 51:497–523
153. Mochida K, Shinozaki K (2011) Advances in omics and bioinformatics tools for systems analyses of plant functions. Plant Cell Physiol 52:2017–2038
154. Baxevanis Andreas D, Davison Daniel B, Page Roderic DM, Petsko Gregory A, Stein Lincoln D, Stormo Gary D (2003) Current protocols in bioinformatics. Vol. 1, John Wiley & Sons Inc, New York
155. Xie C, Mao X, Huang J, Ding Y, Wu J, Dong S et al (2011) KOBAS 2.0: a web server for annotation and identification of enriched pathways and diseases. Nucleic Acids Res 39:W316–W322
156. Du Z, Zhou X, Ling Y, Zhang Z, Su Z (2010) agriGO: a GO analysis toolkit for the agricultural community. Nucleic acids research: W64–70
157. Jensen LJ, Kuhn M, Stark M, Chaffron S, Creevey C, Muller J et al (2009) STRING 8—a global view on proteins and their functional interactions in 630 organisms. Nucleic Acids Res 37:D412–D416
158. Usadel B, Poree F, Nagel A, Lohse M, Czedik-Eysenberg A, Stitt M (2009) A guide to using MapMan to visualize and compare Omics data in plants: a case study in the crop species, *Maize*. Plant Cell Environ 32:1211–1229
159. Maere S, Heymans K, Kuiper M (2005) BiNGO: a Cytoscape plugin to assess overrepresentation of gene ontology categories in biological networks. Bioinformatics 21:3448–3449
160. Bindea G, Mlecnik B, Hackl H, Charoentong P, Tosolini M, Kirilovsky A et al (2009) ClueGO: a cytoscape plug-into decipher functionally grouped gene ontology and pathway annotation networks. Bioinformatics 25:1091–1093
161. Ye J, Fang L, Zheng H, Zhang Y, Chen J, Zhang Z et al (2006) WEGO: a web tool for plotting GO annotations. Nucleic Acids Res 34:W293–W297
162. Pascovici D, Keighley T, Mirzaei M, Haynes PA, Cooke B (2012) PloGO: plotting gene ontology annotation and abundance in multi-condition proteomics experiments. Proteomics 12:406–410
163. Nakagami H, Sugiyama N, Ishihama Y, Shirasu K (2012) Shotguns in the front line: phosphoproteomics in plants. Plant Cell Physiol 53:118–124
164. Cox J, Mann M (2011) Quantitative, high-resolution proteomics for data-driven systems biology. Annu Rev Biochem 80:273–299
165. Yao Q, Ge H, Wu S, Zhang N, Chen W, Xu C et al (2014) P3DB 3.0: from plant phosphorylation sites to protein networks. Nucleic Acids Res 42:D1206–D1213
166. Bessarabova M, Ishkin A, Jebailey L, Nikolskaya T, Nikolsky Y (2012) Knowledge-based analysis of proteomics data. BMC Bioinformatics 13:1

Chapter 2
Seed Proteomics: An Overview

Kanika Narula, Arunima Sinha, Toshiba Haider, Niranjan Chakraborty and Subhra Chakraborty

Abstract Seed is vital for propagation of spermatophytes in biome and as food source for inhabitants of the earth. Studies on seed proteins provide platform for new avenues to explore molecular networks and pathways governing seed filling, maturation, germination, and seedling formation. Protein expression changes of three genetically different sub-regions of angiosperm seeds are reflected in ordered chain of biological events represented from family differences in different taxas. Different families of angiosperm show divergence of seed protein evolution and thus provide insights into seed structure and function. A gamut of information is available on seed proteomic datasets from approximately 3500 proteins that impinge on protein function in diverse plant families. The functional modularity of seed proteins were compared amongst species that span from dicot to moncot and diploid to polyploid. Transitions of protein complement revealed difference between dormancy and germination towards understanding biological check point at translational level. Goal of this chapter is to critically review data available till date on seed proteomic studies and identify family and cross genera knowledge gaps. The information thus obtained would unravel new components and an unparallel understanding of the molecular processes underlying translational and post-translational variations under different conditions that involves histodifferentiation and organogenesis of the seed.

Keywords Crop plant · Seed proteomics · Seed · Dormancy · Germination

Kanika Narula and Arunima Sinha are equally contributed in this work.

K. Narula · A. Sinha · T. Haider · N. Chakraborty · S. Chakraborty (✉)
National Institute of Plant Genome Research, Aruna Asaf Ali Marg,
New Delhi 110067, India
e-mail: subhrac@hotmail.com

© Springer International Publishing Switzerland 2016
G.H. Salekdeh (ed.), *Agricultural Proteomics Volume 1*,
DOI 10.1007/978-3-319-43275-5_2

2.1 Introduction

Evolution of angiosperm has facilitated the transitional dominance of seed producing plants into a terrestrial environment. According to the fossil records their first appearance dated back to 365 million years ago when majority of landscape was successfully and naturally selected for angiosperm survival [1]. The physiological and genetic control of seed is in part responsible for dispersal mechanism which proved to be a spectacular phenomenon responsible for prevalence of seed plants on livable planet, the earth. True seed is fertilized mature ovule possessing an embryonic plant, stored food, and a protective coat. A major factor during angiosperm embryogenesis is the switch from the radial symmetry of the globular embryo to the bilateral symmetry followed by differentiation of cotyledons and embryonal axis [2]. Seed development and germination is a continuous and fine-tuned process with natural circumscription engrossing three phases of embryogenesis recognized as rapid cell division, deposition of reserves, and desiccation. Commonly, mature seeds are classified as albuminous and ex-albuminous depending on the presence or absence of endosperm. Earlier in 1946, Martin studied the internal morphology of seeds belonging to 1287 genera of angiosperm and classified them based on size of the embryo in relation to endosperm and differences in the size, shape, and position of embryo in the seed. The developmental program of monocot and dicot embryogenesis is different but a highly ordered phenomenon. In addition, regulatory pathways, dissection of complex traits, and developmental reprogramming reflects the fundamental differences of the molecular biology of two taxas.

To build new perspectives approach for the physiological and biochemical factors controlling seed traits in two taxas, omics studies of seed development were performed since year 2000 exploiting the availability of genome sequence and related resources [3]. Further, advancement in high-throughput proteome analysis using gel and non-gel based approaches provided detailed protein profile at different developmental stages which might help to elucidate the regulatory network of embryogenesis related proteins. Currently, proteomics is playing an important role in: (i) understanding plant biology, (ii) developing plant biomarkers for human health and food security and (iii) food analysis and bio-safety issues [4]. Seed proteomics research has largely focused on agriculturally important crop plants. Typically, these seeds are obtained from a commercial source. It is reasonable to assume that they are genetically uniform thus it is not biased by the contributions of a "contaminating" proteome. The strategy of using combinatorial-ligand random peptide beads appears to have substantial potential to deplete the supra-abundant SSP from input samples [5]. In addition, proteomic approaches also dissected double fertilization reprogramming at translational and post-translational level in two different clades of angiosperm. This review aims to compile proteomic analyses performed until now to understand evolution of monocot and dicot seed protein patterning, analyzing protein profile of inaccessible regions of seeds, regulatory networks of underlying mechanism of embryogenesis and filling to modulate and outline the strategies to identify candidate seed proteins controlling the regulatory switches.

2.2 Perspect of Seed Biology

Seed is a multifaceted organ which develops from fertilized ovule and is important for plant survival, evolution and agricultural production. Strictly defined, seed development is accompanied with many distinct metabolic, cellular and physiological changes including imbibition, respiration, RNA and protein synthesis, enzyme activities within its surviving structures like endosperm, nucellus, cotyledon, teguments and components including funiculus and integument. Seed biology research has been conducted extensively due to its importance in food industry. Seeds of different genera and family of angiosperm have diverse importance being used for oil, spices, fibre, carbohydrate, fat, secondary metabolites and in brewing industry. They ensure the perpetuation of life forms and spread of the species to new areas by means of autochory, anemochory, hydrochory, and zoochory [6]. Distinct physio-chemical properties of diverse angiosperm seeds compelled the researchers to study their molecular entities including protein complement to understand cellular circuitry correlated to the morphological and physiological adaptations from germination to seedling growth.

2.3 Protein Organization of Seed Parts

In angiosperm, the female gametophyte is seated deep in the ovarian cavity far away from stigma where pollen germinates. There is sufficient evidence to suggest that gametophyte specific proteins were present during fertilization phase [7]. Further, the triploid endosperm, the most common nutritive tissue for the developing seed in angiosperm, maintains a critical protein balance with embryo and maternal tissues. Nuclear, helobial and cellular endosperm shows either symmetric or asymmetric growth that is exemplified by differences in protein patterns [8]. He et al. [9] performed comparative proteomic analysis of wheat embryo and endosperm during seed germination. The most abundant proteins both in the embryo and endosperm were found to be seed storage proteins such as legumins, vicilins and albumins. Housekeeping enzymes, actin-binding profilin, defense-related protein kinases, nonspecific lipid transfer protein and proteins involved in general metabolism were also identified. In monocot, morphologically and biochemically distinct outermost layer of endosperm namely aleurone exhibit differences in protein organization and constitutes an important accumulatory reserve tissue. Following a predetermined mode of development, fertilized egg give rise to embryo which show differences in morphology, anatomy and biochemistry both in monocot and dicots, and protein composition is no exception. Proteomic studies had begun to reveal proteins that are necessary for the events such as pattern formation, cell differentiation and organ development. Up till now approximately 1500 diverse proteins are reported to be active in dicot plants and 928 proteins in monocot plants (Table 2.1). Many of these are expressed in specific cell types and regions of seeds. For modular

Table 2.1 A comprehensive list of angiosperm seed proteome study

Clade	Family	Plant	Organ/tissue	References[a]
Dicot	Brassicaceae	Arabidopsis	Whole seed	[25]
			Whole seed	[23]
			Whole seed	[19]
			Whole seed	[21]
			Whole seed	[27]
			Whole seed	[26]
			Whole seed	[22]
			Whole seed	[17]
			Whole seed	[18]
		Camelina	Whole seed	[48]
		Castor bean	Nucellus	[45]
			Whole seed	[44]
			Whole seed	[42]
			Whole seed	[46]
		Mustard	Whole seed	[30]
			Whole seed	[49]
		Oilseed rape	Embryo	[35]
		Rapeseed	Cotyledon	[31]
			Endosperm	[34]
			Whole seed	[28]
			Whole seed	[33]
			Whole seed	[32]
			Whole seed	[47]
			Whole seed	[29]
		Rapeseed, arabidopsis	Whole seed	[43]
	Euphorbiaceae	Jatropha	Embryo and endosperm	[36]
			Embryo, endosperm	[37]
			Endosperm	[41]
			Inner integument	[39]
			Whole seed	[40]
			Whole seed	[38]
	Leguminoceae	Chickpea	Whole seed	[80]
		Common bean	Whole seed	[78]
		Lentil	Whole seed	[79]
			Cotyledons, hypocotyl	[73]
			Endosperm,	[74]
			Whole seed	[71]
			Whole seed	[72]

(continued)

Table 2.1 (continued)

Clade	Family	Plant	Organ/tissue	References[a]
		Medicago	Whole seed	[68]
			Whole seed	[66]
			Whole seed	[69]
		Medicago, black bean	Whole seed	[70]
		Mungbean	Whole seed	[81]
		Pea	Embryonic axis	[57]
			Whole seed	[75]
			Whole seed	[76]
		Peanut	Whole seed	[82]
			Whole seed and testa	[83]
		Pigeon pea	Whole seed	[84]
		Soybean	Cotyledon	[64]
			Cotyledon	[63]
			Hypocotyl, radicle	[65]
			Seed coat	[13]
			Seed coat	[12]
			Whole seed	[51]
			Whole seed	[52]
			Whole seed	[5]
			Whole seed	[60]
			Whole seed	[61]
			Whole seed	[58]
			Whole seed	[62]
			Whole seed	[59]
			Whole seed	[57]
			Whole seed	[56]
			Whole seed	[55]
			Whole seed	[54]
			Whole seed	[53]
			Whole seed	[59]
	Amaranthceae	Sugarbeet	Whole seed	[85]
	Anacardiaceae	Cashew	Cotyledon	[86]
	Cucurbitaceae	Melon	Whole seed	[89]
	Oleaceae	Olive	Whole seed	[90]
	Rosaceae	Cherry	Cotyledons, embryos, testae	[88]
	Rubiaceae	Coffee	Embryo	[92]
	Solanaceae	Tomato	Embryo, endosperm	[87]
	Theaceae	Tea	Whole seed	[93]
	Vitaceae	Grape	Endosperm	[91]

(continued)

Table 2.1 (continued)

Clade	Family	Plant	Organ/tissue	References[a]
Monocot	Poaceae	Barley	Aleurone	[150]
			Aleurone, endosperm, embryo, whole Seed	[144]
			Endosperm	[143]
			Whole seed	[154]
			Whole seed	[147]
			Whole seed	[142]
			Whole seed	[145]
			Whole seed	[148]
			Whole seed	[149]
			Whole seed	[146]
			Whole seed	[152]
			Whole seed	[151]
			Whole seed	[153]
		Maize	Embryo	[122]
			Embryo	[135]
			Embryo	[134]
			Embryo	[133]
			Embryo	[132]
			Embryo, endosperm	[139]
			Embryo, endosperm	[140]
			Embryo, endosperm	[102]
			Endosperm	[138]
			Scutellum	[136]
			Whole seed	[137]
		Rice	Embryo	[114]
			Embryo	[115]
			Embryo	[116]
			Embryo	[109]
			Embryo	[108]
			Embryo	[107]
			Embryo	[106]
			Endosperm	[110]
			Endosperm	[111]
			Endosperm	[113]
			Whole seed	[105]
			Whole seed	[104]
			Whole seed	[57]
			Whole seed	[101]
			Whole seed	[100]

(continued)

Table 2.1 (continued)

Clade	Family	Plant	Organ/tissue	References[a]
			Whole seed	[99]
			Whole seed	[97]
			Whole seed	[103]
		Wheat	Amyloplast	[128]
			Amyloplast	[127]
			Endosperm	[125]
			Endosperm	[130]
			Endosperm	[129]
			Endosperm	[126]
			Endosperm	[131]
			Whole seed	[124]
			Whole seed	[119]
			Whole seed	[120]
			Whole seed	[118]
			Whole seed	[117]
			Whole seed, embryo	[123]

[a]Based on PubMed search dated November 5, 2015

organization of seeds belonging to different clades of angiosperm, expression of proteins unique to each autonomous region must be regulated at both translational and post-translational level. During development or maturation it is typical to analyze whole seeds as the seed coats and embryonic axes make a relatively small contribution to the total mass [10]. Besides acting as a physical barrier, the seed coat has other multifunctional roles majorly in the metabolic control of seed development and dormancy, disease resistance and in nutrient metabolism from parent plant [11, 12]. In case of soybean seed coat, shotgun proteomic approach was used to identify 1372 seed coat proteins majorly involved in primary and secondary metabolism, cellular structure, stress responses, nucleic acid metabolism, protein synthesis, folding and targeting, hormone synthesis, signaling, and biogenesis of seed storage proteins (SSPs) [13].

2.4 Prototype Extraction of Seed Proteins

Being a sink organ, seeds of angiosperm have reserves of carbohydrate, fat, protein, oil, secondary metabolites, organic acids and cyclic compounds, which make the extraction of protein a daunting task. Total protein content in different angiosperm families vary and show differences in stability, activity and selectivity. Seeds were subjected to protein profiling much before the concept of proteome had emerged. Most of the seed proteomic studies have been carried out using differential biochemical extraction method for fractionation followed by resolution utilizing two dimensional gel electrophoresis (2-DE). Study of seed proteins dates back for over 270 years with the isolation of wheat gluten in 1745 by Beccari. Thereafter, systematic studies on seed proteins were carried out by Osborne in 1924 who classified seed proteins based on their solubility and extraction in a series of solvent, for example, water soluble (albumins), dilute saline soluble (globulins), alcohol soluble (prolamins) and dilute acid or alkali soluble (glutelins). More systematic seed protein extraction was carried out in 1965 that focused on the salt soluble seed proteins from pumkin by extraction through column gradient [14]. Subsequently, seed protein extraction was performed utilizing weak buffer at neutral pH and low ionic strength by Kriz [15]. These conditions were found to enrich low abundant proteins, which are otherwise masked by the presence of high abundant storage proteins. Furthermore, Kreis et al. in 1985 reported extraction of proteins in only deionised water. Improvement in the fractionation procedure underlined of seed sub-regions namely, embryo, endosperm, aleurone, scutellum, cotyledons involved dedicated extraction methods like tris-phenol, acid or alkaline extraction, two-phase separation and urea solubilization [10]. Organellar proteomes from seeds are very less studied due to unavailability of sufficient material.

2.5 Seed Proteomes: A Composite Insight

2.5.1 Seed Proteomics of Angiosperm

Evolutionary relationship between monocot and dicot seeds is an outcome of protein diversity that could be interpreted from the translational landscape obtained from proteomic approaches. Protein signatures in seeds of diverse taxas provide an integrated view based on molecular characteristics. Sub-regions of seed reveal similarities and differences in biological processes among different taxas depicting common themes in diverse genera. Unraveling the molecular basis of embryogenesis in angiosperm at protein level has laid a foundation for the rational improvement of agricultural production. Seed sub-regions have features that appear to be common in angiosperm, including conserved systems for deploying developmentally related proteins and storage strategies [16]. Studies on the mechanism of embryogenesis have revealed how different sub-regions of seed have mimicked fundamental strategies for protein communication in angiosperms. Buildup of precursor metabolites, rapid endoreduplication, growth of pod, testa and endosperm leading to increase in storage protein synthesis and cell division suggests that conserved developmental events exist among angiosperm, although difference lie at molecular, anatomical and physiological levels. These findings highlight the utility of the cross-taxa comparison of seed proteomes for rigorous dissection of fundamental components of seed developmental machinery (Fig. 2.1).

2.6 Assessment of Seed Sub-region Proteomes of Eudicot

2.6.1 Brassicaceae Seed Proteomics: Sub-region Protein Functions

In total, twenty-eight seed proteomes have been reported till date encompassing twenty-four whole seed, one each on nucellus, embryo, endosperm and cotyledon (Table 2.1). Brassicaceae is one of the most assorted angiosperm families whose seeds have varied economic importance. The relationship between most of the allied genera is of importance to understand regulatory responses of seed development among family members (Fig. 2.1).

The seed proteome collinearity of several species of Brassicaceae was compared and three major conclusions emerged. First, more than 2000 proteins, about 800 phosphoproteins, and 200 phosphopeptides were identified in Arabidopsis whole seed proteome pointing towards metabolite fluxes restructuring, polar transport of hormones, proteins related to growth and development, desiccation tolerance, germination, dormancy release, vigor alteration and responses to environmental factors. These

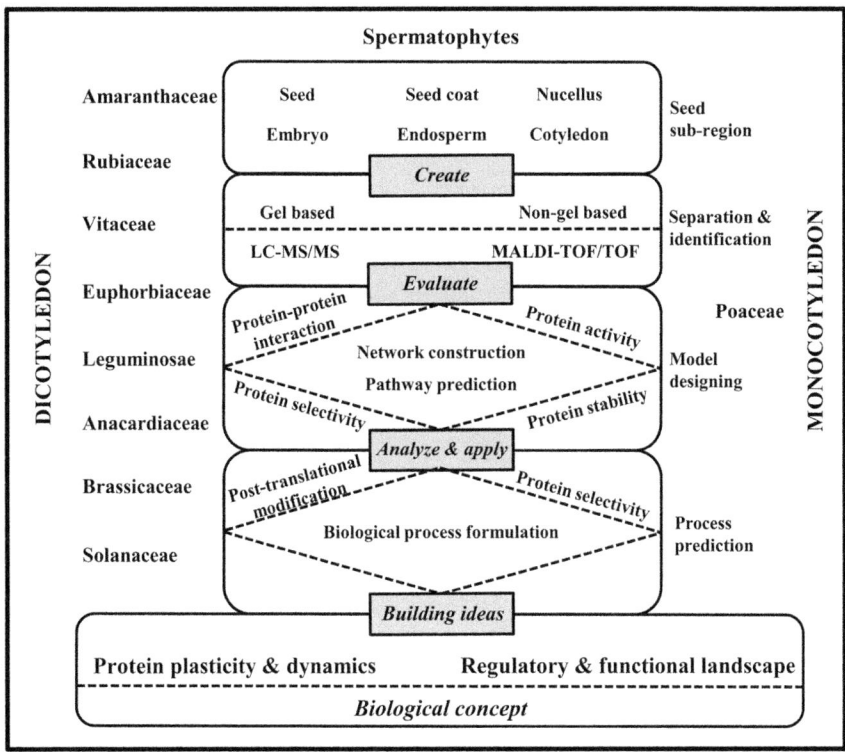

Fig. 2.1 Schematic representation of family based seed proteomics

reports greatly expands knowledge about Brassicaceae seed biology [17–27]. Secondly, around 700 proteins from whole seed, 930 proteins from endosperm and 37 proteins from cotyledons were identified from *Brassica napus* [28–33]. These studies depicted that proteins involved in genetic information processing, carbohydrate metabolism, environmental information processing, energy metabolism, cellular processes and amino acid metabolism were predominant. In endosperm proteome, proteins involved in sugar conversion and recycling, ascorbate metabolism, amino acid biosynthesis and redox balancing were detected confirming the fact that metabolite reallocation and reformation is the major functionality of endosperm during development [34]. Implications of the seed filling process and the function of the embryo were also elaborated in one of the study, where proteins involved in organogenesis, embryogenesis and development were identified [35]. Furthermore, in another study on *Brassica campestris* seed development led to the identification of 260 proteins involved in oxidation/detoxification, energy, defense, transcription, protein synthesis, transport, cell structure, signal transduction, secondary metabolism, transposition, DNA repair and storage [30]. Altogether, these studies revealed conserved protein complexes, processes and functionality amongst the studied species. Nonetheless, diversity lies in signaling and developmental strategies displaying morphological, anatomical, physiological and biochemical variations.

2.6.2 Euphorbiaceae Seed Proteomics: Shared and Distinct Proteins

The *modus operandi* in investigating the seed sub-region proteomes of Euphorbiaceae species was the availability of relevant proteomic studies conducted till date. The outcome suggested that 187 whole seed proteins, approximately 5000 endosperm associated proteins, 28 embryo specific and about 3000 integument derived proteins were detected using gel and non-gel based approaches from *Jatropha curcas* (Table 2.1) [36–41]. Data showed that majority of proteins were unique while some housekeeping proteins were common to specific sub-regions. The identified proteins revealed the predominance of protein inhibitor, metabolism, ROS regulated, transport, development and protein degradation related proteins. The creation of prognostic protein modules were used to identify specific regulators operating in the developmental circuitry responsible for coordinating biological processes. Seeds of Jatropha are prospective resource of biodiesel generation. Seed proteomic studies would laid a foundation to understand basic information on the biosynthetic pathways associated with synthesis of toxic diterpenes, fatty acids and triacylglycerols and deposition of storage proteins during seed development. These studies provide an important glimpse into the enzymatic machinery devoted to the production of carbon (C) and nitrogen (N) sources to sustain seed development and quality. Another family member, *Ricinus communis* was explored for its seed protein dynamics wherein around 2700 proteins from whole seed and 766 proteins from nucellus were identified those involved mainly in fatty acid and amino acid metabolism [35, 42–49].

2.6.3 Leguminosae Seed Proteomics: Common and Contrasting Facets

Renovating Leguminosae family for seed proteomic studies proved to be a descriptive and distinct approach for many agriculturally important legumes including pea, Medicago, soybean, mungbean, peanut, Lens, chickpea, common bean and Lotus. Till date, thirty nine seed proteomic studies comprising of whole seed, cotyledons, embryonic axis and endosperm were conducted on legume species to shed light on the mechanism that regulates seed development, morphogenesis and embryogenesis (Table 2.1). Proteomic studies on soybean seed identified 2472 whole seed proteins involved in nitrogen, carbon and lipid metabolism [50–62]. Also, 472 seed coat proteins identified from soybean showed that cell wall associated bioenergetic pathways were integrated with carbon anabolism and catabolism of fatty acids which might contribute to seed coat formation [12, 13]. Further, 328 cotyledon proteins were identified from different cultivars of soybean those associated with oxidative modification of distinct seed-stored mRNAs having role in oxidative phosphorylation, ribosome biogenesis and nutrient reservoir suggesting the significance of post-transcriptional repression of these biological processes regulating seed dormancy [63–65]. Collectively, soybean datasets provide evidence that several regulatory pathways encompassing metabolic, signal transduction and transport related protein contribute to the seed embryogenesis. Cataloging of seed proteins in another agriculturally important crop of the same family, Medicago revealed an imperative corollary, which shows that overall studies until now have been conducted only on whole seed that led to the identification of 308 proteins [66–70]. To comprehend further, Lotus seed proteomic studies when analysed showed the presence of 1500 proteins and 343 phosphoproteins predominantly involved in metabolism and signal transduction forming a regulatory hub that might be controlled by feedback loops [71–74]. Pea seed proteomics spotlight total repertoire of accumulatory and storage proteins from seed sub-regions. It is apparent that primary metabolism, secondary metabolic processes and ROS associated pathways are activated during seed development [75–77]. Unlike soybean, Medicago, Lotus and pea, seed proteomes in chickpea, common bean, Lens, mungbean and peanut are less studied [78–84]. Metabolic pathways were found to be distinct in subfamilies of legumes. Therefore, decoding seed proteome dynamics in less explored legumes are of utmost importance to understand diversity among the protein complement in this family.

2.6.4 Seed Proteomics of Less Studied Plant Families: A Way Forward

To explore less studied plant families, a defined scheme of seed proteome research were adapted that illustrate some protein variants were functionally and structurally modular and involved in developmental processes. Seed specific proteomes of cashew (Anacardiaceae), sugarbeet (Amaranthaceae), coffee (Rubiaceae), grape (Vitaceae), flax, tomato and potato (Solanaceae), Prunus (Rosaceae) and melon (Cucurbitaceae) involving climacteric and non-climacteric fruits of economic importance have been reported (Table 2.1) [85, 86]. In total seed proteomic studies of climacteric fruit tomato (788 proteins from embryo and endosperm), Prunus (1266 proteins from cotyledon, embryo, and testae), and melon (3 peptides from seed) identified an array of proteins associated with ripening, cell wall strengthening, organ development, storage reserve accumulation and embryogenesis [87–89]. Seed proteins from non-climacteric fruits namely grape (3 proteins from endosperm) and olive (231 proteins from seed) were mostly involved in seed formation and development [90, 91] comprising ascorbate peroxidase, amylase, malate dehydrogenase and triose phosphate isomerase. In contrary, seed proteomic studies from plants used for beverages viz., coffee (10 proteins from embryo) and tea (34 proteins from whole seeds) showed the presence of high levels of ROS related proteins that might alter the redox status and determine seed viability [92, 93]. Till date three reports have been published on flax seed proteomes that identified 1744 proteins involved in reorganization of seed cellular machinery during development promoting primary and secondary metabolites reallocation. In addition, oxidative homeostasis, photosynthesis, fruit quality, embryogenesis and development related proteins were also reported in flax seed proteome (Table 2.1) [94–96].

2.7 Assessment of Monocot Seed Proteomes

2.7.1 Grain Seed Proteomics: Overlapping and Unique Proteins

To address key protein complement of monocot seeds, Poaceae family which represents the most extensively investigated members have been used for assessing grain proteome. Of the total 53 proteomic studies on seed sub-regions, rice and wheat having twenty and seventeen reports formed the predominantly studied members whereas fifteen and eleven reports were from barley and maize, respectively (Table 2.1). When rice seed proteomes were explored, twenty one proteomic studies including whole seed (ten reports), embryo (three reports) and endosperm (eight reports) revealed reorganization of protein pool during various developmental stages. 2186 differentially expressed whole seed proteins were found to be involved in central metabolic or regulatory pathways, including carbohydrate metabolism (especially cell wall synthesis) and protein synthesis, folding and degradation. These provide proteomic confirmation of the notion that seed formation and development involves diverse but delicately regulated pathways [97–105]. Our inclusive analysis of endosperm (408 proteins) and embryo (1656 proteins) proteomics of rice, a valuable proteomic resource highlight characterization of pathways contributing to organ development and embryogenesis at molecular and biochemical levels [106–116]. Phosphorylation, a well-studied post-translational modification is fundamental in the signal transduction cascades during histodifferentiation and embryogenesis. We observed that till date only one report illustrates the role of phosphorylation in embryo formation in rice. The study identified 168 phosphoproteins elucidating the involvement of biomolecular signaling and hormonal interplay during cell division, differentiation and delineation of rice embryo. A glance at wheat seed proteomics yielded a total of 2327 whole seed proteins from ten separate studies (Table 2.1) [3, 46, 117–124]. Wheat seed sub-region proteome have identified 2789 endosperm specific proteins and 63 embryo related proteins [9, 125–131]. Investigating functional and regulatory context of wheat seed proteome research paved the way to understand sink tissue biology of a polyploid crop. Data showed that differentially expressed proteins from embryo were mainly related to carbohydrate metabolism, amino acid metabolism, nucleic acid metabolism and stress-related proteins; whereas those from the endosperm were mainly involved in protein storage, carbohydrate metabolism, protein inhibitors, stress response, and protein synthesis. Translational changes of wheat whole seed display distinct differentially expressed proteins and their synergistic expression provide a mechanistic basis for the normal germination of wheat seeds. Analysis of maize seed proteomes yielded eleven reports including whole seed (one report), embryo (five reports), embryo and endosperm (three reports), endosperm (one report) and scutellum (one report) till date (Table 2.1) [102, 122, 132–140]. Furthermore, 2809 embryo associated proteins involved in organ development, transportation, amino acid metabolism, defense response, molecular chaperone function, protein synthesis, proteolysis, secondary metabolism and signal transduction have been catalogued using gel and non-gel based proteome analysis. Additionally, 183 endosperm related proteins were identified involved in storage, C and N recycling and biogenesis. Molecular basis of seed development of maize was elucidated from whole seed proteotypes that resolved 4511 proteins. Proteins involved in turgor pressure generation, energy metabolism, secondary metabolism, protein synthesis and oxidative burst were identified in different maize genotypes. Further, fifteen seed proteomic studies of barley has identified 423 proteins from whole seed associated with photosynthesis and energy metabolism, carbohydrate metabolism, protein degradation and defense (Table 2.1). Also, 168 proteins from aleurone, embryo and endosperm belongs to diverse functional categories such as metabolite allocation, carbohydrate metabolism, amino acid metabolism, defense response, protein folding and stabilization and oxidative stress tolerance [141–154]. A vast array of monocot seed proteins add diversity to the seed biological properties leading to exclusivity and specificity in cellular processes. Therefore, characterization of the seed sub-region proteome holds promise of increasing understanding about the regulation of genes and their function in diverse monocot genera.

2.8 Conclusion

Protein expression of recalcitrant or orthodox seed and climacteric or non-climacteric seed belonging to diverse families of angiosperm revealed that the enzymatic machinery initiate germination during the maturation phase is a common theme. The extent to which the functionally active translational machinery would synthesize protein at different stages of seed development differs between different family, genera and species. Protein content and profiles of the seed tissues (cotyledon, embryonic axis and endosperm) were reasonably diverse in different families. The embryonic axis showed proteins related to cell division, histodifferentiation, organogenesis and embryogenesis. Majority of endosperm related proteins, being classified according to their function into major group primarily involved in macronutrient metabolism, metabolite accumulation and assimilation. Cotyledon showed higher number of metabolic and storage related proteins compared to embryonic axis. The tegument, aleurone, nucellus and scutellum presented the largest number of the transport, signaling and cytoskeleton proteins. Protein patterning is in agreement with the biological role of the tissues. Studies on seed tissue and sub-region proteome confirmed a compartmentalization of biological pathways and a partition of metabolic fluxes between different regions of seed. This partition and compartmentalization uncovered the divergence and particularities of the seeds from different families.

Acknowledgments This research work was supported by grants from the Department of Biotechnology (DBT), Ministry of Science and Technology, Govt. of India (Grant No. BT/PR12919/AGR/02/676/2009, BT/AGR/CG-Phase II/01/2014) and the National Institute of Plant Genome Research, India. K. N. and A. S. are the recipient of pre-doctoral fellowship from Council of Scientific and Industrial Research (CSIR), Govt. of India. T. H. is the recipient of senior research associate fellowship from Council of Scientific and Industrial Research (CSIR), Govt. of India.

References

1. Lord JM, Westoby M (2012) Accessory costs of seed production and the evolution of angiosperms. Evolution 66:200–210
2. Dekkers BJ, Pearce S, Van Bolderen-Veldkamp R, Marshall A, Widera P, Gilbert J et al (2013) Transcriptional dynamics of two seed compartments with opposing roles in Arabidopsis seed germination. Plant Physiol 163:205–215
3. Young ND, Debelle F, Oldroyd GE, Geurts R, Cannon SB, Udvardi MK et al (2011) The Medicago genome provides insight into the evolution of rhizobial symbioses. Nature 480:520–524

4. Narula K, Pandey A, Gayali S, Chakraborty N, Chakraborty S (2015) Birth of plant proteomics in India: a new horizon. J Proteomics 127:34–43
5. Krishnan HB, Oehrle NW, Natarajan SS (2009) A rapid and simple procedure for the depletion of abundant storage proteins from legume seeds to advance proteome analysis: a case study using Glycine max. Proteomics 9:3174–3188
6. Nathan R, Schurr FM, Spiegel O, Steinitz O, Trakhtenbrot A, Tsoar A (2008) Mechanisms of long-distance seed dispersal. Trends Ecol Evol 23:638–647
7. Mascarenhas JP (1975) The biochemistry of angiosperm pollen development. Bot Rev 41:259–314
8. Tebbji F, Nantel A, Matton DP (2010) Transcription profiling of fertilization and early seed development events in a solanaceous species using a 7.7 K cDNA microarray from Solanum chacoense ovules. BMC Plant Biol 10:174
9. He M, Zhu C, Dong K, Zhang T, Cheng Z, Li J et al (2015) Comparative proteome analysis of embryo and endosperm reveals central differential expression proteins involved in wheat seed germination. BMC Plant Biol 15:1
10. Miernyk JA, Hajduch M (2011) Seed proteomics. J Proteomics 74:389–400
11. Weber H, Borisjuk L, Wobus U (2005) Molecular physiology of legume seed development. Annu Rev Plant Biol 56:253–279
12. Gupta R, Min CW, Kim SW, Wang Y, Agrawal GK, Rakwal R et al (2015) Comparative investigation of seed coats of brown-versus yellow-colored soybean seeds using an integrated proteomics and metabolomics approach. Proteomics 15:1706–1716
13. Miernyk JA, Johnston ML (2013) Proteomic analysis of the testa from developing soybean seeds. J Proteomics 89:265–272
14. Alekseeva M (1964) [Investigation of salt-soluble proteins of pumpkin seeds (*Cucurbita pepo* L.) by a column gradient extraction method]. Biokhimiia (Moscow, Russia) 30:60–66
15. Kriz AL (1989) Characterization of embryo globulins encoded by the maize Glb genes. Biochem Genet 27:239–251
16. Deng ZY, Gong CY, Wang T (2013) Use of proteomics to understand seed development in rice. Proteomics 13:1784–1800
17. Gallardo K, Job C, Groot SP, Puype M, Demol H, Vandekerckhove J et al (2001) Proteomic analysis of Arabidopsis seed germination and priming. Plant Physiol 126:835–848
18. Gallardo K, Job C, Groot SP, Puype M, Demol H, Vandekerckhove J et al (2002) Proteomics of Arabidopsis seed germination. A comparative study of wild-type and gibberellin-deficient seeds. Plant Physiol 129:823–837
19. Chibani K, Ali-Rachedi S, Job C, Job D, Jullien M, Grappin P (2006) Proteomic analysis of seed dormancy in Arabidopsis. Plant Physiol 142:1493–1510
20. Fait A, Angelovici R, Less H, Ohad I, Urbanczyk-Wochniak E, Fernie AR et al (2006) Arabidopsis seed development and germination is associated with temporally distinct metabolic switches. Plant Physiol 142:839–854
21. Higashi Y, Hirai MY, Fujiwara T, Naito S, Noji M, Saito K (2006) Proteomic and transcriptomic analysis of Arabidopsis seeds: molecular evidence for successive processing of seed proteins and its implication in the stress response to sulfur nutrition. Plant J 48:557–571
22. Rajjou L, Belghazi M, Huguet R, Robin C, Moreau A, Job C et al (2006) Proteomic investigation of the effect of salicylic acid on Arabidopsis seed germination and establishment of early defense mechanisms. Plant Physiol 141:910–923
23. Chen M, Mooney BP, Hajduch M, Joshi T, Zhou M, Xu D et al (2009) System analysis of an Arabidopsis mutant altered in de novo fatty acid synthesis reveals diverse changes in seed composition and metabolism. Plant Physiol 150:27–41
24. Hajduch M, Hearne LB, Miernyk JA, Casteel JE, Joshi T, Agrawal GK et al (2010) Systems analysis of seed filling in Arabidopsis: using general linear modeling to assess concordance of transcript and protein expression. Plant Physiol 152:2078–2087
25. Arc E, Chibani K, Grappin P, Jullien M, Godin B, Cueff G et al (2012) Cold stratification and exogenous nitrates entail similar functional proteome adjustments during Arabidopsis seed dormancy release. J Proteome Res 11:5418–5432

26. Galland M, Huguet R, Arc E, Cueff G, Job D, Rajjou L (2014) Dynamic proteomics emphasizes the importance of selective mRNA translation and protein turnover during Arabidopsis seed germination. Mol Cell Proteomics 13:252–268
27. Nguyen TP, Cueff G, Hegedus DD, Rajjou L, Bentsink L (2015) A role for seed storage proteins in Arabidopsis seed longevity. J Exp Bot 66:6399–6413
28. Hajduch M, Casteel JE, Hurrelmeyer KE, Song Z, Agrawal GK, Thelen JJ (2006) Proteomic analysis of seed filling in Brassica napus. Developmental characterization of metabolic isozymes using high-resolution two-dimensional gel electrophoresis. Plant Physiol 141:32–46
29. Jolivet P, Boulard C, Bellamy A, Valot B, D'andréa S, Zivy M et al (2011) Oil body proteins sequentially accumulate throughout seed development in *Brassica napus*. J Plant Physiol 168:2015–2020
30. Li W, Gao Y, Xu H, Zhang Y, Wang J (2012) A proteomic analysis of seed development in *Brassica campestri* L. PLoS ONE 7:e50290
31. Garg H, Li H, Sivasithamparam K, Barbetti MJ (2013) Differentially expressed proteins and associated histological and disease progression changes in cotyledon tissue of a resistant and susceptible genotype of *Brassica napus* infected with *Sclerotinia sclerotiorum*. PLoS ONE 8: e65205
32. Kubala S, Garnczarska M, Wojtyla L, Clippe A, Kosmala A, Zmienko A et al (2015) Deciphering priming-induced improvement of rapeseed (*Brassica napus* L.) germination through an integrated transcriptomic and proteomic approach. Plant Sci 231:94–113
33. Yin X, He D, Gupta R, Yang P (2015) Physiological and proteomic analyses on artificially aged *Brassica napus* seed. Front Plant Sci 6:112
34. Lorenz C, Rolletschek H, Sunderhaus S, Braun HP (2014) Brassica napus seed endosperm—metabolism and signaling in a dead end tissue. J Proteomics 108:382–426
35. Borisjuk L, Neuberger T, Schwender J, Heinzel N, Sunderhaus S, Fuchs J et al (2013) Seed architecture shapes embryo metabolism in oilseed rape. Plant Cell 25:1625–1640
36. Liu H, Liu YJ, Yang MF, Shen SH (2009) A comparative analysis of embryo and endosperm proteome from seeds of *Jatropha curcas*. J Integr Plant Biol 51:850–857
37. Liu H, Yang Z, Yang M, Shen S (2011) The differential proteome of endosperm and embryo from mature seed of *Jatropha curcas*. Plant Sci 181:660–666
38. Liu H, Wang C, Komatsu S, He M, Liu G, Shen S (2013) Proteomic analysis of the seed development in Jatropha curcas: from carbon flux to the lipid accumulation. J Proteomics 91:23–40
39. Soares EL, Shah M, Soares AA, Costa JH, Carvalho P, Domont GB et al (2014) Proteome analysis of the inner integument from developing *Jatropha curcas* L. seeds. J Proteome Res 13:3562–3570
40. Liu H, Wang C, Chen F, Shen S (2015) Proteomic analysis of oil bodies in mature *Jatropha curcas* seeds with different lipid content. J Proteomics 113:403–414
41. Shah M, Soares EL, Carvalho PC, Soares AA, Domont GB, Nogueira FC et al (2015) Proteomic analysis of the endosperm ontogeny of *Jatropha curcas* L. seeds. J Proteome Res 14:2557–2568
42. Houston NL, Hajduch M, Thelen JJ (2009) Quantitative proteomics of seed filling in castor: comparison with soybean and rapeseed reveals differences between photosynthetic and nonphotosynthetic seed metabolism. Plant Physiol 151:857–868
43. Meyer LJ, Gao J, Xu D, Thelen JJ (2012) Phosphoproteomic analysis of seed maturation in Arabidopsis, rapeseed, and soybean. Plant Physiol 159:517–528
44. Nogueira FC, Palmisano G, Schwammle V, Campos FA, Larsen MR, Domont GB et al (2012) Performance of isobaric and isotopic labeling in quantitative plant proteomics. J Proteome Res 11:3046–3052
45. Nogueira FC, Palmisano G, Soares EL, Shah M, Soares AA, Roepstorff P et al (2012) Proteomic profile of the nucellus of castor bean (*Ricinus communis* L.) seeds during development. J Proteomics 75:1933–1939

46. Nogueira FC, Palmisano G, Schwammle V, Soares EL, Soares AA, Roepstorff P et al (2013) Isotope labeling-based quantitative proteomics of developing seeds of castor oil seed (*Ricinus communis* L.). J Proteome Res 12:5012–5024
47. Puumalainen TJ, Puustinen A, Poikonen S, Turjanmaa K, Palosuo T, Vaali K (2015) Proteomic identification of allergenic seed proteins, napin and cruciferin, from cold-pressed rapeseed oils. Food Chem 175:381–385
48. Alvarez S, Roy Choudhury S, Sivagnanam K, Hicks LM, Pandey S (2015) Quantitative proteomics analysis of Camelina sativa seeds overexpressing the AGG3 gene to identify the proteomic basis of increased yield and stress tolerance. J Proteome Res 14:2606–2616
49. Hummel M, Wigger T, Brockmeyer J (2015) Characterization of mustard 2S albumin allergens by bottom-up, middle-down, and top-down proteomics: a consensus set of isoforms of Sin a 1. J Proteome Res 14:1547–1556
50. Senescence L, Grabau LJ, Blevins G, Minor C (1986) P nutrition during seed development. Plant Physiol 82:1008–1012
51. Hajduch M, Ganapathy A, Stein JW, Thelen JJ (2005) A systematic proteomic study of seed filling in soybean. Establishment of high-resolution two-dimensional reference maps, expression profiles, and an interactive proteome database. Plant Physiol 137:1397–1419
52. Natarajan SS, Xu C, Bae H, Caperna TJ, Garrett WM (2006) Characterization of storage proteins in wild (*Glycine soja*) and cultivated (*Glycine max*) soybean seeds using proteomic analysis. J Agric Food Chem 54:3114–3120
53. Danchenko M, Skultety L, Rashydov NM, Berezhna VV, Mátel LU, Salaj TZ et al (2009) Proteomic analysis of mature soybean seeds from the Chernobyl area suggests plant adaptation to the contaminated environment. J Proteome Res 8:2915–2922
54. Natarajan SS, Krishnan HB, Lakshman S, Garrett WM (2009) An efficient extraction method to enhance analysis of low abundant proteins from soybean seed. Anal Biochem 394:259–268
55. Brandão A, Barbosa H, Arruda M (2010) Image analysis of two-dimensional gel electrophoresis for comparative proteomics of transgenic and non-transgenic soybean seeds. J Proteomics 73:1433–1440
56. Kim HT, Choi U-K, Ryu HS, Lee SJ, Kwon O-S (2011) Mobilization of storage proteins in soybean seed (*Glycine max* L.) during germination and seedling growth. Biochimica et Biophysica Acta (BBA)-Proteins and Proteomics 1814:1178–1187
57. Wang WQ, Moller IM, Song SQ (2012) Proteomic analysis of embryonic axis of *Pisum sativum* seeds during germination and identification of proteins associated with loss of desiccation tolerance. J Proteomics 77:68–86
58. Gomes LS, Senna R, Sandim V, Silva-Neto MRA, Perales JE, Zingali RB et al (2014) Four conventional soybean [*Glycine max* (L.) Merrill] seeds exhibit different protein profiles as revealed by proteomic analysis. J Agric Food Chem 62:1283–1293
59. Han C, Yin X, He D, Yang P (2013) Analysis of proteome profile in germinating soybean seed, and its comparison with rice showing the styles of reserves mobilization in different crops. PLoS ONE 8:e56947
60. Capriotti AL, Caruso G, Cavaliere C, Samperi R, Stampachiacchiere S, Zenezini Chiozzi R et al (2014) Protein profile of mature soybean seeds and prepared soybean milk. J Agric Food Chem 62:9893–9899
61. Mataveli LR, Arruda MA (2014) Expanding resolution of metalloprotein separations from soybean seeds using 2D-HPLC-ICP-MS and SDS-PAGE as a third dimension. J Proteomics 104:94–103
62. Smith-Hammond CL, Swatek KN, Johnston ML, Thelen JJ, Miernyk JA (2014) Initial description of the developing soybean seed protein Lys-N(epsilon)-acetylome. J Proteomics 96:56–66
63. Komatsu S, Makino T, Yasue H (2013) Proteomic and biochemical analyses of the cotyledon and root of flooding-stressed soybean plants. PLoS ONE 8:e65301

64. Kamal AHM, Rashid H, Sakata K, Komatsu S (2015) Gel-free quantitative proteomic approach to identify cotyledon proteins in soybean under flooding stress. J Proteomics 112:1–13
65. Yin Y, Yang R, Gu Z (2014) Organ-specific proteomic analysis of NaCl-stressed germinating soybeans. J Agric Food Chem 62:7233–7244
66. Gallardo K, Le Signor C, Vandekerckhove J, Thompson RD, Burstin J (2003) Proteomics of *Medicago truncatula* seed development establishes the time frame of diverse metabolic processes related to reserve accumulation. Plant Physiol 133:664–682
67. Djemel N, Guedon D, Lechevalier A, Salon C, Miquel M, Prosperi J-M et al (2005) Development and composition of the seeds of nine genotypes of the *Medicago truncatula* species complex. Plant Physiol Biochem 43:557–566
68. Boudet J, Buitink J, Hoekstra FA, Rogniaux H, Larré C, Satour P et al (2006) Comparative analysis of the heat stable proteome of radicles of *Medicago truncatula* seeds during germination identifies late embryogenesis abundant proteins associated with desiccation tolerance. Plant Physiol 140:1418–1436
69. Gallardo K, Firnhaber C, Zuber H, Héricher D, Belghazi M, Henry C et al (2007) A combined proteome and transcriptome analysis of developing *Medicago truncatula* seeds evidence for metabolic specialization of maternal and filial tissues. Mol Cell Proteomics 6:2165–2179
70. Delahaie J, Hundertmark M, Bove J, Leprince O, Rogniaux H, Buitink J (2013) LEA polypeptide profiling of recalcitrant and orthodox legume seeds reveals ABI3-regulated LEA protein abundance linked to desiccation tolerance. J Exp Bot 64:4559–4573
71. Dam S, Laursen BS, Ørnfelt JH, Jochimsen B, Stærfeldt HH, Friis C et al (2009) The proteome of seed development in the model legume *Lotus japonicus*. Plant Physiol 149:1325–1340
72. Nautrup-Pedersen G, Dam S, Laursen BS, Siegumfeldt AL, Nielsen K, Goffard N et al (2010) Proteome analysis of pod and seed development in the model legume *Lotus japonicus*. J Proteome Res 9:5715–5726
73. Ino Y, Ishikawa A, Nomura A, Kajiwara H, Harada K, Hirano H (2014) Phosphoproteome analysis of *Lotus japonicus* seeds. Proteomics 14:116–120
74. Moro CF, Fukao Y, Shibato J, Rakwal R, Timperio AM, Zolla L et al (2015) Unraveling the seed endosperm proteome of the lotus (Nelumbo nucifera Gaertn.) utilizing 1DE and 2DE separation in conjunction with tandem mass spectrometry. Proteomics 15:1717–1735
75. Schiltz S, Gallardo K, Huart M, Negroni L, Sommerer N, Burstin J (2004) Proteome reference maps of vegetative tissues in pea. An investigation of nitrogen mobilization from leaves during seed filling. Plant Physiol 135:2241–2260
76. Bourgeois M, Jacquin F, Savois V, Sommerer N, Labas V, Henry C et al (2009) Dissecting the proteome of pea mature seeds reveals the phenotypic plasticity of seed protein composition. Proteomics 9:254–271
77. Barac M, Cabrilo S, Pesic M, Stanojevic S, Zilic S, Macej O et al (2010) Profile and functional properties of seed proteins from six pea (*Pisum sativum*) genotypes. Int J Mol Sci 11:4973–4990
78. De La Fuente M, Borrajo A, Bermúdez J, Lores M, Alonso J, López M et al (2011) 2-DE-based proteomic analysis of common bean (*Phaseolus vulgaris* L.) seeds. J Proteomics 74:262–267
79. Ialicicco M, Viscosi V, Arena S, Scaloni A, Trupiano D, Rocco M et al (2012) Lens culinaris Medik. Seed proteome: analysis to identify landrace markers. Plant Sci 197:1–9
80. Vessal S, Siddique KH, Atkins CA (2012) Comparative proteomic analysis of genotypic variation in germination and early seedling growth of chickpea under suboptimal soil-water conditions. J Proteome Res 11:4289–4307
81. Kazłowski B, Chen M-R, Chao P-M, Lai C-C, Ko Y-T (2013) Identification and roles of proteins for seed development in mungbean (*Vigna radiata* L.) seed proteomes. J Agric Food Chem 61:6650–6659

82. Kottapalli KR, Zabet-Moghaddam M, Rowland D, Faircloth W, Mirzaei M, Haynes PA et al (2013) Shotgun label-free quantitative proteomics of water-deficit-stressed midmature peanut (*Arachis hypogaea* L.) seed. J Proteome Res 12:5048–5057
83. White BL, Gokce E, Nepomuceno AI, Muddiman DC, Sanders TH, Davis JP (2013) Comparative proteomic analysis and IgE binding properties of peanut seed and testa (skin). J Agric Food Chem 61:3957–3968
84. Swathi M, Lokya V, Swaroop V, Mallikarjuna N, Kannan M, Dutta-Gupta A et al (2014) Structural and functional characterization of proteinase inhibitors from seeds of *Cajanus cajan* (cv. ICP 7118). Plant Physiol Biochem 83:77–87
85. Catusse J, Strub J-M, Job C, Van Dorsselaer A, Job D (2008) Proteome-wide characterization of sugarbeet seed vigor and its tissue specific expression. Proc Natl Acad Sci 105:10262–10267
86. Ponte LFA, Silva ALCD, Carvalho FEL, Maia JM, Voigt EL, JaG Silveira (2014) Salt-induced delay in cotyledonary globulin mobilization is abolished by induction of proteases and leaf growth sink strength at late seedling establishment in cashew. J Plant Physiol 171:1362–1371
87. Sheoran IS, Olson DJ, Ross AR, Sawhney VK (2005) Proteome analysis of embryo and endosperm from germinating tomato seeds. Proteomics 5:3752–3764
88. Lee CS, Chien CT, Lin CH, Chiu YY, Yang YS (2006) Protein changes between dormant and dormancy-broken seeds of *Prunus campanulata* maxim. Proteomics 6:4147–4154
89. Priyanto AD, Doerksen RJ, Chang CI, Sung WC, Widjanarko SB, Kusnadi J et al (2015) Screening, discovery, and characterization of angiotensin-I converting enzyme inhibitory peptides derived from proteolytic hydrolysate of bitter melon seed proteins. J Proteomics 128:424–435
90. Esteve C, D'amato A, Marina ML, García MC, Citterio A, Righetti PG (2012) Identification of olive (*Olea europaea*) seed and pulp proteins by nLC-MS/MS via combinatorial peptide ligand libraries. J Proteomics 75:2396–2403
91. Gazzola D, Vincenzi S, Gastaldon L, Tolin S, Pasini G, Curioni A (2014) The proteins of the grape (*Vitis vinifera* L.) seed endosperm: fractionation and identification of the major components. Food Chem 155:132–139
92. Franco OL, Pelegrini PB, Gomes CP, Souza A, Costa FT, Domont G et al (2009) Proteomic evaluation of coffee zygotic embryos in two different stages of seed development. Plant Physiol Biochem 47:1046–1050
93. Chen Q, Yang L, Ahmad P, Wan X, Hu X (2011) Proteomic profiling and redox status alteration of recalcitrant tea (*Camellia sinensis*) seed in response to desiccation. Planta 233:583–592
94. Klubicová KN, Danchenko M, Skultety L, Miernyk JNA, Rashydov NM, Berezhna VV et al (2010) Proteomics analysis of flax grown in Chernobyl area suggests limited effect of contaminated environment on seed proteome. Environ Sci Technol 44:6940–6946
95. Barvkar VT, Pardeshi VC, Kale SM, Kadoo NY, Giri AP, Gupta VS (2012) Proteome profiling of flax (*Linum usitatissimum*) seed: characterization of functional metabolic pathways operating during seed development. J Proteome Res 11:6264–6276
96. Renouard S, Cyrielle C, Lopez T, Lamblin F, Laine E, Hano C (2012) Isolation of nuclear proteins from flax (*Linum usitatissimum* L.) seed coats for gene expression regulation studies. BMC Res Notes 5:15
97. Yang P, Li X, Wang X, Chen H, Chen F, Shen S (2007) Proteomic analysis of rice (*Oryza sativa*) seeds during germination. Proteomics 7:3358–3368
98. Xu SB, Li T, Deng ZY, Chong K, Xue Y, Wang T (2008) Dynamic proteomic analysis reveals a switch between central carbon metabolism and alcoholic fermentation in rice filling grains. Plant Physiol 148:908–925
99. Doroshenk KA, Crofts AJ, Morris RT, Wyrick JJ, Okita TW (2009) Proteomic analysis of cytoskeleton-associated RNA binding proteins in developing rice seed. J Proteome Res 8:4641–4653

100. He D, Han C, Yao J, Shen S, Yang P (2011) Constructing the metabolic and regulatory pathways in germinating rice seeds through proteomic approach. Proteomics 11:2693–2713
101. Sano N, Permana H, Kumada R, Shinozaki Y, Tanabata T, Yamada T et al (2012) Proteomic analysis of embryonic proteins synthesized from long-lived mRNAs during germination of rice seeds. Plant Cell Physiol 53:687–698
102. Wang YD, Wang X, Ngai SM, Wong YS (2013) Comparative proteomics analysis of selenium responses in selenium-enriched rice grains. J Proteome Res 12:808–820
103. Liao JL, Zhou HW, Zhang HY, Zhong PA, Huang YJ (2014) Comparative proteomic analysis of differentially expressed proteins in the early milky stage of rice grains during high temperature stress. J Exp Bot 65:655–671
104. Lin Z, Zhang X, Yang X, Li G, Tang S, Wang S et al (2014) Proteomic analysis of proteins related to rice grain chalkiness using iTRAQ and a novel comparison system based on a notched-belly mutant with white-belly. BMC Plant Biol 14:1–17
105. Qian D, Tian L, Qu L (2015) Proteomic analysis of endoplasmic reticulum stress responses in rice seeds. Sci Rep 5:14255
106. Komatsu S, Abbasi F, Kobori E, Fujisawa Y, Kato H, Iwasaki Y (2005) Proteomic analysis of rice embryo: An approach for investigating Gα protein-regulated proteins. Proteomics 5:3932–3941
107. Kim ST, Kang SY, Wang Y, Kim SG, Hwang DH, Kang KY (2008) Analysis of embryonic proteome modulation by GA and ABA from germinating rice seeds. Proteomics 8:3577–3587
108. Wang W, Meng B, Ge X, Song S, Yang Y, Yu X et al (2008) Proteomic profiling of rice embryos from a hybrid rice cultivar and its parental lines. Proteomics 8:4808–4821
109. Kim ST, Wang Y, Kang SY, Kim SG, Rakwal R, Kim YC et al (2009) Developing rice embryo proteomics reveals essential role for embryonic proteins in regulation of seed germination. J Proteome Res 8:3598–3605
110. Xu SB, Yu HT, Yan LF, Wang T (2010) Integrated proteomic and cytological study of rice endosperms at the storage phase. J Proteome Res 9:4906–4918
111. Yu HT, Xu SB, Zheng CH, Wang T (2012) Comparative proteomic study reveals the involvement of diurnal cycle in cell division, enlargement, and starch accumulation in developing endosperm of *Oryza sativa*. J Proteome Res 11:359–371
112. Xu H, Zhang W, Gao Y, Zhao Y, Guo L, Wang J (2012) Proteomic analysis of embryo development in rice (*Oryza sativa*). Planta 235:687–701
113. Nallamilli BR, Zhang J, Mujahid H, Malone BM, Bridges SM, Peng Z (2013) Polycomb group gene OsFIE2 regulates rice (*Oryza sativa*) seed development and grain filling via a mechanism distinct from Arabidopsis. PLoS Genet 9:e1003322
114. Han C, He D, Li M, Yang P (2014) In-depth proteomic analysis of rice embryo reveals its important roles in seed germination. Plant Cell Physiol pcu114
115. Han C, Wang K, Yang P (2014) Gel-based comparative phosphoproteomic analysis on rice embryo during germination. Plant Cell Physiol 55:1376–1394
116. Han C, Yang P, Sakata K, Komatsu S (2014) Quantitative proteomics reveals the role of protein phosphorylation in rice embryos during early stages of germination. J Proteome Res 13:1766–1782
117. Laino P, Shelton D, Finnie C, De Leonardis AM, Mastrangelo AM, Svensson B et al (2010) Comparative proteome analysis of metabolic proteins from seeds of durum wheat (cv. Svevo) subjected to heat stress. Proteomics 10:2359–2368
118. Nadaud I, Girousse C, Debiton C, Chambon C, Bouzidi MF, Martre P et al (2010) Proteomic and morphological analysis of early stages of wheat grain development. Proteomics 10:2901–2910
119. Bykova NV, Hoehn B, Rampitsch C, Banks T, Stebbing JA, Fan T et al (2011) Redox-sensitive proteome and antioxidant strategies in wheat seed dormancy control. Proteomics 11:865–882

120. Tasleem-Tahir A, Nadaud I, Girousse C, Martre P, Marion D, Branlard G (2011) Proteomic analysis of peripheral layers during wheat (*Triticum aestivum* L.) grain development. Proteomics 11:371–379
121. Guo G, Lv D, Yan X, Subburaj S, Ge P, Li X et al (2012) Proteome characterization of developing grains in bread wheat cultivars (*Triticum aestivum* L.). BMC Plant Biol 12:1
122. Guo B, Chen Y, Zhang G, Xing J, Hu Z, Feng W et al (2013) Comparative proteomic analysis of embryos between a maize hybrid and its parental lines during early stages of seed germination. PLoS ONE 8:e65867
123. Fercha A, Capriotti AL, Caruso G, Cavaliere C, Samperi R, Stampachiacchiere S et al (2014) Comparative analysis of metabolic proteome variation in ascorbate-primed and unprimed wheat seeds during germination under salt stress. J Proteomics 108:238–257
124. Ma C, Zhou J, Chen G, Bian Y, Lv D, Li X et al (2014) iTRAQ-based quantitative proteome and phosphoprotein characterization reveals the central metabolism changes involved in wheat grain development. BMC Genom 15:1029
125. Wong JH, Balmer Y, Cai N, Tanaka CK, Vensel WH, Hurkman WJ et al (2003) Unraveling thioredoxin-linked metabolic processes of cereal starchy endosperm using proteomics. FEBS Lett 547:151–156
126. Vensel WH, Tanaka CK, Cai N, Wong JH, Buchanan BB, Hurkman WJ (2005) Developmental changes in the metabolic protein profiles of wheat endosperm. Proteomics 5:1594–1611
127. Balmer Y, Vensel WH, Dupont FM, Buchanan BB, Hurkman WJ (2006) Proteome of amyloplasts isolated from developing wheat endosperm presents evidence of broad metabolic capability. J Exp Bot 57:1591–1602
128. Dupont FM (2008) Metabolic pathways of the wheat (*Triticum aestivum*) endosperm amyloplast revealed by proteomics. BMC Plant Biol 8:1
129. Merlino M, Bousbata S, Svensson B, Branlard G (2012) Proteomic and genetic analysis of wheat endosperm albumins and globulins using deletion lines of cultivar Chinese Spring. Theor Appl Genet 125:1433–1448
130. Suliman M, Chateigner-Boutin AL, Francin-Allami M, Partier A, Bouchet B, Salse J et al (2013) Identification of glycosyltransferases involved in cell wall synthesis of wheat endosperm. J Proteomics 78:508–521
131. Tasleem-Tahir A, Nadaud I, Chambon C, Branlard G (2012) Expression profiling of starchy endosperm metabolic proteins at 21 stages of wheat grain development. J Proteome Res 11:2754–2773
132. Campo S, Carrascal M, Coca M, Abián J, San Segundo B (2004) The defense response of germinating maize embryos against fungal infection: a proteomics approach. Proteomics 4:383–396
133. Lu TC, Meng LB, Yang CP, Liu GF, Liu GJ, Ma W et al (2008) A shotgun phosphoproteomics analysis of embryos in germinated maize seeds. Planta 228:1029–1041
134. Fu Z, Jin X, Ding D, Li Y, Fu Z, Tang J (2011) Proteomic analysis of heterosis during maize seed germination. Proteomics 11:1462–1472
135. Huang H, Møller IM, Song S-Q (2012) Proteomics of desiccation tolerance during development and germination of maize embryos. J Proteomics 75:1247–1262
136. Tnani H, Lopez I, Jouenne T, Vicient CM (2012) Quantitative subproteomic analysis of germinating related changes in the scutellum oil bodies of *Zea mays*. Plant Sci 191–192:1–7
137. Walley JW, Shen Z, Sartor R, Wu KJ, Osborn J, Smith LG et al (2013) Reconstruction of protein networks from an atlas of maize seed proteotypes. Proc Natl Acad Sci U S A 110: E4808–4817
138. Silva-Sanchez C, Chen S, Li J, Chourey PS (2014) A comparative glycoproteome study of developing endosperm in the hexose-deficient miniature1 (mn1) seed mutant and its wild type Mn1 in maize. Front Plant Sci 5:63
139. Decourcelle M, Perez-Fons L, Baulande S, Steiger S, Couvelard L, Hem S et al (2015) Combined transcript, proteome, and metabolite analysis of transgenic maize seeds engineered

for enhanced carotenoid synthesis reveals pleotropic effects in core metabolism. J Exp Bot 66:3141–3150
140. Wu X, Gong F, Yang L, Hu X, Tai F, Wang W (2014) Proteomic analysis reveals differential accumulation of small heat shock proteins and late embryogenesis abundant proteins between ABA-deficient mutant vp5 seeds and wild-type Vp5 seeds in maize. Front Plant Sci 5:801
141. Finnie C, Melchior S, Roepstorff P, Svensson B (2002) Proteome analysis of grain filling and seed maturation in barley. Plant Physiol 129:1308–1319
142. Ostergaard O, Melchior S, Roepstorff P, Svensson B (2002) Initial proteome analysis of mature barley seeds and malt. Proteomics 2:733–739
143. Borén M, Larsson H, Falk A, Jansson C (2004) The barley starch granule proteome—internalized granule polypeptides of the mature endosperm. Plant Sci 166:617–626
144. Finnie C, Svensson B (2003) Feasibility study of a tissue-specific approach to barley proteome analysis: aleurone layer, endosperm, embryo and single seeds. J Cereal Sci 38:217–227
145. Bak-Jensen KS, Laugesen S, Roepstorff P, Svensson B (2004) Two-dimensional gel electrophoresis pattern (pH 6–11) and identification of water-soluble barley seed and malt proteins by mass spectrometry. Proteomics 4:728–742
146. Finnie C, Steenholdt T, Noguera OR, Knudsen S, Larsen J, Brinch-Pedersen H et al (2004) Environmental and transgene expression effects on the barley seed proteome. Phytochemistry 65:1619–1627
147. Ostergaard O, Finnie C, Laugesen S, Roepstorff P, Svennson B (2004) Proteome analysis of barley seeds: identification of major proteins from two-dimensional gels (pI 4–7). Proteomics 4:2437–2447
148. Alexander RD, Morris PC (2006) A proteomic analysis of 14-3-3 binding proteins from developing barley grains. Proteomics 6:1886–1896
149. Finnie C, Bak-Jensen KS, Laugesen S, Roepstorff P, Svensson B (2006) Differential appearance of isoforms and cultivar variation in protein temporal profiles revealed in the maturing barley grain proteome. Plant Sci 170:808–821
150. Hynek R, Svensson B, Jensen ON, Barkholt V, Finnie C (2006) Enrichment and identification of integral membrane proteins from barley aleurone layers by reversed-phase chromatography, SDS-PAGE, and LC-MS/MS. J Proteome Res 5:3105–3113
151. Witzel K, Surabhi GK, Jyothsnakumari G, Sudhakar C, Matros A, Mock HP (2007) Quantitative proteome analysis of barley seeds using ruthenium(II)-tris-(bathophenanthroline-disulphonate) staining. J Proteome Res 6:1325–1333
152. Laugesen S, Bak-Jensen KS, Hägglund P, Henriksen A, Finnie C, Svensson B et al (2007) Barley peroxidase isozymes: expression and post-translational modification in mature seeds as identified by two-dimensional gel electrophoresis and mass spectrometry. Int J Mass Spectrom 268:244–253
153. Geddes J, Eudes F, Laroche A, Selinger LB (2008) Differential expression of proteins in response to the interaction between the pathogen *Fusarium graminearum* and its host, *Hordeum vulgare*. Proteomics 8:545–554
154. Eggert K, Pawelzik E (2011) Proteome analysis of Fusarium head blight in grains of naked barley (*Hordeum vulgare* subsp. nudum). Proteomics 11:972–985

Chapter 3
Fruit Development and Ripening: Proteomic as an Approach to Study *Olea europaea* and Other Non-model Organisms

Linda Bianco and Gaetano Perrotta

Abstract Boosted by the development of cutting-edge "omics" technologies, powerful tools have been developed to support traditional fruit crop research. Comparative "omics" studies have been extensively applied to investigate complex biological processes, such as fruit development and ripening, pointing out unique pathways, genes and proteins involved in these processes. Due to the availability of new technologies, reduced experimental costs, and optimized protein extraction protocols for recalcitrant plant tissues, proteomics is rapidly expanding, reaching fruit species regarded as non-model plant systems. *Olea europaea* can be undoubtedly ranked as a non-model plant species, thus suffering from a dearth of proteomic investigation when compared to other fruit species. In this chapter, we will briefly travel through the proteomic history of olives as an example of a non-model tree crop, characterized by a proteomic investigation still in its infancy but appearing to be promising. We will highlight what has been already done and we will draw the attention of the reader especially on what can be still done.

Keywords Fruit · *Olea europaea* · Non-model organisms

3.1 Introduction

The consumption of fruit is an important part of a healthy diet as it has been associated with a reduced risk of developing degenerative diseases, like cancer and cardiovascular diseases. Fresh fruits are in fact characterized by high levels of relevant nutrients, such vitamins, fibres, minerals, and anti-oxidant compounds, including polyphenolic flavonoids, vitamin-C, and anthocyanins.

Nutritional and sensory qualities of fruits are largely determined by the catabolic and anabolic processes taking place during fruit development and ripening. The

---

L. Bianco · G. Perrotta (✉)
ENEA, Trisaia Research Center, S.S 106 Jonica, Km 419, 5, 75026 Rotondella, MT, Italy
e-mail: gaetano.perrotta@enea.it

developmental process requires biochemical pathways that are unique to plants and may vary between species. In fleshy fruit, it involves three distinct stages, namely fruit set, fruit development and fruit ripening. During the first phases, very active cell division promotes fruit growth, followed by fruit enlargement due to considerable cell expansion, driven by cell wall extension as well as synthesis of new cell wall material. Ripening initiates after fruit development, when seed maturation has been completed; it is characterized by deep metabolic alterations in biochemistry, physiology and gene expression of the fruit, leading to changes in colour, texture, aroma and nutritional qualities [1].

The transition from immature to mature stage occur when fruits acquire their final colour, edible traits and organoleptic quality through a progression of events accompanied by an intense metabolism. As a matter of fact, chlorophyll is progressively degraded and the photosynthetic apparatus is dismantled; new pigments start being synthetized as well as different types of anthocyanins and carotenoids, such as β-carotene, xanthophyll esters, xanthophylles, and lycopene. Complex mixtures of volatile organic compounds are produced, whereas bitter substances, such as flavonoids, tannins and related compounds, are hydrolysed [2]. Fruit taste is shaped by an increase in sweetness, due to augmented gluconeogenesis, hydrolysis of polysaccharides, especially starch, and a parallel decrease in acidity. Changes in texture also occur, leading to a gradual loss of firmness. Fruit softening is a developmentally programmed process that requires a large number of hydrolytic enzymes, synergistically working to disassembly the cell wall. Tissue softening is strictly related to an increased fruit susceptibility to pathogens; in addition, it is responsible for the reduced tolerance to mechanical damage that might quickly render the fruit unmarketable. At the late stages of ripening, some senescence-related physiological changes also take place, leading to membrane deterioration and initiating programmed cell death.

According to the regulatory mechanisms underlying their ripening process, fruits can be broadly classified in climacteric and non-climacteric. Climacteric fruits, such as tomatoes, apples and pears are characterised by an increased ethylene production and a rise in cellular respiration. By contrast, non-climacteric fruits, such as strawberry, olives and oranges, show no dramatic changes and ethylene production remains at a very low level. Interestingly, this physiological behaviour is not related to taxonomy. Species belonging to the same family can display divergent responses to ethylene. For example, tomato and pepper belonging to *Solanaceae*, are classified as climacteric and non-climacteric, respectively. Although ethylene plays a pivotal role in climacteric fruit ripening, both ethylene-dependent and ethylene-independent gene regulation pathways coexist to coordinate this process [3]. The cross-talk between ethylene and auxin represents, in fact, a critical point in this regulatory network [4].

The complexity of ethylene action during ripening is confirmed by the activation of multiple receptors and signal transduction components. The presence of several active ethylene receptors has been demonstrated even in non-climacteric fruits, where ethylene seems to play a regulatory function, dependent on the interaction with abscissic acid (ABA).

Overall, fruit development is a very complex phenomenon, which displays deep biochemical and physiological changes in response to different hormonal inputs. In addition to ethylene, auxin, and ABA, significant involvement has also been proposed for jasmonates and brassinosteroids in climacteric [5] and non-climacteric fruits [6], respectively. All the hormone categories seem to be directly or indirectly involved in the ripening of climacteric and non-climacteric fruits, supporting the hypothesis of a model common for both fruit classifications [7].

As a general consideration, the developmental program is under strict genetic control and driven by the coordinated expression of fruit-related genes, coding for enzymes directly involved in biochemical and physiological changes as well as for regulatory proteins participating in signalling and transcriptional machineries. All these biochemical and physiological events are influenced by several environmental conditions, as well.

Comparative "omics" studies have been extensively applied to study complex biological processes, such as fruit development and ripening, pointing out unique pathways, genes and proteins involved in this process [8]. An important number of data are nowadays available from large-scale analysis of gene expression during climacteric and non-climacteric fruit development. In recent years, triggered by NGS revolution, comparative proteomics has become increasingly attractive to plant biologists as the avalanche of genomic information provided new opportunities for protein identification and functionality [9]. It represents an extremely informative approach, examining gene expression end products: the proteins. As a matter of fact, a possible divergence between messenger (transcript) and its final effector (mature protein) can occur. As most biological functions in a cell are executed by proteins rather than by mRNA, transcript expression profiling provides partial information for the description of a biological system, such as fruit development and ripening. Several post-transcriptional and post-translational control mechanisms such as the translation rate, the half-lives of mRNAs and proteins, protein modifications and intercellular protein trafficking, have an important influence on the phenotype [10]. Due to the availability of new technologies, reduced experimental costs, and optimized protein extraction protocols for recalcitrant plant tissues, proteomics is rapidly expanding, reaching fruit species regarded as non-model plant systems [10, 11]. Up until some years ago, only very few data on fruit development proteomics were available [12]; nowadays papers on this tricky topic are progressively increasing. We can think about papaya, banana, mango, avogado, and apricot proteomics [9, 11]. Overall, they show a great effort to understand the molecular mechanisms affecting the development and ripening in a large variety of fruits, based on the hypothesis that proteomic-driven knowledge can effectively help to improve their quality traits.

3.2 Olive Drupe Proteomics: A History to Be Completed

Fruit proteomics is a relevant as well as complex and arduous research tool; it represents one of the major challenges, especially when applied to orphan, unsequenced and non-model organisms. The power of all proteomic methods tends to be lost in non-model species due to the lack of genomic information, the complexity of the genome (protein inference problem) or due to the sequence divergence to a related sequenced reference variety or to a related model organism [13].

To date, a very limited number of papers have dealt with olive fruit proteome. Due to its particular size and long generation time, *Olea europaea* can be undoubtedly ranked as a non-model plant species, thus suffering from a dearth of proteomic investigation when compared to the grass and major horticultural crops, like tomato, citrus or grape, or even to other non-model organisms, the aforementioned papaya, banana, mango, avogado tree crops.

In this chapter, we will briefly travel through the proteomic history of olives as an example of a non-model tree crop, characterized by a proteomic investigation still in its infancy but appearing to be promising. We will highlight what has been already done and we will draw the attention of the reader especially on what can be still done.

Olea europaea is one of the most economically relevant tree crops in the Mediterranean basin. Olive fruits, classified as drupes, can be directly consumed as table olives or subjected to mechanical extraction of oil. Olive oil is a predominant component of the so-called "Mediterranean diet", worldwide known for its beneficial effects on human health; its consumption has been associated to a reduced risk of cardiovascular diseases and cancer [14]. Olive oil is particularly enriched in the monounsaturated fatty acid oleate (18:1), reaching percentages up to 75–80 % of total fatty acids, followed by linoleate (C18:2), palmitate (C16:0), stearate (C18:0) and linolenate (C18:3). The final acyl composition enormously varies throughout olive fruit development, according to genotype and environmental conditions. Drupe mesocarp can accumulate important metabolites, including polyphenols, carotenoids, chlorophylls, sterols, terpenoids and a wide range of volatile compounds, all directly or indirectly affecting the olive oil quality and aroma [15]. Given the importance of the olive fruit and the nutritional value of its oil, it would be of great interest the comprehension of metabolic changes leading to the biosynthesis of compounds relevant for the quality of both, fruit and oil.

From a proteomic angle, olive drupes exemplify a recalcitrant plant material. Similarly to other plant tissues, drupe tissues contain very low amount of proteins (approximately 2 %) [16] and display a high content of proteases and metabolites such as phenolics, organic acids, lipids, pigments and polysaccharides, making them a challenging source for protein extraction. In addition, they accumulate oil in the mesocarp, reaching up to 28–30 % of the total pulp fresh weight, with an accumulation peak after the onset of ripening. Such interfering compounds are responsible for irreproducible results, such as proteolytic breakdown, charge heterogeneity and streaking on traditional 2-D gels. This probably explains why scarce research has been devoted to the identification and characterization of olive pulp proteins so far (Fig. 3.1).

As matter of fact, literature is very rich in studies aimed at determining the main components of olives and olive oils; by contrast, minor fruit components, such as proteins, have been scarcely investigated, despite their putative role in oil stability,

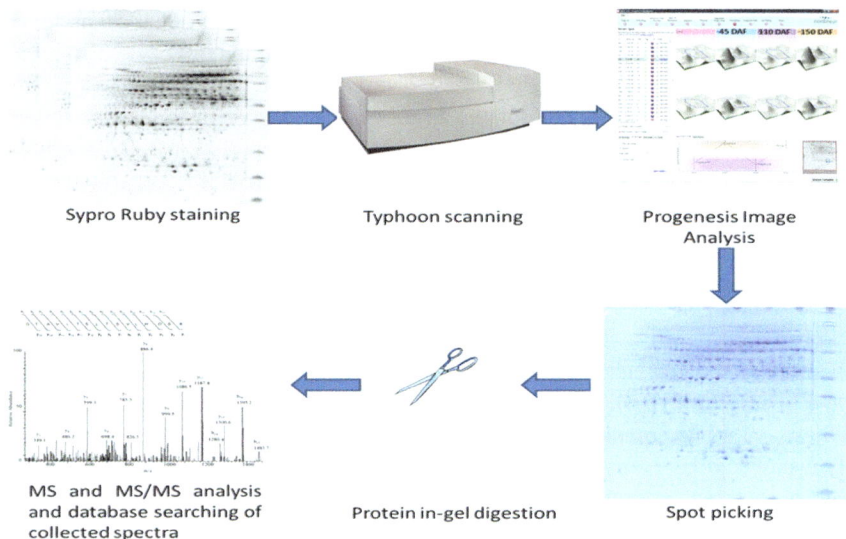

Fig. 3.1 Schematic workflow of the proteomic investigations in *Olea europaea*

as suggested by different authors [17]. It has been hypothesized that peculiar proteins of drupe mesocarp can be transferred to the oil during oil extraction.

Before 2011, the only protein described in olive pulp and oil was the 4.6 kDa oleosin-like polypeptide [16, 18]. In the latest years, efforts to deeply characterize olive drupe protein have been focused on the development of an extraction procedure form both olive pulp and seed, yielding very few bands on Coomassie blue stained gels [19, 20]. The introduction of combinatorial peptide ligand libraries (CPPLs), commercially available under the trade-name ProteoMiner (Bio-rad), has been proven to be an useful tool for protein extraction in difficult matrix [21]. Some authors applied conventional extraction methods followed by CPLLs, identifying 231 olive drupe proteins [22]. Among them, only 9 were identified as proteins corresponding to the species *Olea europea*, due to the little information available in protein database for this species. The remaining identifications were carried out by homology with different sequenced organisms.

Despite the absolute novelty represented by these works in the context of olive drupe proteomics, they are very far from providing significant information about drupe developmental process, as they are focused on technical traits and provide a small sub-set sof proteins. In 2013, Capriotti and colleagues proposed a gel-free proteomic platform for the identification of proteins in ripe olives [23]. In contrast with aforementioned investigations, that were established on mono-dimensional gel-based technique combined to different types of mass spectrometric instrumentation, these authors proposed a shotgun proteomic approach. In particular, they used a high resolving power LTQ-Orbitrap XL coupled to an improved, miniaturized liquid chromatography system, delivering a significant increase in the number of identified proteins compared to previous works. A total of 1456 proteins were identified in fruits of "Caninese" *cultivar*, collected during olive harvesting for oil production. This represents the largest dataset of proteins identified so far in olive drupes. However, only 22 proteins were identified as proteins belonging to the *O. europaea* species. Of these, 7 had been previously reported [19, 22], in particular, the thaumatin-like protein [19], the acyl-[acyl-carrierprotein] desaturase, involved in oil biochemistry [24], the beta-1.3-glucanase, commonly found in higher plants but also recently proven to be a significant allergen [25], and the Cu/Zn superoxide dismutase, widely described in olive pollen as the allergen Ole e 5 [23]. Among the remaining 15 proteins, it is noteworthy to mention the hydroperoxide lyase, which catalyses the cleavage of hydroperoxides from polyunsaturated fatty acids and is responsible for the major components of the aroma of virgin olive oil [26]; the secologanin synthase, enzyme involved in the synthesis of monoterpenic moiety of secoiridoids and not yet fully known [27] polyphenol oxidase, involved in olive fruit browning [28]; the isopentenyl diphosphate isomerase, involved in the phenolic metabolism in plants, thus it is an enzyme of particular interest as olive fruit possesses a wide range of secondary metabolites, like phenols and secoiridoids [27]. Of course, the remaining protein identifications were carried out by sequence homology to known plant genomes [23].

The study proposed by Capriotti and colleagues suffers from the main drawbacks of shotgun proteomics. 1236 out of 1465 proteins were identified through assignments based on single peptide. Although a unique peptide might identify proteins with a high score, the peptide-centric nature of shotgun proteomics complicates the analysis and biological interpretation of the data [29]. The same peptide sequence can in fact be present in multiple different proteins or protein isoforms. Therefore, such shared peptides can lead to ambiguities in determining the identities of sample proteins, especially when the information accessible in databases is far from being complete and assignments are based on sequence homology. This study can be definitely referred as the first descriptive proteomic investigation of olive drupes, providing the systematic analysis of mesocarp proteins. Nevertheless, it does not provide any information about protein fluctuations during drupe development and ripening. So far, only one paper in literature monitors the proteome variations associated with olive fruit development by using comparative proteomics coupled to mass spectrometry [30] and it has been developed in our labs. In details, we investigated the *cultivar* "Coratina", because of its very high phenolic content. In order to monitor major proteome changes during fruit development and to reveal modulations in the biosynthesis of compounds related to major quality traits of olives and oil, we extracted the total proteome content from drupe mesocarp and epicarp, after pit removal. Three different developmental stages were taken into account, corresponding to 45, 110 and 150 DAF (days after flowering). Proteins were extracted by using a classical phenol extraction protocol [31], introducing some modifications to remove major contaminants and to obtain extracts suitable for 2-D electrophoresis. Our investigation, in fact, is based on 2-D gels. The use of electrophoretic protein separation, spot excision and in-gel digestion of each gel spot before mass spec analysis is an approach commonly employed for large-scale proteomic studies, fruit proteomics included.

The main advantage of this approach is the separation of proteins in protein spots, reducing sample complexity prior to digestion. Moreover, different isoforms can be simultaneously displayed on large gels, information that is completely lost by using peptide-centric shotgun approach. To take advantages of 2-D gel over shotgun proteomics, we first optimized a protocol for protein extraction from recalcitrant matrixes. In particular, we applied the protocol developed by Isaacson et al. [31] based on phenol, introducing some washing steps before the extraction. Olive drupes were finely grinded in liquid nitrogen and the resulting powder was washed in 20 % TCA in water, for protein precipitation and removal of phenolics. Precipitated proteins were successively washed twice with 20 % TCA in 80 % acetone for oil removal. Then, proteins were extracted by using phenol, as described before [19, 31]. By applying this extraction technique, approximately 1600 protein spots were detected on 2-D gels, *per* each developmental stage, proving to be an effective protocol for such recalcitrant plant material and providing the first 2-D map of olive drupes realized until now. In order to detect protein spots changing in abundance during olive drupe development and ripening, 2-D gels were subjected to image analysis, revealing 247 differential accumulated protein spots [30]. Of them, 170 were manually excised from the gel, while the remaining 77 differentially accumulated spots were too faint for being manually picked up. 121 out of 170 spots were successfully identified. To get this high identification rate, we performed database searching against the olive fruit EST database [32] and against in an in-house *Olea europaea* flower EST database, both generated in our labs. Only when identifications failed, spectra were searched against *Viridiplantae* subset of the non-redundant NCBI protein database.

Among identified protein spots, we found a large number of proteins strictly related to fruit development. As mentioned before, we analyzed fruit protein content at 45, 110 and 150 DAF, corresponding to well-defined drupe developing phases. 45 DAF corresponds to a period of rapid fruit growth, due to both development of endocarp and intense cell division; 110 DAF is a phase marked by a mesocarp development, mainly due to the expansion of pre-existing flesh cells, whereas at 150 DAF oil accumulation reaches the completion. During fruit development, cells first divide, as supported by the identification of several proteins playing a role in cell division; all of them showed an increase in abundance from 45 to 110 DAF and remained approximately stable during the transition from 110 to 150 DAF. Then, they expand, as justified by the accumulation of protein spots corresponding to subunits of vacuolar H^+-ATPase. The proton electrochemical gradient generated by this multi-subunit enzyme might represent a driving force for cell expansion during development [33, 34]. The changes undergone by fruit cells, which first multiply and then enlarge, must be supported by massive structural remodelling of the cell wall and changes in cytoskeleton structure. As a matter of fact, we identified annexins, playing a role in cell expansion [35, 36], and alpha/beta-tubulins, supporting changes in cytoskeleton.

Interestingly, several spots corresponding to methionine synthase and S-adenosylmethionine synthetase were also identified. Their abundance intriguingly decreased during olive drupe development. Both enzyme belong to the ethylene biosynthetic pathway, but they are also involved in the biosynthesis of polyamines. These latter are required for cell growth and cell division. In this context, their decreasing levels could be related to the transition from early developmental stages, characterized by intense cell division, to later stages, where their decrease could be driven by both the cessation of cell division and the non-climacteric nature of olive drupe ripening, as changes in the availability of soluble methionine limit ethylene production.

As expected, during fruit development and ripening, photosynthetic apparatus is progressively dismantled. However, olive drupes seem to retain, for a considerable period of time, active chloroplasts, that are responsible for photosynthetic activities [37]. As a matter of fact, different protein spots related to fruit photosynthesis were identified; many of them showed a similar accumulation pattern at 45 and 110 DAF, to finally decrease at 150 DAF. The only exception is represented by RuBisCO large subunit-binding protein subunit alpha, which showed an increase in spot intensity during fruit development. Photosynthesis occur in fruits with characteristics different from either C3 or C4/CAM plants. The intense metabolism occurring during development is responsible for increased level of CO_2, which accumulates in high concentration in the fruit cell-free space, due to the impermeability of fruit cuticle. Inorganic carbon is fixed into oxalacetate, converted into malate by malate-dehydrogenase. Malate can be decarboxylated by cytosolic or mitochondrial malic enzyme to yield pyruvate and CO_2. The latter can further be photosynthetically fixed into triose phosphate in the fruit chloroplasts. It has been demonstrated that fruit photosynthesis contributes to the carbon economy of developing fruits and hence to olive oil biogenesis [38, 39]. Remarkably, NAD-malic enzyme (ME) involved in C4 photosynthesis showed a differential accumulation in our investigation, increasing during the transition from 45 to 110 DAF, and remaining approximately stable from 110 to 150 DAF. The accumulation of ME might represent the proof that in *Olea*, during fruit photosynthesis, refixing of CO_2 occurs. Besides, this might explain the contribution of fruit photosynthesis toward the biogenesis of olive oil, as well. As a matter of fact, the reaction catalyzed by ME yields pyruvate, which is the precursor of fatty acid biosynthesis. The malic enzyme accumulates during developmental stages, where an intense oil accumulation is expected to occur [15] suggesting a pivotal role for this enzyme in oleogenesis. In this context, the discussed increase of RuBisCO large subunit-binding protein subunit alpha might find a new possible explanation: this protein could work as chaperonin, stabilizing RuBisCO and thus its activity in fixing CO_2, yielded by the malic enzyme beside pyruvate production.

Of course, many proteins related to the metabolism of fatty acids, phenolics and aroma compounds were also detected; their trend is consistent with the accumulation pattern of oil, phenolic content and volatile compounds.

Olive drupes, and in particular drupes of *cultivar* "Coratina", contains high levels of phenolics as well as aroma compounds, which can be transferred to the oil, during mechanical extraction. Thus, quality traits of oil are strictly correlated to the quality of olives. The olive fruit fly *Bactrocera oleae* can attack olive drupes, significantly affecting their quantity and quality as well as the nutritional and sensory profile of olive oil. Gel-based comparative proteomic has been applied to investigate specific proteomic changes in drupes with larval feeding tunnels [40]. 26 protein spots exhibited differential accumulation in infested fruits. Among them, 8 spots increased in abundance in insect-attacked olives, while the remaining spots showed the opposite trend. 23 out 26 spots were successfully identified by database search against *Olea europea* ESTs available over the WEB, or a plant non-redundant sequence database. Beta-glucosidase, major latex proteins, phosphogluconolactonase and 6-phosphogluconate dehydrogenase showed increased accumulation. All of them have been previously reported as putative defensive proteins [41–43]. Interestingly, protein involved in the regulation of redox status were detected as differentially accumulated. Some of them exhibited decreasing levels, indicative of plant efforts to maintain homeostasis under stress conditions, preventing the risk associated to the production of highly reactive cytotoxic ROS.

3.3 Perspectives on Olive Drupe Proteomics

In terms of olive and olive oil production, Italy ranks second in the world, after Spain. Nowadays, great attention has been paid to olive tree crops, due to *Xylella fastidiosa* infection in southern Italy. The fear of spread in the Mediterranean basin brought out the importance of molecular tools to understand and face the challenges posed by pests and biotic stresses. The establishment of these molecular tools cannot prescind from the sequencing of olive genome, which is currently under investigation. The International Olive Genome Consortium (IOGC http://olivegenome.karatekin.edu.tr) has the purpose to sequence the whole genome of olive, whereas the Italian project OLEA developed genomic resources aimed at identifying, isolating and determining the function of genes associated with both vegetative and reproductive phenotype [44]. However, the information currently accessible in databases for olive is far from being complete. While waiting for the genome sequences, proteomic approaches can be in any case exploited, as mass spectra can be collected and searched as soon as genomic resources will become available. The traditional proteomics workflow for non-model organisms is based on the interpretation of data against protein databases constructed from annotated genomes, either from the non-model organism itself or from the most closely related organisms, or from expressed sequence tag and transcripts information obtained by sequencing cDNA.

With the advent of high-accuracy tandem mass spectrometers, experimental proteomics data can be used to refine genome annotations, by using an expanding approach known as proteogenomics, enabling the molecular studies of non-model organisms, as *Olea europea*, at an unprecedented depth [45, 46]. Custom databases can be in fact generated from genomic and transcriptomic information using NGS technology in a rather straightforward way and at a reasonable cost for any organism. Such database can be used to identify novel peptides from MS-based proteomic data and, in a iterative process, mass spec data can be used to provide evidences of gene expression at protein level and to help manual re-annotation/curation.

Olive fruit proteomics is still in its infancy, thus there is huge room for investigation of olive drupe ripening process. The only published work dealing with proteome variation during drupe development does not investigate ripening [30]. Information in this context is totally lacking, but absolutely necessary to dissect the molecular mechanisms underlying quality traits related to oil production and phenol accumulation. Set of high-quality EST can be easily generated from fully ripe drupes, as described by Parra et al. [47]. The combination of "omics" data in proteogenomic-flavored approach is expected to provide relevant information on such relevant process.

Proteogenomics could be also applied to shed light on complex biosynthetic pathways, for example oleuropein biosynthesis. Olive drupes contain variable amounts of phenolic content, where secoiridoids represent the most important class. Oleuropein is the main secoiridoid, representing up to 82 % of the total bio-phenols and is responsible for the characteristic bitter and pungent taste of the olive drupes and oil. Proteomic investigation carried out so far revealed no traces of the enzymes involved in oleuropein biosynthetic pathway on the traditional 2-D maps, likely due to the lack of sequence information as well as to the high dynamic range characterizing olive drupes. All biological samples, olive drupe included, show a small set of proteins, often as few as 20–30, present in a large excess, which can leave very little room for sampling and detection of all other species present therein [48]. Olive drupe proteomics could take advantages from the development of protein enrichment strategies, such as CPLL, and couple it to comparative or quantitative proteomics techniques, to detect low-abundance proteins and their variations during ripening [49–51].

Many other biological questions remain open talking about olive drupe proteomics. At the veraison, when fruit change colour and start ripening, olives start losing their firmness, while increasing the concentration and quality of their oil. This process greatly depends on the *cultivar*. From a proteomic point of view, the comprehension of the mechanisms by which fatty acid composition varies from *cultivar* to *cultivar* during olive fruit development and ripening would represent an important step toward the ultimate goal of regulating these processes in a directed and predictable manner.

References

1. Barry CS, Giovannoni JJ (2007) Ethylene and fruit ripening. J Plant Growth Regul 26:143–159
2. Defilippi BG, Manriquez D, Luengwilai K, González-Agüero M (2009) Aroma volatiles: biosynthesis and mechanisms of modulation during fruit ripening. Adv Bot Res 50:1–37

3. Pech J-C, Bouzayen M, Latché A (2008) Climacteric fruit ripening: ethylene-dependent and independent regulation of ripening pathways in melon fruit. Plant Sci 175:114–120
4. Trainotti L, Tadiello A, Casadoro G (2007) The involvement of auxin in the ripening of climacteric fruits comes of age: the hormone plays a role of its own and has an intense interplay with ethylene in ripening peaches. J Exp Bot 58:3299–3308
5. Ziosi V, Bonghi C, Bregoli AM, Trainotti L, Biondi S, Sutthiwal S et al (2008) Jasmonate-induced transcriptional changes suggest a negative interference with the ripening syndrome in peach fruit. J Exp Bot 59:563–573
6. Symons GM, Davies C, Shavrukov Y, Dry IB, Reid JB, Thomas MR (2006) Grapes on steroids. Brassinosteroids are involved in grape berry ripening. Plant Physiol 140:150–158
7. Giovannoni J (2001) Molecular biology of fruit maturation and ripening. Annu Rev Plant Physiol Plant Mol Biol 52:725–749
8. Gapper NE, Giovannoni JJ, Watkins CB (2014) Understanding development and ripening of fruit crops in an 'omics' era. Hortic Res 1:14034
9. Palma JM, Corpas FJ, Luís A (2011) Proteomics as an approach to the understanding of the molecular physiology of fruit development and ripening. J Proteomics 74:1230–1243
10. Carpentier SC, Panis B, Vertommen A, Swennen R, Sergeant K, Renaut J et al (2008) Proteome analysis of non-model plants: a challenging but powerful approach. Mass Spectrom Rev 27:354–377
11. Righetti PG, Esteve C, D'amato A, Fasoli E, Luisa Marina M, Concepcion Garcia M (2015) A sarabande of tropical fruit proteomics: avocado, banana, and mango. Proteomics 15:1639–1645
12. Rocco M, D'ambrosio C, Arena S, Faurobert M, Scaloni A, Marra M (2006) Proteomic analysis of tomato fruits from two ecotypes during ripening. Proteomics 6:3781–3791
13. Carpentier SC, America T (2014) Proteome analysis of orphan plant species, fact or fiction? Plant Proteomics: Meth Protocols 333–346
14. Coni E, Di Benedetto R, Di Pasquale M, Masella R, Modesti D, Mattei R et al (2000) Protective effect of oleuropein, an olive oil biophenol, on low density lipoprotein oxidizability in rabbits. Lipids 35:45–54
15. Conde C, Delrot S, Geros H (2008) Physiological, biochemical and molecular changes occurring during olive development and ripening. J Plant Physiol 165:1545–1562
16. Zamora R, Alaiz M, Hidalgo FJ (2001) Influence of cultivar and fruit ripening on olive (*Olea europaea*) fruit protein content, composition, and antioxidant activity. J Agric Food Chem 49:4267–4270
17. Koidis A, Boskou D (2006) The contents of proteins and phospholipids in cloudy (veiled) virgin olive oils. Eur J Lipid Sci Technol 108:323–328
18. Hidalgo FJ, Alaiz M, Zamora R (2001) Determination of peptides and proteins in fats and oils. Anal Chem 73:698–702
19. Esteve C, Canas B, Moreno-Gordaliza E, Del Rio C, Garcia MC, Marina ML (2011) Identification of olive (*Olea europaea*) pulp proteins by matrix-assisted laser desorption/ionization time-of-flight mass spectrometry and nano-liquid chromatography tandem mass spectrometry. J Agric Food Chem 59:12093–12101
20. Esteve C, Del Rio C, Marina ML, Garcia MC (2011) Development of an ultra-high performance liquid chromatography analytical methodology for the profiling of olive (*Olea europaea* L.) pulp proteins. Anal Chim Acta 690:129–134
21. Boschetti E, Righetti PG (2014) Plant proteomics methods to reach low-abundance proteins. Plant Proteomics: Meth Protocols 111–129
22. Esteve C, D'amato A, Marina ML, Garcia MC, Citterio A, Righetti PG (2012) Identification of olive (*Olea europaea*) seed and pulp proteins by nLC-MS/MS via combinatorial peptide ligand libraries. J Proteomics 75:2396–2403
23. Capriotti AL, Cavaliere C, Foglia P, Piovesana S, Samperi R, Stampachiacchiere S et al (2013) Proteomic platform for the identification of proteins in olive (*Olea europaea*) pulp. Anal Chim Acta 800:36–42

24. Banilas G, Moressis A, Nikoloudakis N, Hatzopoulos P (2005) Spatial and temporal expressions of two distinct oleate desaturases from olive (*Olea europaea* L.). Plant Sci 168:547–555
25. Palomares O, Villalba M, Quiralte J, Polo F, Rodriguez R (2005) 1, 3-β-glucanases as candidates in latex–pollen–vegetable food cross-reactivity. Clin Exp Allergy 35:345–351
26. Salas JNJ, Sánchez J (1999) Hydroperoxide lyase from olive (*Olea europaea*) fruits. Plant Sci 143:19–26
27. Alagna F, Mariotti R, Panara F, Caporali S, Urbani S, Veneziani G et al (2012) Olive phenolic compounds: metabolic and transcriptional profiling during fruit development. BMC Plant Biol 12:162
28. Ortega-García F, Blanco S, Peinado MÁ, Peragón J (2008) Polyphenol oxidase and its relationship with oleuropein concentration in fruits and leaves of olive (*Olea europaea*) cv. 'Picual'trees during fruit ripening. Tree Physiol 28:45–54
29. Nesvizhskii AI, Aebersold R (2005) Interpretation of shotgun proteomic data: the protein inference problem. Mol Cell Proteomics 4:1419–1440
30. Bianco L, Alagna F, Baldoni L, Finnie C, Svensson B, Perrotta G (2013) Proteome regulation during *Olea europaea* fruit development. PLoS ONE 8:e53563
31. Isaacson T, Damasceno CM, Saravanan RS, He Y, Catalá C, Saladié M et al (2006) Sample extraction techniques for enhanced proteomic analysis of plant tissues. Nat Protoc 1:769–774
32. Alagna F, D'agostino N, Torchia L, Servili M, Rao R, Pietrella M et al. (2009) Comparative 454 pyrosequencing of transcripts from two olive genotypes during fruit development. BMC Genom 10:399
33. Amemiya T, Kanayama Y, Yamaki S, Yamada K, Shiratake K (2006) Fruit-specific V-ATPase suppression in antisense-transgenic tomato reduces fruit growth and seed formation. Planta 223:1272–1280
34. Faurobert M, Mihr C, Bertin N, Pawlowski T, Negroni L, Sommerer N et al (2007) Major proteome variations associated with cherry tomato pericarp development and ripening. Plant Physiol 143:1327–1346
35. Clark GB, Sessions A, Eastburn DJ, Roux SJ (2001) Differential expression of members of the annexin multigene family in Arabidopsis. Plant Physiol 126:1072–1084
36. Konopka-Postupolska D (2007) Annexins: putative linkers in dynamic membrane-cytoskeleton interactions in plant cells. Protoplasma 230:203–215
37. Proietti P, Nasini L, Famiani F (2006) Effect of different leaf-to-fruit ratios on photosynthesis and fruit growth in olive (*Olea europaea* L.). Photosynthetica 44:275–285
38. Sánchez J (1995) Olive oil biogenesis. Contribution of fruit photosynthesis. In: Plant lipid metabolism. Springer, Berlin, pp 564–566
39. Sánchez J, Harwood JL (2002) Biosynthesis of triacylglycerols and volatiles in olives. Eur J Lipid Sci Technol 104:564–573
40. Corrado G, Alagna F, Rocco M, Renzone G, Varricchio P, Coppola V et al (2012) Molecular interactions between the olive and the fruit fly Bactrocera oleae. BMC Plant Biol 12:1
41. Van De Ven WT, Levesque CS, Perring TM, Walling LL (2000) Local and systemic changes in squash gene expression in response to silverleaf whitefly feeding. Plant Cell 12:1409–1423
42. Xiong Y, Defraia C, Williams D, Zhang X, Mou Z (2009) Characterization of Arabidopsis 6-phosphogluconolactonase T-DNA insertion mutants reveals an essential role for the oxidative section of the plastidic pentose phosphate pathway in plant growth and development. Plant Cell Physiol 50:1277–1291
43. Konno K (2011) Plant latex and other exudates as plant defense systems: roles of various defense chemicals and proteins contained therein. Phytochemistry 72:1510–1530
44. Muleo R, Cavallini A, Perrotta G, Baldoni L, Morgante M, Velasco R (2012) Olive tree genomic. INTECH Open Access Publisher, Rijeka
45. Agrawal GK, Pedreschi R, Barkla BJ, Bindschedler LV, Cramer R, Sarkar A et al (2012) Translational plant proteomics: a perspective. J Proteomics 75:4588–4601
46. Armengaud J, Trapp J, Pible O, Geffard O, Chaumot A, Hartmann EM (2014) Non-model organisms, a species endangered by proteogenomics. J Proteomics 105:5–18

47. Parra R, Paredes MA, Sanchez-Calle IM, Gomez-Jimenez MC (2013) Comparative transcriptional profiling analysis of olive ripe-fruit pericarp and abscission zone tissues shows expression differences and distinct patterns of transcriptional regulation. BMC Genom 14:1
48. Righetti PG, Fasoli E, D'amato A, Boschetti E (2014) The "Dark side" of food stuff proteomics: the CPLL-marshals investigate. Foods 3:217–237
49. Roux-Dalvai F, De Peredo AG, Simó C, Guerrier L, Bouyssié D, Zanella A et al (2008) Extensive analysis of the cytoplasmic proteome of human erythrocytes using the peptide ligand library technology and advanced mass spectrometry. Mol Cell Proteomics 7:2254–2269
50. Hartwig S, Czibere A, Kotzka J, Paßlack W, Haas R, Eckel J et al (2009) Combinatorial hexapeptide ligand libraries (ProteoMiner™): an innovative fractionation tool for differential quantitative clinical proteomics. Arch Physiol Biochem 115:155–160
51. Boschetti E, Righetti PG (2013) Low-abundance proteome discovery: state of the art and protocols. Newnes

Chapter 4
Proteomics in Detection of Contaminations and Adulterations in Agricultural Foodstuffs

Javad Gharechahi, Mehrshad Zeinolabedini
and Ghasem Hosseini Salekdeh

Abstract Proteins are essential components of our diet and are found in almost all foodstuffs. Food proteomics is a broad term used to describe technological and methodological approaches used to characterize protein constituents of a particular food product. In recent years, incremental advances in mass spectrometry (MS)-based proteomics have resulted in the development of robust, sensitive, and versatile analytical tools that can be used to describe safety, quality, traceability, and originality of different food products. Interestingly, MS-based proteomics has now become the method of choice for rapid, targeted, and cost-effective analysis of foodstuffs of different origin for possible adulteration and contamination. In addition, proteomics has well-performed in characterization of allergen proteins in foodstuffs as well as in safety assessment of processed food products in terms of the presence of food allergens, food-borne microbes or microbial toxins. In this chapter, we will review achievements obtained by proteomics with an especial emphasis on the sensitivity and detection limit of currently-available MS-based proteomics approaches used for the detection of food contaminations and adulterations.

Keywords Proteomics · Food contamination · Food adulteration · MALDI-TOF/TOF · ESI-Q/TOF · HPLC-MS/MS · MRM

J. Gharechahi
Chemical Injuries Research Center, Baqiyatallah University of Medical Sciences,
Tehran, Iran

M. Zeinolabedini · G.H. Salekdeh (✉)
Department of Systems Biology, Agricultural Biotechnology Research Institute of Iran,
Agricultural Research, Education, and Extension Organization, Karaj, Iran
e-mail: hsalekdeh@yahoo.com

4.1 Proteins: A Major Constituent of Our Foods

Proteome is defined as a total set of proteins expressed by a genome at a definite point in time and under a particular cellular state. In contrast to genome that is largely static, proteome is highly dynamic and significantly differs in between different cells, tissues, and organs of an organism and also changes in response to environmental stimuli. Proteins are critical for almost every biological process in living organisms including catalysis, signaling, motility, immunity, and sensing. Indeed, proteins are the main building blocks of living organisms, found in nearly all biological samples and therefore, constitute the major part of our foods. Proteins largely define nutritional and physicochemical characteristics of foodstuffs including viscosity, thermal conductivity, and vapor pressure while contribute to the formation and stabilization of foams, gels, and fibrillary structures in foodstuffs as well [1, 2]. As an important component of foodstuffs, proteins and their amino acid constituents, contribute to the color, flavor, and aroma formed during thermal or enzymatic reactions carried out for the production, processing, and storage of foodstuffs too [2]. For instance, in wheat flour-derived products, gluten proteins contribute to the physical properties by providing visco-elasticity to the resultant dough [3]. In other food products such as milk, meat, vegetables, and fruits, which display complex protein composition, changes in physical properties during processing are largely determined by their protein contents [1, 4]. The protein composition of foodstuffs varies depending on their origin (animal or plant), species, tissue or organ used as food, and the extent of processing including storage, fermentation, cooking, etc. The huge diversity of proteins marks them as suitable candidate markers for the detection of food contamination and adulteration as well as for acknowledging food authenticity, safety, and traceability (Fig. 4.1). On such basis, proteomics is now providing the power, specificity, and precision required for detailed exploration of food proteins to satisfy food forensics requirements.

4.2 Proteomics Tools and Techniques

Proteomics enables detailed characterization of the proteins present in a biological sample in terms of type, abundance, post-translational modification (PTMs), interactions, and cellular localization [5]. Proteomics analysis involves a combination of different analytical methodologies for high resolution protein separation, quantification, and identification as well as bioinformatics for data management and analysis. Indeed, the term proteomics came into practice through the improvements achieved in two-dimensional gel electrophoresis (2-DE) for protein separation. In 2-DE based proteomics analysis, the signal intensity of protein features is compared between two biological conditions and newly synthesized, disappeared, and up/down regulated proteins are identified. The candidate spots are subsequently gel-recovered and subjected to mass spectrometry (MS) for protein identification

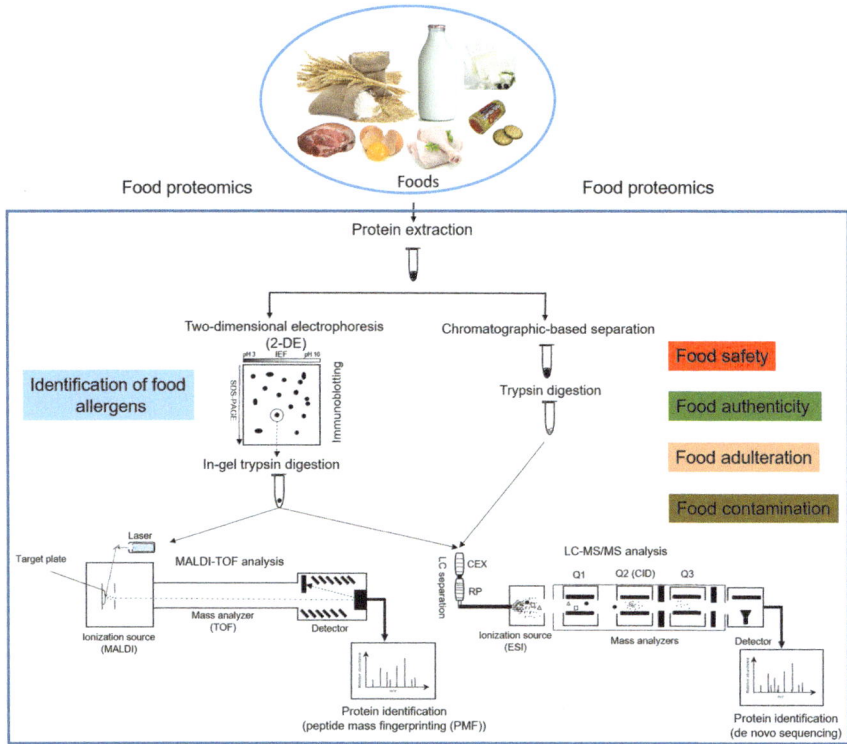

Fig. 4.1 A general workflow for food proteomics. Food proteins are extracted and separated using 2-DE or chromatography based approaches. 2-DE has been extensively utilized for the identification of food allergens. In this approach, 2-DE separated proteins are transferred to a nylon membrane and immunoblotted using sera of patients displaying immunogenic reaction to that food. Candidate spots are then gel recovered and subjected to in-gel trypsin digestion and analyzed for protein identification by either MALDI-TOF MS or LC-ESI-MS/MS instruments. In chromatography based separation, extracted proteins are trypsin digested and the resulting peptides are separated using a combination of cation exchange (*CEX*) and reverse phase (*RP*) chromatography. In this approach, the separated peptides are directly injected into the ionization source of MS for tandem MS (MS/MS) based protein identification. MS-based proteomics in combination with multiple reaction monitoring (MRM) enabled sensitive, accurate, and targeted analysis of a specific set of protein markers for the detection of food adulteration, contamination, as well as for acknowledging food authenticity

(Fig. 4.1). Although 2-DE has extensively been utilized for systematic proteome analysis in diverse biological systems, the limitations associated with this approach including difficulty in detection of low abundant, poor soluble, as well as very small or large proteins and those with extreme p*I*, have largely restricted its applicability for high throughput proteome analysis specially for the detection of food contaminations and adulterations. However, 2-DE based proteomics has been successfully applied to the detection of allergen proteins in foodstuffs [6].

It is worth quoting that proteomics has been largely enhanced with the advances made in the application of MS for protein identification. The introduction of soft ionization techniques, namely electrospray ionization (ESI) [7] and matrix-assisted laser desorption ionization (MALDI) [8], for gentile ionization of proteins and peptides, was a breakthrough for proteomics. A typical MS instrument consists of an ion source (MALDI or ESI) for peptide ionization, a mass analyzer (quadrupoles (Q), time-of-flight (TOF), ion trap (IT), Orbitrap, and ion cyclotron resonance (ICR) for separation of peptide ions according to their mass to charge ratio (m/z), and a detector for ion detection and quantification. Depending on their mode of action, MS instruments most commonly used in proteomics analysis fall into two categories: single stage MS instruments and tandem mass spectrometers (MS/MS) [9]. The most common single stage MS instrument is configured by coupling the MALDI ionization source with TOF mass analyzer (MALDI-TOF), an instrument widely used for protein identification using peptide mass fingerprinting (PMF) [9]. The PMF is particularly suitable for the identification of proteins whose sequences are available in the databases, proteins from species for which genome sequences are available, as well as for the identification of proteins from 2-DE separated spots [10, 11]. The MS/MS instruments are typically equipped with two mass analyzers which are arranged in tandem and are separated by a collision-induced dissociation (CID) cell. In order to reduce sample complexity prior to MS analysis, most modern tandem mass spectrometers are coupled on-line with a multidimensional chromatography-based separation. A different combination of chromatography-based separation including size exclusion chromatography, cation or anion exchange chromatography, reverse phase (RP) chromatography, and affinity chromatography is used for protein and peptide separation through high performance liquid chromatography (HPLC) interface. In a typical HPLC-MS/MS system, peptides eluted from the last chromatography column (usually RP) are ionized in the ionization source (ESI) and are scanned for m/z ratios in the first mass analyzer (TOF, Q, or ion trap). Subsequently, a specific set of ions is selected (user defined, usually most intense ions) and the ions are allowed to sequentially enter into the CID (based on their m/z ratios), in which they are subjected to fragmentation through collision with inert gas molecules. Fragmented peptide ions are then scanned by the second mass analyzer (TOF, Q, ion trap, Orbitrap, or FTICR) and their m/z ratios are recorded as MS/MS spectra. The MS/MS spectral data are then used for protein identification using de novo sequencing, peptide mass tagging, and/or in silico matching of MS/MS spectral data with the theoretical spectra calculated for all peptides in the databases (Fig. 4.1) [12]. New MS instruments with different MS/MS configurations including MALDI or ESI-Q/TOF, MALDI-TOF/TOF, ESI-triple quadrupole (TQ), ESI-LIQ, ESI-LTQ/Orbitrap, and ESI-LTQ-FTICR have been developed which have varying resolution, sensitivity, mass accuracy, dynamic range, and applicability in proteome analysis [13]. Advances in the MS-based proteomics has now enabled researchers to specifically monitor the abundance of a single protein across multiple samples with high reproducibility and accuracy through multiple reactions monitoring (MRM) [14,

15]. This approach is particularly useful for monitoring the abundances of a specific set of marker proteins for the detection of food contamination and adulteration.

In recent years, 2-DE and to a larger extent, MS-based proteomics has been extensively utilized for the analysis of food composition, authentication, and safety [16, 17], and also to the search for protein markers for the detection of food contamination [16, 18–22], adulteration [23–32], and allergens [6, 33]. MS-based proteomics approaches have proven to be highly sensitive, specific, accurate, and dynamic compared with the other methodologies used for the detection of food contamination and adulteration [34]. In this chapter, we will review achievements obtained by proteomics with an emphasis on the sensitivity and large scale applicability of currently-available MS-based proteomics approaches for the detection of food contaminations and adulterations.

4.3 Proteomics in Assessment of Food Safety

Food safety is a broad term used to describe concerns regarding food quality or composition, origin, and the presence of allergens, pathogens, and other contaminations [35]. Food safety is therefore one of the most important food-related issues which requires novel analytical tools for sensitive, accurate, and robust analysis of food composition for the detection of biotic and abiotic contaminations that might compromise the health of the end user consumers. In this respect, the characterization of food-born allergens and pathogens represents a growing challenge in the field of food safety.

4.3.1 Food Allergens

Food allergens are naturally occurring compounds, found in diverse food products as main ingredients or included as additive to foodstuffs during processing and storage. These compounds are capable of causing adverse immune reactions in sensitized individuals. The huge diversity of food allergens and different immune reactions of individuals to these allergens make it difficult to achieve an ultimate cure. Therefore, allergic patients are forbidden to consume potentially allergenic foods, even at small quantities [36]. Proteins are the major allergens found in foodstuffs derived from diverse agricultural products including milk, egg, wheat, soybeans, peanuts, tree nuts (e.g., walnuts, hazelnut, almonds, pecans, and cashews), and fish [37]. Allergic reactions to milk, egg, and soybean are more common in children, whereas fresh fruit, nut, and seafood are more allergic in adults [38]. For example, peanuts and tree nuts cause anaphylaxis reactions which are among the most life-threatening allergic reactions in humans [39]. It is estimated that allergy to food products affects 5 % of adults and 8 % of children worldwide [38]. This highlights the importance and the urgent need for the development of high-throughput proteomics approaches for the detection and characterization of potential food protein allergens, PTMs that may give allergic properties to food proteins, and changes that may occur in immunogenic potential of food proteins during processing, storage, and cooking. Identification of food allergens allows for recombinant production of allergen proteins for use in immunotherapeutic approaches and also for the generation of genetically-modified plants knocked out for allergen proteins or epitopes. Interestingly, genetically-modified soybean and tomato plants with reduced allergen proteins content have been generated and some of them have been approved for use as safe non-allergic foods by the U.S. food and drug administration (FDA).

The most widely-used proteomics approach for the detection and identification of food allergen proteins involves a combination of 2-DE and IgE immunoblotting analysis with the sera of allergic patients as probe. Potential candidate allergen proteins are then gel-recovered and subjected to MS or MS/MS for protein identification. The immunological responses to food allergens can be either IgE-mediated or cell-mediated. As an essential constituent of human diet, milk is one of the main allergic foods particularly in the early childhood. It has been estimated that between 2–3 % of infants younger than 2 years of age are allergic to cow's milk [40]. More than 25 different proteins are found in cow's milk, out of which only a few are known as allergens [41]. Caseins and β-Lactoglobulins are among the highly-abundant milk proteins eliciting an IgE-mediated immune response in patients suffering from cow's milk intolerance. Proteomics studies have also found that less abundant proteins such as lactoferrin, IgG, bovine serum albomin may also elicit allergic reactions [40].

Peanuts or peanut derived food products are among the most frequently reported allergens in the U.S. [39]. Using a combined 2-DE, immunoblotting, and MS analysis, Chassaigne et al. were able to identify several isoforms of storage proteins as main peanut allergens, namely Ara h 1, Ara h 2 and Ara h 3/4 [42]. In addition, proteomics analyses have also shown that different varieties of peanuts display variable amounts of these allergens [43]. 2-DE was also applied to search for sesame seed allergens. Probing 2-DE resolved sesame proteins using sera of patients with sesame seed allergy resulted in the identification of a storage protein named Ses i 3 [44]. Interestingly, Ses i 3 showed 80 % sequence similarity with one of the IgE-binding epitopes of peanut allergen Ara h 1. Similarly, 2-DE analysis coupled to immunoblotting resulted in the identification of an allergen protein, Cor a 9, as hazelnut food allergen which belonged to the 11S globulin family of seed storage proteins [45]. Interestingly, an IgE-binding epitope of peanut allergen protein Ara h 3 also showed 67 % amino acid sequence similarity with a corresponding region in Cor a 9, suggesting that IgE-binding epitopes in different food allergens are evolutionary conserved. A similar approach also showed that all three subunits of beta-conglycinin protein are capable of mounting allergic reactions in patients with soybean allergy [46]. Targeted quantification using tandem MS in MRM mode represents as promising approach in the detection of food allergens. Using this methodology, Houston et al. [47] monitored the concentration of 10 allergen proteins in 20 commercially-available soybean varieties and demonstrated that seed concentration of soybean allergens was quite similar among different genotypes and ranged from 0.5 to 5.7 µg/mg of total protein. MS-based proteomics has also been used for the detection of trace amount of peanut allergens in food products that might be unintentionally contaminated with peanut. In this respect, Careri et al. [48] applied LC-MS/MS analysis to investigate the presence of peanut allergens of Ara h 2 and Ara h 3/4 in selected foodstuffs. The method allowed identification of as low as 1 µg/g peanut allergens in rice crisp and chocolate-based snacks.

Celiac disease (CD) is a non-IgE or cell-mediated allergic reaction to gluten proteins found in cereal crops such as wheat (*Triticum aestivum*), barley (*Hordeum vulgare*), and rye (*Secale cereale*). Consumption of food products containing gluten by patients suffering from the CD elicits a chronic inflammatory reaction and causes damages to small intestine mucosa followed by severe consequences for the patient. CD is known to associate with the presence of HLA-DQ2/8 and the generation of circulating autoantibodies to the enzyme tissue transglutaminase [49]. Indeed, in patients suffering from the CD, the immune system reacts to the peptides derived from gluten proteins. The immune response is associated with increased population of a subset of lymphocytes and the generation of antibodies that attack the lining of the small intestine causing damages to the intestinal villi and consequently leading to reduced nutrient absorption [21, 50]. The prevalence of CD is estimated to be 0.6–1 % of population worldwide [51].

The consumption of as little as 1 mg of gluten, which is roughly equivalent to the amount of gluten found in a half of a grain of barley is sufficient to elicit the immune response and to compromise the health of the patients suffering from CD [52]. Since gluten is the only known trigger for CD, gluten-free diet is its ultimate scientifically proven treatment. Therefore, gluten has to be eliminated from all food products and medications obtained from wheat, barley, and rye because even trace amounts of gluten in dietary foods has severe consequences for gluten intolerant patients. Clear and accurate labeling of food products is therefore critical for the detection of gluten-free foods and the protection of gluten intolerance patients. It has now been well-documented that the consumption of wheat, barley, and rye is harmful to patient suffering from the CD. However, a recent long-term feeding study suggested that oat (*Avena sativa*) was safe for CD suffers [53]. Pure oat can therefore be used as safe gluten-free nutrient for CD patients. However, oats are frequently contaminated with other celiacogenic cereals such as wheat, barley, and rye during farming, transport, storage, and processing [54]. Accurate detection of wheat, barley, and rye contamination in oat and soybean flour and other food products is critical for protecting consumers with the CD. According to the United States FDA, food products labeled as gluten-free must contain less than 20 ppm gluten from wheat, barley, rye, and crossbreds cereals like triticale [55]. Although enzyme-linked immunosorbent assay (ELISA) is known as sensitive and reliable method for the detection and quantification of gluten, this method suffers from high false-positivity and inadequate quantification. ELISA has been successfully applied to the detection of wheat gluten and barley hordeins contaminations in 109 out of 134 tested oats and oat-derived food products [56].

MS-based proteomics represents a complementary approach with unique capability in targeted as well as sequence and species-specific detection and quantification of gluten proteins in different food products derived from cereal crops. This approach has been used for the detection of immunogenic gluten peptides eliciting immune and inflammatory responses in CD patients [57]. MS-based proteomics has been successfully applied to detect wheat gluten contamination in various food products. In a study, Sealey-Voyksner et al. applied LC-MS/MS analysis to monitor the presence of trace amounts of wheat gluten peptides in a list of gluten-free and gluten-containing food products [58]. ESI-triple quadrupole MS/MS instrument (ESI-QQQ) enabled the detection and quantification of six gluten peptides in different foodstuffs over a range of 10 pg/mg–100 ng/mg. The accuracy of detection was estimated to be 90 % for the lower concentration level and 98 % for concentrations within the range of 30–60 ng/mg. In a recent study, targeted MS-based proteomics enabled highly sensitive and accurate detection of close to 1 ppm wheat gluten contamination in oat flour [54]. Colgrave et al. also developed a MS-based proteomics approach involving a combination of LC-MS/MS and MRM for targeted quantification of wheat gluten in commercially sourced flours, including rye, millet, oats, sorghum, buckwheat, and three varieties of soy [21]. The method also enabled the rapid, sensitive, and reproducible detection of wheat gluten peptides in intentionally-contaminated soy flour at concentrations down to 15 mg/kg. These studies have provided evidences showing that MS-based proteomics methodologies are extremely robust and sensitive in the detection of food contaminations in a wide dynamic range.

4.3.2 Food Borne Microbes

The detection of food-borne pathogenic microorganisms, mostly bacteria and fungi, and their toxins in food products is one of the main challenges of food safety. Food-borne microbes impose a serious health risk to humans and thus, are the major causes of food-borne illnesses (commonly known as food poisoning) worldwide. With the concurrent development of antibiotic resistance, the health risk associated with food-borne microbes is increasing. According to the US Department of Agriculture's Food Safety and Inspection Service, the most common pathogens for food-borne illnesses are Salmonella, Campylobacter, Shigella, Cryptosporidium, Shiga-toxin producing *Escherichia coli*, Vibrio, Listeria, Yersina, and Cyclospora [59]. Food-borne pathogens usually release proteinaceous factors (toxins) into the food matrix without changing appearance, odor or flavor of the food product. These toxic proteins are produced to facilitate the infection and multiplication of the microbe itself in the host cells. Some food borne bacteria pathogens secrete the most powerful human poisons known including enterotoxins, neurotoxins, leukocidins and hemolysins [60]. The majority of food borne microbial toxins are heat stable and are able to escape and remain active in thermally processed food products. With respect to the sensitivity and robustness of the MS-based proteomic methodologies, proteomics can therefore be used for the exploration of food products for the presence of microbial pathogens or their toxin proteins.

MALDI-TOF MS represents a fast, sensitive, and cost-effective tool for the identification of microbial toxins in diverse food products. Among food borne pathogens, *Staphylococcus aureus* is an important enteric bacteria commonly detected in foodstuffs of animal origin such as meat and milk. The pathogenicity of *S. aureus* largely depends on enterotoxins produced and secreted by the bacterium into the food matrix. The secreted toxins are resistant to heat, freezing, and irradiation and therefore, may remain active even after sterilization and pasteurization. The safety concern regarding the presence of *S. aureus* in food products has motivated food scientists to search for novel technologies for targeted detection of secreted enterotoxins. In a preliminary study, SDS-PAGE fractionation of milk proteins precipitated by a mixture of dichloromethane and acidified water followed by MALDI-TOF MS analysis led to the detection of Staphylococcal enterotoxin A in contaminated milk samples [61]. Recently, highly sensitive and targeted LC-MS/MS in MRM mode was successfully applied to quantify Staphylococcal enterotoxins A and B in milk samples [62]. The method also allowed successful discrimination of A and B type toxins at a detection limit of 8 and 4 ng/g, respectively, in a single run, proving the power of the targeted MS analysis for the identification of contaminated food products.

MS-based proteomics has also well performed in the identification of food borne pathogens such as *Listeria monocytogenes, E. coli, and Yersinia pestis*. Consumption of food products contaminated by *L. monocytogenes* leads to the development of listeriosis, which has a high fatality rate (20–30 %) in high-risk individuals [63]. *L. monocytogenes* was conventionally detected by culture-based methods and biochemical tests, which are generally highly expensive and time-consuming. Recently, Jadhav et al. [64] applied a MALDI–TOF MS approach for direct detection of *L. monocytogenes* in food enrichment broth taken from three different solid foods including chicken pâté, fresh cantaloupe, and Camembert cheese with a relatively high sensitivity (down to 10 colony-forming unit (cfu) per mL) and limited time (30 h). In another study, a combination of intact cell immunocapture and LC-MS/MS in targeted SRM mode was used for direct detection of *Y. pestis* in contaminated milk samples [65]. In this approach, bacterial cells in milk samples were enriched using immobilized monoclonal antibodies specifically binding to the *Y. pestis*-specific plasmid encoded surface proteins (pFra and pPla) followed by trypsin digestion of protein mixture and targeted identification of peptide markers. The method allowed for the rapid detection of *Y. pestis* in contaminated milk or tap water at a detection limit of 20,000 cfu/mL, which was comparable to the sensitivity of the conventional immunoassay tests. Ochoa et al. [66] employed a similar methodology for the identification of enterohemorrhagic *E. coli* serotype O157:H7 in ground beef samples. Through a pre-enrichment step, authors were able to monitor meat contamination for pathogenic *E. coli* at a detection limit of 2 million cells per mL.

Contamination of food products with microbes and their toxins are not limited to bacteria. Fungi are also important source of biological contaminations, especially in cereal derived products, compromising the safety of food products and imposing health hazards by the secretion of mycotoxins. Mycotoxins are toxic secondary metabolites commonly detected in foods contaminated by molds of genera Aspergillus, Penicillium and Fusarium [67]. Aflatoxins, zearalenone, fumonisins, ochratoxins, trichothecenes, tremorgenic toxins, and ergot alkaloids are among common food borne mycotoxins detected in contaminated food products. Currently, LC-MS/MS is one of the technically robust and sensitive tools for rapid and simultaneous detection and quantification of non-proteinaceous toxins in contaminated foods [68]. For example, this method has been successfully applied by Capriotti et al. for the detection of a range of mycotoxins including thricotecenes A and B, zearalenone, fumonisins, ochratoxin A, enniatins and beauvericin in biscuits [69]. Using a combination of solid phase extraction procedure and LC-MS/MS analysis, the authors were able to quantify as low as 0.04 µg/kg mycotoxin in commercially sourced biscuits samples.

Proteomics has also been used to overcome food safety concerns regarding the presence of prion proteins in animal derived products. Prions are unique class of transmittable infectious proteins causing a group of transmissible spongiform encephalopathies diseases in animals and humans [70]. Prion proteins are usually transmitted by the consumption of meat products derived from infected animals [71]. As an example, the consumption of meat products from cows infected with bovine spongiform encephalopathy (mad cow disease) in the United Kingdom resulted in transmission of the disease to humans and the development of a new variant of Creutzfeldt-Jakob Disease (nvCJD) [70]. Sensitive and accurate detection of prion proteins are therefore necessary for the assessment of the safety of meat products and also for the early detection of diseased animals and their elimination from food chain. To date, the detection of prion protein (PrP^{Sc}) has largely relied on western blotting (WB), ELISA, and the conformation dependent immunoassay (CDI) with the detection limit ranging from 10 to 20 pmol for WB down to the 0.1 pmol for the CDI immunoassay [72]. However, prion proteins are present in attomol quantities in biological samples of infected animals, necessitating the improvement of the detection limit for the assurance of the safety of animal derived food products. To this aim, Onisko and coworkers applied targeted nLC-MS/MS analysis to quantify PrP protein in the brains of terminally ill Syrian hamsters at a detection limit of 27–30 amol [73], proving the power of the MS-based proteomics in the detection of biologically-contaminated food products at a detection limit far below what could be detected by the other analytical procedures.

4.4 Proteomics in Detection of Food Adulterations

Food adulteration is defined as any intentional or unintentional partial or complete substitution of a food product with inferior or forbidden materials or the removal of some valuable ingredients from the main food article, which both may lead to decreased food quality. Food adulteration is now increasingly being practiced in various consumer sectors. Most agricultural foodstuffs including dairy and meat products, cereals, legumes, beverages, eggs, and fruit products are commonly subjected to adulteration for short-term economic profit, without any concerns regarding the potential health risks. Consumption of food products containing undeclared ingredients may impose health risks such as food toxicity, allergy, and intolerance in sensitized individuals [74]. Authentication of raw materials and processed food products are therefore of utmost importance from both consumer's and industrie's points of view. The detection of food adulteration is a technical challenge since adulterated food products have almost the same chemical composition compared with their original counterparts. This has forced food standard agencies and control laboratories to search for novel technologies to define the molecular composition of different food products and to differentiate genuine products from similar but adulterated ones [74]. It is worth to note that the definition of a food product at the molecular level is not a trivial task, since food products are usually complex in nature and are composed of different biological materials, each with unique chemical and molecular compositions [74]. Numerous analytical methods based on chemical and physical properties of foodstuffs have been developed and applied to verify the identity of different food products and to check for possible illegal adulteration. Since protein composition of foodstuffs changes due to adulteration, proteomics can therefore be used for the detection of such food frauds. In addition, proteomics can also be employed for the identification of unique protein markers for rapid, sensitive, and high-throughput detection of adulterations commonly practiced in different food products.

4.4.1 Adulteration in Dairy Products

As a major constituent of human diet and main source of nutrients, proteins, and microelements for the newborns and adults, raw milk is frequently subjected to adulteration. Ovine, caprine, and buffalo milk are usually adulterated with bovine milk, because of their limited availability and higher prices. Interestingly, caprine milk has high nutritional value compared with bovine milk and has been recommended as a nutritious food for physically weak people [31]. It also displays limited allergic reactions in infants who are intolerant to cow's milk [75]. To demonstrate the applicability of the MS-based proteomics in detection of milk adulteration, Chen and coworkers [31] applied HPLC/ESI-MS to detect bovine milk adulteration in caprine milk. Using a combination of solvent fractionation and MS analysis, they were able to separate, quantify, and identify beta-lactoglobulin as marker for cow's milk adulteration in caprine milk at levels as low as 5 %. Recently, Girolamo et al. [24] reported the identification of bovine, buffalo, and ovine milk adulteration in caprine milk at an adulteration level down to the 0.5 % using a combination of MALDI-TOF MS and principle component analysis (PCA). Interestingly, MALDI-TOF was represented as fast, reliable, robust, and sensitive analytical instrument that could be adapted for routine analysis of dairy products without any needs for laborious pre-analytical sample separation steps. In another study, Sassi et al. [23] developed a peptidomic profiling strategy based on MALDI-TOF MS instrument to explore cow's milk adulteration in ovine, caprine, and water buffalo milk as well as the addition of powdered cow's milk to its fresh counterpart. In addition, MALDI-TOF analysis also enabled rapid detection of thermal treatment markers in different types of commercially sourced milks. MALDI-TOF MS-based proteomic has also been used for the determination of cow milk adulteration in donkey milk with a detection limit of 0.5 % [24]. Donkey milk can be used as the best substitution for mother's milk in newborns that are intolerant to cow's milk. Indeed, donkey milk has high nutritional and health promoting characteristics as well as limited allergic properties compared to cow's milk [76].

4.4.2 Adulteration in Meat Products

Clear and informative labeling of meat products are critical from both economic and religious points of view. For example, the consumption of meat products derived from species such as pork, horse, donkey, and many other non-ruminant and carnivorous animals is forbidden by Islam marked as non-Halal or Haram foodstuffs. Therefore, the presence of trace amounts of meat from such species in food products make them Haram and/or Makrouh for Muslims. Nevertheless, raw meats or processed meat products are frequently subjected to adulteration by fraudulent addition or substitution with meat from lower priced or forbidden species. This necessitates the development of species-specific markers for the authentication of meat products. In this regard, MS-based proteomics approaches have been shown to provide the required specificity and precision. To this aim, Sentandreu and coworkers applied a MS-based proteomics approach to substantiate the presence of chicken in meat mixes using species-specific peptide biomarkers derived from myofibrillar proteins [18]. Myofibrillar proteins are of particular interest in this respect because they are quite resistant to food processing such as cooking and heat treatment making them ideal candidate markers for meat authentication. The developed method allowed the detection of as low as 0.5 % w/w contaminating chicken in pork meat with a high sensitivity and precision, even after cooking. Von Bargen et al. [77] applied a combination of fractionation for the myofibrillar and sarcoplasmic protein fractions from unprocessed meat samples and targeted MRM-based method for quantitative monitoring of species-specific peptides for rapid and sensitive detection of horse and pork meat contamination in beef. Single MRM transition allowed the detection of as low as 0.55 % horse and pork meat contamination in beef. However, triple MRM transition using a QTRAP instrument extended the limit of detection to as low as 0.13 % pork meat contamination in beef, suggesting that triple MRM significantly improved the specificity and sensitivity of detection. In a complementary study, von Bargen and coworkers applied the same approach for the detection of pork and horse meat contamination in processed beef and several commercially sourced food products

[17]. The method resulted in the detection of as low as 0.24 % horse and pork meat in processed beef meat matrix. They demonstrated the applicability of targeted MS-based proteomics approach for the rapid and sensitive detection of meat contamination in extensively processed (boiled or fried) food products. Recently, Montowska and coworkers applied the ambient liquid extraction surface analysis mass spectrometry (LESA-MS) methodology for the identification of peptide markers for authentication of thermally processed meat products [78]. Interestingly, the method allowed the detection of 10 % (w/w) of pork, horse, and turkey meat and 5 % (w/w) of chicken meat in beef, relatively faster and simpler than previously used approaches. Indeed, the LESA-MS technique requires minimal sample preparation and does not need any sample pre-fractionation steps, enabling rapid and sensitive detection of species-specific peptide biomarkers for the authentication of raw and processed food products. These results suggest that peptide biomarkers are sufficiently resistant during thermal treatments and are therefore best suited for the authentication of processed food products.

The adulteration of meat products is not just limited to the fraudulent mixing of meat samples of different origins, but it also involves the use of less valuable components of animal origin such as offal and connective tissues or the addition of vegetable proteins in place or in combination with meat in the production of processed meat products. For example, food products manufactured from meat are sometimes supplemented with limited amount of soybean proteins as emulsifiers in order to avoid fat coalescence during heat treatment. Soybean proteins have excellent nutritional qualities and functional properties which have promoted their use in a variety of food products [79]. However, the lower cost of soybean proteins compared with proteins of animal origin promotes their fraudulent use in quantities exceeding the permitted values as emulsifiers. Because of the allergic properties of soybean proteins, undeclared addition of these proteins may impose health risk to patients who have allergy to soybean proteins [79]. This has forced standard agencies to develop analytical methodologies enabling sensitive and accurate quantification of soybean proteins in diverse food products that might be at risk of soybean adulteration. In this regard, MS-based proteomics approaches have the required speed and sensitivity for monitoring soybean-specific proteins in processed food products, as exemplified by Leitner and coworkers [29]. In this work, authors were able to specifically monitor the presence of soybean proteins in processed meat products. Using a combination of chromatographic separation and MS/MS analysis, different subunits of glycinin A protein (more specifically glycinin G4 subunit A4) were identified as markers for discriminating soybean-containing foods from soybean-free counterparts.

4.4.3 Adulteration in Cereals

Proteomics can also be used for the identification of adulterations commonly practiced in cereals. The detection of common wheat contamination or adulteration in durum wheat is a well-known example. Durum wheat (*Triticum turgidum* ssp. *durum* L.) is usually cultivated for use in the pasta and spaghetti manufacturing industries, while common wheat (*Triticum astivum*) is largely used for bread, biscuits, and bakery production. The presence of common wheat in durum wheat changes the rheological properties of durum wheat flour, making it unsuitable for pasta production [80]. The higher price of durum wheat (by about 25 %) compared with common wheat has led to fraudulent addition of common wheat to durum wheat, necessitating the development of sensitive approaches for the detection of possible adulterations [19]. In this regards, many studies have exploited the inherent differences that arise from different ploidy levels of durum wheat (AABB) and common wheat (AABBDD) for the detection of common wheat in durum wheat samples. Targeted amplification of DNA sequences belonging to the DD genome by polymerase chain reaction (PCR) can be performed with a relatively high sensitivity and specificity using a pair of species-specific primers. This approach has been successfully used for the detection of common wheat contamination in durum wheat at a detection limit down to the 0.2 % [81]. Although this approach has been proved to be sensitive enough for the detection of common wheat contamination in durum wheat, the requirement for DNA extraction and sensitivity of DNA to food processing make it unsuitable for high throughput screenings. Recently, Prandi et al. [19] exploited the power of LC-MS for the detection of common wheat contamination in durum wheat. Detection was made possible with a single common wheat specific peptide resulted from the co-digestion of proteins with pepsin and chymotrypsin. In addition, they were also able to accurately detect common wheat contamination in diverse commercially-sourced durum wheat flour on the Italian market, indicating that common wheat contamination in durum wheat was more common. In a complementary study, Russo and coworkers applied ultra-performance liquid chromatography (UPLC)-ESI-MS/MS based on MRM for quantification of common wheat [22]. Targeted analysis of a single tryptic peptide from puroindoline a (Pin-a) and a cysteine-rich amphiphilic lipid binding protein from common wheat, allowed the quantification of common wheat in durum wheat at a detection and quantification limit of 0.01 and 0.03 %, respectively. In addition, the method allowed accurate monitoring of common wheat contamination in both raw materials (kernels) and processed durum wheat-derived products (pasta).

4.5 Conclusion

Proteomics has now become the method of choice for safety and quality assessment of many agricultural foodstuffs, thanks to the advances made in MS analysis. Methods based on 2-DE proteomics in combination with immunoblotting have been extensively utilized for the identification of food allergens in a diverse range of food products. Proteomics has also been well-performed in the detection of microbial contamination and their toxins. Interestingly, microbial toxins including both proteins and metabolites in contaminated food products can now be detected and quantified at attomol quantities in targeted and multiplexed mode using novel LC-MS/MS instruments. Recent progress in MS including the introduction of MRM holds great promises for rapid, targeted, and cost-effective detection and quantification of marker proteins for food contamination and adulteration. With respect to food adulterations, proteomics has now provided the sensitivity and robustness required for the detection of an adulterated food item from its genuine counterpart, even if both have almost the same chemical compositions.

References

1. Mamone G, Picariello G, Caira S, Addeo F, Ferranti P (2009) Analysis of food proteins and peptides by mass spectrometry-based techniques. J Chromatogr A 1216:7130–7142
2. Josić D, Kovac S, Gaso-Sokac D (2013) Nutritionally Relevant Proteins. In: Toldrá F, Nollet LLM (eds) Proteomics in Foods: principles and applications. Springer, US, Boston, MA, pp 425–446
3. Veraverbeke WS, Delcour JA (2002) Wheat protein composition and properties of wheat glutenin in relation to breadmaking functionality. Crit Rev Food Sci Nutr 42:179–208
4. Gašo-Sokač D, Kovač S, Josić D (2010) Application of proteomics in food technology and food biotechnology: process development, quality control and product safety. Food Technol Biotechnol 48
5. Ong SE, Mann M (2005) Mass spectrometry-based proteomics turns quantitative. Nat Chem Biol 1:252–262
6. Picariello G, Mamone G, Addeo F, Ferranti P (2011) The frontiers of mass spectrometry-based techniques in food allergenomics. J Chromatogr A 1218:7386–7398
7. Fenn JB, Mann M, Meng CK, Wong SF, Whitehouse CM (1989) Electrospray ionization for mass spectrometry of large biomolecules. Science 246:64–71
8. Hillenkamp F, Karas M, Beavis RC, Chait BT (1991) Matrix-assisted laser desorption/ionization mass spectrometry of biopolymers. Anal Chem 63:1193A–1203A
9. Gygi SP, Aebersold R (2000) Mass spectrometry and proteomics. Curr Opin Chem Biol 4:489–494
10. Yates JR 3rd (2000) Mass spectrometry. From genomics to proteomics. Trends Genet 16:5–8
11. Aebersold R, Goodlett DR (2001) Mass spectrometry in proteomics. Chem Rev 101:269–295
12. Zhang G, Annan RS, Carr SA, Neubert TA (2014) Overview of peptide and protein analysis by mass spectrometry. Curr Protoc Mol Biol 108:10 21 11–10 21 30
13. Yates JR, Ruse CI, Nakorchevsky A (2009) Proteomics by mass spectrometry: approaches, advances, and applications. Annu Rev Biomed Eng 11:49–79
14. Lange V, Picotti P, Domon B, Aebersold R (2008) Selected reaction monitoring for quantitative proteomics: a tutorial. Mol Syst Biol 4:222
15. Gillette MA, Carr SA (2013) Quantitative analysis of peptides and proteins in biomedicine by targeted mass spectrometry. Nat Methods 10:28–34
16. Cunsolo V, Muccilli V, Saletti R, Foti S (2013) MALDI-TOF mass spectrometry for the monitoring of she-donkey's milk contamination or adulteration. J Mass Spectrom 48:148–153
17. Von Bargen C, Brockmeyer J, Humpf HU (2014) Meat authentication: a new HPLC-MS/MS based method for the fast and sensitive detection of horse and pork in highly processed food. J Agric Food Chem 62:9428–9435
18. Sentandreu MA, Fraser PD, Halket J, Patel R, Bramley PM (2010) A proteomic-based approach for detection of chicken in meat mixes. J Proteome Res 9:3374–3383

19. Prandi B, Bencivenni M, Tedeschi T, Marchelli R, Dossena A, Galaverna G et al (2012) Common wheat determination in durum wheat samples through LC/MS analysis of gluten peptides. Anal Bioanal Chem 403:2909–2914
20. Calvano CD, Monopoli A, Loizzo P, Faccia M, Zambonin C (2013) Proteomic approach based on MALDI-TOF MS to detect powdered milk in fresh cow's milk. J Agric Food Chem 61:1609–1617
21. Colgrave ML, Goswami H, Byrne K, Blundell M, Howitt CA, Tanner GJ (2015) Proteomic profiling of 16 cereal grains and the application of targeted proteomics to detect wheat contamination. J Proteome Res 14:2659–2668
22. Russo R, Cusano E, Perissi A, Ferron F, Severino V, Parente A et al (2014) Ultra-high performance liquid chromatography tandem mass spectrometry for the detection of durum wheat contamination or adulteration. J Mass Spectrom 49:1239–1246
23. Sassi M, Arena S, Scaloni A (2015) MALDI-TOF-MS platform for integrated proteomic and peptidomic profiling of milk samples allows rapid detection of food adulterations. J Agric Food Chem 63:6157–6171
24. Di Girolamo F, Masotti A, Salvatori G, Scapaticci M, Muraca M, Putignani L (2014) A sensitive and effective proteomic approach to identify she-donkey's and goat's milk adulterations by MALDI-TOF MS fingerprinting. Int J Mol Sci 15:13697–13719
25. Campos Motta TM, Hoff RB, Barreto F, Andrade RB, Lorenzini DM, Meneghini LZ et al (2014) Detection and confirmation of milk adulteration with cheese whey using proteomic-like sample preparation and liquid chromatography-electrospray-tandem mass spectrometry analysis. Talanta 120:498–505
26. Huang S, Zhang CP, Li GQ, Sun YY, Wang K, Hu FL (2014) Identification of catechol as a new marker for detecting propolis adulteration. Molecules 19:10208–10217
27. Addeo F, Pizzano R, Nicolai MA, Caira S, Chianese L (2009) Fast isoelectric focusing and antipeptide antibodies for detecting bovine casein in adulterated water buffalo milk and derived mozzarella cheese. J Agric Food Chem 57:10063–10066
28. Cozzolino R, Passalacqua S, Salemi S, Malvagna P, Spina E, Garozzo D (2001) Identification of adulteration in milk by matrix-assisted laser desorption/ionization time-of-flight mass spectrometry. J Mass Spectrom 36:1031–1037
29. Leitner A, Castro-Rubio F, Marina ML, Lindner W (2006) Identification of marker proteins for the adulteration of meat products with soybean proteins by multidimensional liquid chromatography-tandem mass spectrometry. J Proteome Res 5:2424–2430
30. Luykx DM, Cordewener JH, Ferranti P, Frankhuizen R, Bremer MG, Hooijerink H et al (2007) Identification of plant proteins in adulterated skimmed milk powder by high-performance liquid chromatography—mass spectrometry. J Chromatogr A 1164:189–197
31. Chen RK, Chang LW, Chung YY, Lee MH, Ling YC (2004) Quantification of cow milk adulteration in goat milk using high-performance liquid chromatography with electrospray ionization mass spectrometry. Rapid Commun Mass Spectrom 18:1167–1171
32. Cordewener JH, Luykx DM, Frankhuizen R, Bremer MG, Hooijerink H, America AH (2009) Untargeted LC-Q-TOF mass spectrometry method for the detection of adulterations in skimmed-milk powder. J Sep Sci 32:1216–1223
33. Kuppannan K, Albers DR, Schafer BW, Dielman D, Young SA (2011) Quantification and characterization of maize lipid transfer protein, a food allergen, by liquid chromatography with ultraviolet and mass spectrometric detection. Anal Chem 83:516–524
34. Ellis DI, Brewster VL, Dunn WB, Allwood JW, Golovanov AP, Goodacre R (2012) Fingerprinting food: current technologies for the detection of food adulteration and contamination. Chem Soc Rev 41:5706–5727
35. D'alessandro A, Zolla L (2012) We are what we eat: food safety and proteomics. J Proteome Res 11:26–36
36. Herrero M, Simo C, Garcia-Canas V, Ibanez E, Cifuentes A (2012) Foodomics: MS-based strategies in modern food science and nutrition. Mass Spectrom Rev 31:49–69
37. Di Girolamo F, Muraca M, Mazzina O, Lante I, Dahdah L (2015) Proteomic applications in food allergy: food allergenomics. Curr Opin Allergy Clin Immunol 15:259–266

38. Sicherer SH, Sampson HA (2014) Food allergy: epidemiology, pathogenesis, diagnosis, and treatment. J Allergy Clin Immunol 133(291–307):e295
39. Stevenson SE, Chu Y, Ozias-Akins P, Thelen JJ (2009) Validation of gel-free, label-free quantitative proteomics approaches: applications for seed allergen profiling. J Proteomics 72:555–566
40. Natale M, Bisson C, Monti G, Peltran A, Garoffo LP, Valentini S et al (2004) Cow's milk allergens identification by two-dimensional immunoblotting and mass spectrometry. Mol Nutr Food Res 48:363–369
41. Hochwallner H, Schulmeister U, Swoboda I, Spitzauer S, Valenta R (2014) Cow's milk allergy: from allergens to new forms of diagnosis, therapy and prevention. Methods 66:22–33
42. Chassaigne H, Tregoat V, Norgaard JV, Maleki SJ, Van Hengel AJ (2009) Resolution and identification of major peanut allergens using a combination of fluorescence two-dimensional differential gel electrophoresis, Western blotting and Q-TOF mass spectrometry. J Proteomics 72:511–526
43. Schmidt H, Gelhaus C, Latendorf T, Nebendahl M, Petersen A, Krause S et al (2009) 2-D DIGE analysis of the proteome of extracts from peanut variants reveals striking differences in major allergen contents. Proteomics 9:3507–3521
44. Beyer K, Bardina L, Grishina G, Sampson HA (2002) Identification of sesame seed allergens by 2-dimensional proteomics and Edman sequencing: seed storage proteins as common food allergens. J Allergy Clin Immunol 110:154–159
45. Beyer K, Grishina G, Bardina L, Grishin A, Sampson HA (2002) Identification of an 11S globulin as a major hazelnut food allergen in hazelnut-induced systemic reactions. J Allergy Clin Immunol 110:517–523
46. Krishnan HB, Kim WS, Jang S, Kerley MS (2009) All three subunits of soybean beta-conglycinin are potential food allergens. J Agric Food Chem 57:938–943
47. Houston NL, Lee DG, Stevenson SE, Ladics GS, Bannon GA, Mcclain S et al (2011) Quantitation of soybean allergens using tandem mass spectrometry. J Proteome Res 10:763–773
48. Careri M, Costa A, Elviri L, Lagos JB, Mangia A, Terenghi M et al (2007) Use of specific peptide biomarkers for quantitative confirmation of hidden allergenic peanut proteins Ara h 2 and Ara h 3/4 for food control by liquid chromatography-tandem mass spectrometry. Anal Bioanal Chem 389:1901–1907
49. Schuppan D, Junker Y, Barisani D (2009) Celiac disease: from pathogenesis to novel therapies. Gastroenterology 137:1912–1933
50. Tye-Din JA, Stewart JA, Dromey JA, Beissbarth T, Van Heel DA, Tatham A et al (2010) Comprehensive, quantitative mapping of T cell epitopes in gluten in celiac disease. Sci Transl Med 2:41ra51
51. Biagi F, Klersy C, Balduzzi D, Corazza GR (2010) Are we not over-estimating the prevalence of coeliac disease in the general population? Ann Med 42:557–561
52. Biagi F, Campanella J, Martucci S, Pezzimenti D, Ciclitira PJ, Ellis HJ et al (2004) A milligram of gluten a day keeps the mucosal recovery away: a case report. Nutr Rev 62:360–363
53. Kaukinen K, Collin P, Huhtala H, Maki M (2013) Long-term consumption of oats in adult celiac disease patients. Nutrients 5:4380–4389
54. Fiedler KL, Mcgrath SC, Callahan JH, Ross MM (2014) Characterization of grain-specific peptide markers for the detection of gluten by mass spectrometry. J Agric Food Chem 62:5835–5844
55. Anonymous (2013) Food labeling; gluten-free labeling of foods. In: Federal Register, pp 47154–47179
56. Hernando A, Mujico JR, Mena MC, Lombardía M, Mendez E (2008) Measurement of wheat gluten and barley hordeins in contaminated oats from Europe, the United States and Canada by Sandwich R5 ELISA. Eur J Gastroenterol Hepatol 20:545–554
57. Shan L, Molberg O, Parrot I, Hausch F, Filiz F, Gray GM et al (2002) Structural basis for gluten intolerance in celiac sprue. Science 297:2275–2279

58. Sealey-Voyksner JA, Khosla C, Voyksner RD, Jorgenson JW (2010) Novel aspects of quantitation of immunogenic wheat gluten peptides by liquid chromatography-mass spectrometry/mass spectrometry. J Chromatogr A 1217:4167–4183
59. American Medical A, American Nurses Association-American Nurses F, Centers for Disease C, Prevention, Center for Food S, Applied Nutrition F et al (2004) Diagnosis and management of foodborne illnesses: a primer for physicians and other health care professionals. MMWR Recomm Rep 53:1–33
60. Rajkovic A (2014) Microbial toxins and low level of foodborne exposure. Trends Food Sci Technol 38:149–157
61. Sospedra I, Soler C, Manes J, Soriano JM (2011) Analysis of staphylococcal enterotoxin A in milk by matrix-assisted laser desorption/ionization-time of flight mass spectrometry. Anal Bioanal Chem 400:1525–1531
62. Andjelkovic M, Tsilia V, Rajkovic A, De Cremer K, Van Loco J (2016) Application of LC-MS/MS MRM to determine *Staphylococcal enterotoxins* (SEB and SEA) in Milk. Toxins (Basel) 8
63. Ramaswamy V, Cresence VM, Rejitha JS, Lekshmi MU, Dharsana KS, Prasad SP et al (2007) Listeria–review of epidemiology and pathogenesis. J Microbiol Immunol Infect 40:4–13
64. Jadhav S, Sevior D, Bhave M, Palombo EA (2014) Detection of *Listeria monocytogenes* from selective enrichment broth using MALDI-TOF mass spectrometry. J Proteomics 97:100–106
65. Chenau J, Fenaille F, Simon S, Filali S, Volland H, Junot C et al (2014) Detection of *Yersinia pestis* in environmental and food samples by intact cell immunocapture and liquid chromatography-tandem mass spectrometry. Anal Chem 86:6144–6152
66. Ochoa ML, Harrington PB (2005) Immunomagnetic isolation of enterohemorrhagic *Escherichia coli* O157:H7 from ground beef and identification by matrix-assisted laser desorption/ionization time-of-flight mass spectrometry and database searches. Anal Chem 77:5258–5267
67. Zinedine A, Brera C, Elakhdari S, Catano C, Debegnach F, Angelini S et al (2006) Natural occurrence of mycotoxins in cereals and spices commercialized in Morocco. Food Control 17:868–874
68. Capriotti AL, Caruso G, Cavaliere C, Foglia P, Samperi R, Lagana A (2012) Multiclass mycotoxin analysis in food, environmental and biological matrices with chromatography/mass spectrometry. Mass Spectrom Rev 31:466–503
69. Capriotti AL, Cavaliere C, Foglia P, Samperi R, Stampachiacchiere S, Ventura S et al (2014) Multiclass analysis of mycotoxins in biscuits by high performance liquid chromatography–tandem mass spectrometry. Comparison of different extraction procedures. J Chromatogr A 1343:69–78
70. Aguzzi A, Polymenidou M (2004) Mammalian prion biology: one century of evolving concepts. Cell 116:313–327
71. Gough KC, Maddison BC (2010) Prion transmission: prion excretion and occurrence in the environment. Prion 4:275–282
72. Lee DC, Stenland CJ, Hartwell RC, Ford EK, Cai K, Miller JL et al (2000) Monitoring plasma processing steps with a sensitive Western blot assay for the detection of the prion protein. J Virol Methods 84:77–89
73. Onisko B, Dynin I, Requena JR, Silva CJ, Erickson M, Carter JM (2007) Mass spectrometric detection of attomole amounts of the prion protein by nanoLC/MS/MS. J Am Soc Mass Spectrom 18:1070–1079
74. Mamone G, Picariello G, Nitride C, Addeo F, Ferranti P (2013) The role of proteomics in the discovery of marker proteins of food adulteration. In: Toldrá F, Nollet LLM (eds) Proteomics in foods: principles and applications. Springer, US, Boston, MA, pp 465–501
75. Roncada P, Gaviraghi A, Liberatori S, Canas B, Bini L, Greppi GF (2002) Identification of caseins in goat milk. Proteomics 2:723–726
76. Salimei E, Fantuz F (2012) Equid milk for human consumption. Int Dairy J 24:130–142

77. Von Bargen C, Dojahn J, Waidelich D, Humpf HU, Brockmeyer J (2013) New sensitive high-performance liquid chromatography-tandem mass spectrometry method for the detection of horse and pork in halal beef. J Agric Food Chem 61:11986–11994
78. Montowska M, Alexander MR, Tucker GA, Barrett DA (2014) Rapid detection of peptide markers for authentication purposes in raw and cooked meat using ambient liquid extraction surface analysis mass spectrometry. Anal Chem 86:10257–10265
79. Saz JM, Marina ML (2007) High performance liquid chromatography and capillary electrophoresis in the analysis of soybean proteins and peptides in foodstuffs. J Sep Sci 30:431–451
80. Dexter JE, Matsuo RR (1980) Relationship between durum wheat protein properties and pasta dough rheology and spaghetti cooking quality. J Agric Food Chem 28:899–902
81. Arlorio M, Coïsson J, Cereti E, Travaglia F, Capasso M, Martelli A (2013) Polymerase chain reaction (PCR) of puroindoline b and ribosomal/puroindoline b multiplex PCR for the detection of common wheat (*Triticum aestivum*) in Italian pasta. Eur Food Res Technol 216:253–258

Chapter 5
Holistic Sequencing: Moving Forward from Plant Microbial Proteomics to Metaproteomics

Behnam Khatabi, Neda Maleki Tabrizi and Ghasem Hosseini Salekdeh

Abstract In natural environments, plant and microbial communities continuously interact with each other as well as with the environment. Plant-associated microbial communities are critical to plant growth and health. However, the biological impact of microbial communities within natural habitats is not only attributed to the biological activities of a specific microbe, but are also impacted by the microbial communities as a whole. Hence, a deep understanding of the complexity of microbial systems cannot be achieved solely by monitoring certain microbes in isolation. The integration of omics data provides a unique opportunity to tackle long-term problems in the area of plant and microbial ecology. Environmental proteomics or metaproteomics provides a practical tool for a better understanding of the function, structure, dynamics and significance of plant-associated microbial communities in both natural and man-made environments. To begin this chapter, we will present the importance of plant associated microbial communities and their impact on plant growth and health. We will then progress to the application of different omics approaches, especially proteomics for large-scale protein analysis, in order to dissect the molecular basis of plant-microbial interactions in a post genomic era. Next we will examine the advantages of integrating omics data to give a comprehensive understanding of plant microbial communities toward the development of efficient management strategies that reduce the impact of environmental stress and control plant disease epidemiology. We will conclude with a real application of metaproteomics to manage a plant phytobiome, thereby promoting agricultural sustainability.

B. Khatabi
Department of Biological Sciences, Delaware State University, Dover, Delaware, USA

N.M. Tabrizi
Department of Agronomy and Plant Breeding, College of Agriculture and Natural Resources, University of Tehran, Karaj, Iran

N.M. Tabrizi · G.H. Salekdeh (✉)
Department of Systems Biology, Agricultural Biotechnology Research Institute of Iran, Agricultural Research, Education, and Extension Organization, Karaj, Iran
e-mail: hsalekdeh@yahoo.com

Keywords Mass spectrometry · Microbial communities · Plant-environment interactions

5.1 The Second Green Revolution May Depend on Applications of Beneficial Microbes

Plants often establish associations with beneficial microorganisms to cope with unfavorable conditions. They receive benefits from symbiosis, which include increased nutrient uptake from the soil, improved pest and disease resistance and enhanced tolerance to abiotic stresses. Several studies have focused on the biological impact of beneficial plant microbes on plant hosts. Scientists aim at enhancing plant tolerance to environmental stresses by adapting the host plant through the introduction of beneficial microorganisms and initiating a state of symbiosis. They have actively pursued research in applications of beneficial microorganisms as an environmentally-friendly way to help plants to defend themselves from environmental stresses. Instead of using chemical fertilizers or pesticides to combat harmful pest and diseases, plant-microorganism symbiosis can be employed as an alternative way to harness these benefits of microorganisms. Making optimal usage of this symbiosis is important for the sustainability of agriculture, especially for organic farmers, who do not use chemical fertilizers or pesticides.

The biological impact of plant-microbial association on the growth, development and health of plants is substantial. Plant-microbe symbiotic interactions increase biomass production, improve plant health, boost stress tolerance and can enable bioremediation of crop species. An important consideration moving forward is that plant microbial communities are very dynamic and interactive in that both the plant and microbes contribute to the outcome of their relationships. However, as a basis, we know that microorganisms serving together as the plant microbiome produce considerable improvements to plant traits and special functions. Therefore, improving plant microbial communities can boost agricultural productivity. A detailed understanding of how the symbiotic relationships are capable of enhancing plant adaption is crucial and can lead to sustainable agriculture.

5.2 An Omics-Based Approach to Study Microbial Communities

Microbial systems biology aims to understand the basis of multifaceted plant microbial associations by incorporating the analyses of various aspects. The Omics approach is one of the most robust and fastest growing areas within modern biology. The integration of omics data creates a new link between computational analyses and the systems biology of microbial communities such as genomics, population genomics and metagenomics. We now can gain a better understanding about the interactive partners affecting the complexity of plant microbial communities. Highly sensitive metagenomics techniques connect the genetics of each member while simultaneously considering their diversity and how these factors interact within an environment as seen when studying the microbial community system which must consider their individual genetics and this interplay as well as their heterogeneity [1–3]. However, understanding the composition and activity of microbial communities is more than just studying the biodiversity of microbial communities; therefore, there are multiple facets to consider regarding metagenomics data. Importantly, the structure and function of the phytobiome, which is driven by both plant species and environmental conditions, cannot be defined by metagenomics analysis alone [4].

5.3 Plant Microbial Systems Biology; from a Single Microbe to Microbial Communities

Several recent endeavors in microbial systems biology have focused on plant microbial communities to accomplish a whole-system understanding of a dynamics of microbial communities rather than focusing on a single microbe at a certain time point. All of these interacting microbial communities present in the inner and outer of parts of a plant are described in total as the phytobiome [5]. Therefore, integrating several layers of information obtained from computational and experimental techniques including omics data are necessary to construct a complete picture of the metabolic network active during plant growth and adaptation. Metagenomics contributes through high-throughput sequencing data to provide the complete genomes of many plants and microbes. On the other side, with the current advances in mass spectrometry, we have dramatically increased the depth of proteomic discoveries. In addition, the recent boosts in genome and proteome data allow the scientific communities to explore different aspects of plant microbial communities in natural and man-made environments.

In this context, a new subfield of proteomics, termed metaproteomics, or environmental proteomics, has been established which provides substantial knowledge about the structure and function of microbial communities in real environments. As a result of metaproteomics techniques, several novel proteins involved in the regulation of metabolic and signaling pathways have been identified. Specifically, metaproteomics characterizes the function and structure of plant microbial communities, which help us to have a better understanding about the interactive metabolic response and microbe-derived signaling molecules between plants and associated microbial communities. Although the knowledge about the proteins produced and secreted by microbes provides a valuable resource to the microbiological communities, this is just the beginning.

5.4 Moving Forward from Plant Microbial Proteomics to Metaproteomics

Environmental proteomics, or metaproteomics, provides a practical tool for better understanding the function, structure, dynamics and significance of plant-associated microbial communities in both natural and man-made environments. Proteomics data provides a better understanding of the different aspects of gene expression including the proteome and processes affected by modifications of the genome (epigenomics), as well as metabolism (metabolomics). As a complementary technique, metaproteomics not only provides substantial knowledge about the structure and function of microbial communities in real environments, but also covers metagenomics data as it represents the functional molecules active within natural habitats. The metaproteome was first described as "the large scale characterization of the entire protein complement of environmental microbiota at a given point in time" by Wilmes and Bond [6]. Through metaproteomics research we are able to cover different aspects of plant-microbe associations including the function and structure of microbial communities as well as explore plant signaling molecules during pathogenic and mutualistic relationships.

During the past few years, several good reviews have been published highlighting the potential of a proteomics technique in the study of plant microbial communities [7]. However, due to the added value in the application of metaproteomics, the current book chapter has been included to update information concerning environmental proteomics to represent advances in the field of plant microbial ecology.

5.5 Metaproteomics Reveals the Structure and Function of Microbial Communities

Many different factors including the diversity, structure, function, and population dynamics of plant microbial communities are altered in association with different plant species and environmental conditions. It has been shown that the alteration in the structure and dynamics of plant-associated microbial communities has a direct impact on plant health and growth, which enables us to select effective strategies to improve plant health [8]. A number of researchers focus on the phytobiome profile of the plant microbial communities associated with healthy and diseased plants using metaproteomics [9, 10]. There is a clear difference in rhizosphere plant-associated microbial populations between plants grown with or without

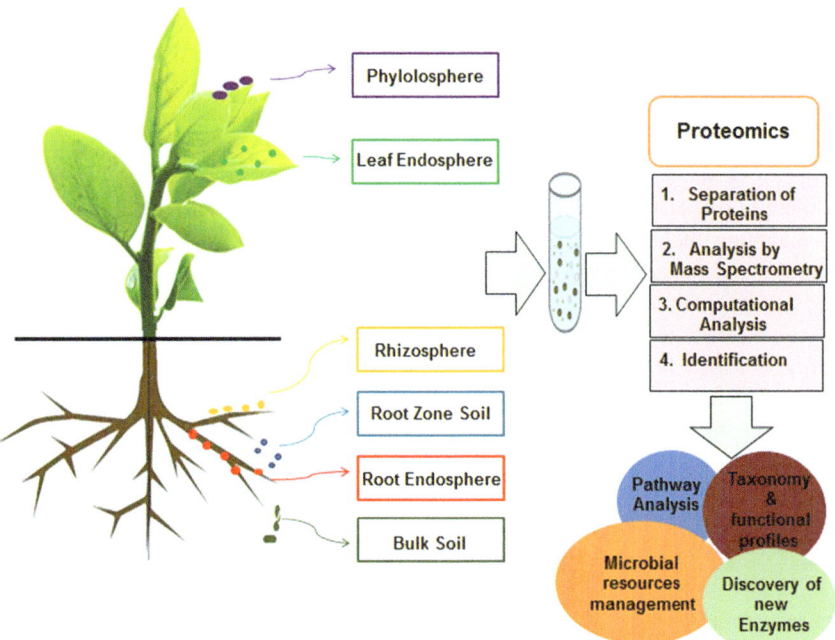

Fig. 5.1 Overview of metaproteomics in plant-microbiome interaction. Plant metaproteomics is the study of proteins in microbial community and present in different parts of a plant such as phyllosphere, leaf endosphere, rhizosphere, root endosphere, root zone soil and bulk soil. Current methods in proteomics were used for making omic data for different targets such as identification of novel enzyme and protein pathways. Also high accuracy identification of microbial functions can lead to management of microbial resources for sustainable agriculture

establishing mutualistic and pathogenic interactions with plant roots. In Fig. 5.1, we show the overview of metaproteomics techniques used in environmental studies.

5.6 Using Metaproteomics to Dissect Plant-Microbe Communications

Several scientists have tried to answer the question of how plants and plant microbial communities interact with each other in real environments. Metaproteomics is a helpful tool to study the active compounds and the secreted proteins released by plants and microbes during the actual interplay seen in nature [11]. However, the exchange between plant signaling molecules and microbial communities is not fully understood. Only recently, the links between the composition of root exudates released into the rhizosphere and the fluctuations of the microbial communities have been studied [12]. The structure and composition of plant-associated microbial communities exhibit differences during incompatible and compatible interactions with a pathogen, which leads to disease resistance or susceptibility [13].

Several attempts have been made to show the metaproteomics profile of rhizosphere-associated microbes under different environment conditions. Comparative metaproteomics data analyses collected from samples from the rhizosphere of different crops have led to the identification of several secreted proteins and signaling molecules [11]. Metaproteomics represents the molecular basis of plant stress adaptation responses induced by plant-associated microbes [14]. We hope that by applying metaproteomics techniques secreted proteins participating in the communication taking place during plant microbial interactions that are important for pathogen development can be identified (unless you are currently working in this area). For instance, by blocking the plant-pathogen crosstalk, we hope to be able to generate broad-spectrum disease resistance in plants.

5.7 Technical Overview of Metaproteomics Approach

Metaproteomics as a novel approach for studying functional microbial ecology investigations involves several key steps, each of them with its own challenges. These factors should be carefully considered such as: experimental design, sampling method, protein purification, protein isolation, liquid chromatography (LC)-tandem mass spectrometry (MS/MS), bioinformatics, and protein identification. Additionally, reproducible scientific studies require exact programing and well-documented experimentation. To achieve higher resolution, the practical endeavors should be outlined before a study's launch and potential challenges should be noted.

Before facing other challenges, providing optimized samples with sufficient copy number is essential. Hence, the preparation of high quality protein extracts is a very important issue, which can be challenging when dealing with environmental samples. This is clearly shown in the way most proteome research is limited only to isolated microbes in axenic culture. In fact, protein isolation methods need a noticeable improvement of protein extraction protocols and sample preparation [15, 16].

During cell lysis and protein extraction, the greatest problem is the presence of a high amount of impurities, especially the high amount of humic acid in soil samples, and also the various presentations of microbes in a population where some are sporadic and some form a biofilm given a special substrate [17]. The process of dealing with impurities needs improvement when performing protein extraction and fractionation methods for metaproteomics. Thus, optimization of sample preparation should be emphasized because the challenging computational analyses that take place at the end of the study are critically affected by these primary stages. A complicated metaproteome analysis depends on these three stages.

5.7.1 Protein Extraction

In order to characterize the whole protein profile of an entire microbial community, methods of extraction need to be employed using the most reproducible and efficient methods. Appropriate protocols currently being used lyse the cells by using buffers and include one or more detergents (SDS, CHAPS, Triton X-100) [6, 18, 19] chaotropic agents (urea, guanidine hydrochloride) [6], reducing agents (dithiothreitol (DTT), tributylphosphine) [20], other organic/inorganic compounds (phenol, NaOH) [21, 22] and temperature treatment (such as boiling, freezing, thawing) [23, 24] or mechanical destruction (grinding, bead beating and sonication) [25–27]. Since Gram positive and Gram negative bacteria and fungi have huge differences in their structures and each of them respond differently to protein extraction, an optimized protocol should be used for each type to reach the maximum yield. Using phenol extraction to remove the humic acid that sometimes impacts highly complex samples is another example of tailoring the protocol to the situation which is an essential step for metaproteome analysis.

5.7.2 Sample Preparation

The next step is included because all detergents used during extraction and any other interfering compounds should be removed before digesting the proteins. After microbial population protein extraction, those components that might interfere in enzyme activities during isolation or MS analysis should be eliminated. To do this, a classic action trichloroacetic acid, acetone, or ammonium acetate/methanol [19, 22, 28] is added to the protein extract. The protein plate can be solved in a buffer, which is used in the later steps, but a considerable portion of protein will be lost due to aggregation during this procedure. Another suitable method pursued for gel digestion of the extracted protein is one-dimensional electrophoresis (1-DE) that causes the interferer components to be trapped in the gel matrix and allows for categorization of samples in gel slices [22]. This method is efficient, but it is also time consuming and has a low level of reproducibility. The novel alternative is represented through the filter-aided sample preparation (FASP) method in which enzymatic cleavage and clean up occurs in a centrifugal filter with a high molecular weight cut off [29]. This method is very suitable for environmental samples, especially for those with low amounts of protein [30].

5.7.3 Pre-fractionation

Pre-fractionation should be done on peptides and proteins before MS detection to decrease sample complexity and increase analysis depth. This improves the extent of information achievable by shotgun MS analysis. This has been accomplished in previous metaproteomic studies by protein fractionation (mainly by 1-DE and GELFrEE approaches) [28, 31] and/or peptide level (most commonly by means of two-dimensional liquid chromatography (2D-LC). However, all of these useful methods required extensive laboratory effort, as well as increasing challenges in analytical reproducibility. In particular, 2D-LC tandem mass spectrometry (MS/MS) has very remarkable analysis depth and is technically demanding, but, above all, requires long processing times for a single sample (22 h in a typical experimental setting for metaproteome samples) [32, 33]. Recently, an approach has been described as single-run nanoLC-MS/MS, enabling the identification of several thousands of proteins per run from different types of samples [34–37]. After separation, computational analysis should be done on the acquired data. In the case of a fully sequenced organism's proteins identification, it is done with comparison to the peptide fingerprints in data banks. But in species without a sequenced genome, protein identification can be done with peptides and their MS/MS fragments if there is high homology to known proteins, or in the case of low homology, with peptides which are sequenced with MS and show good homology with published sequences. Routinely, this identification is done with algorithms like XI tandem [38, 39].

Standard banks for protein identification include the National Center for Biotechnology Information (NCBI) [40], UniProt/SwissPort and UniProtkB/TrEMBL [41]. Depending on the metaproteome, more specific databases or searches against metagenomes of the same samples can result in further protein identification [42]. The difficulty in characterizing the metaproteome in comparison to simpler proteomic investigations that use defined culture media is that the complex metaproteome's taxonomic composition is unknown and can result in increases in false positives. In this regard, decreasing the size to only the known sequences provides a viable option [43]. In this way, some of the false positives will be reduced and a wider range of proteins will be identified. The next step is protein identification based on the homology of two peptides [44], however, in the case of high resolution MS data, using one peptide is also acceptable [45]. By using multiple algorithms, we increase the number of correctly identified peptides and proteins. Even with the best algorithm, only the proteins with corresponding sequences are covered in a single database. To solve this matter, we can use de novo sequencing of peptides using acquired spectra and identify proteins using a basic local alignment tool based in Ms. A complication arises because de novo results' evaluation requires manual inspection. Consequently, samples can be more directly analyzed using metagenome sequencing. The likelihood of a single identification can be improved by acquiring meta information concerning taxonomy and function from repositories after successful protein identification. Additionally, engaging similar peptide sets [33], one shared peptide [46] or sequence similarity comparisons can be the basis of redundant protein identifications due to similar peptides from homologous proteins. At last, taxonomy of protein can be redefined by the common ancestor taxonomy of all proteins in one group that permits a reliable phylogenetic assignment of metaproteins avoiding risky assignments of species or strains. To ensure an even more reliable study, we can visualize taxonomic composition in a Krona plot, which is based on known peptides or spectra from the National Center for Biotechnology Information taxonomy [47].

Species richness, the overall community organization and their interacting dynamics can be calculated in a taxonomic profile comparison from different samples or time points. This has been extensively discussed in the concept of Microbial Resource Management. It is beneficial to shift to protein functions using overview plots, such as a Voronoi Treemap [48] or a common pie chart, based on gene ontologies or UniProt keywords [41]. Assignment of identified proteins to biochemical pathways demands even more importance. By using KEGG (Kyoto Encyclopedia of Genes and Genomes) ontologies or enzyme commission numbers a direct mapping to MetaCyc patH pathway [49] or Kyoto Encyclopedia of Genes and Genomes (KEGG) pathways can be achieved. Proteome studies often result in long lists of up regulated and down regulated proteins confirmed by statistical tests.

5.8 Novel Applications of Metaproteomics

In an agricultural context, microorganisms in the plant rhizosphere and phyllosphere can be artificially manipulated to increase agricultural production in an environmentally sound manner [50]. Plant microbial communities are predicted based on the knowledge obtained from plant-microbial interactions. Therefore, a comprehensive understanding of the phytobiome enables us to answer several fundamental questions in the areas of plant pathology, crop physiology and microbial biology. Not surprisingly, understanding the mechanisms by which the phytobiome supports plant health and productivity have gained a lot of interest. These studies highlight the power and potential of metaproteomics to characterize the function and structure of plant-associated microbial communities [51]. Despite a limited number of publications on plant microbial metagenomics, metatranscriptomics, and metaproteomics, these data have changed our understanding of the function, structure, and population dynamics of the phytobiome.

It is known that the structure and populations of plant-associated microbial communities have *effects on plant* growth and *health* by *influencing* the *outcome* of *pathogen* infection. Metaproteomics provides a useful tool for a comprehensive understanding of the phytobiome by utilizing plant-associated microbial communities in support of plant health against insects and diseases. Metaproteomics also sheds light on the impact of humans on the environment and agriculture. For instance, the release of *transgenic plants* into fields has a *direct impact* on the properties and functions of soil and plant-associated microbial communities can be *monitored by metaproteomics* [21]. *Some of these recent studies are shown in* Table 5.1.

Table 5.1 Recent applications of metaproteomics approach in the field of plant sciences

Plants	Comment	References
Leaves of soybean, clover and *Arabidopsis thaliana*	Identification of community members that are responsible for providing energy and transport processes	[66]
Different crop rhizospheric soils including rice, sugarcane, *Pseudostellariae heterophylla*, *Rehmanniae* sp., tobacco	Detection of several metabolic pathways such as the defense machinery, energy production, protein biosynthesis and turnover, and secondary metabolism by developing an optimized extraction method named C/S-P-M	[67]
Phyllosphere plus rhizoshere microbiome of rice	Detection of 4600 proteins that were similar in phyllosphere in different geographical conditions but in the rhizosphere observed various proteins that interfered in stress responses and one carbon compound cycle	[68]
Soil of *Rehmannia glutinosa* (herbal medicine) under consecutive monoculture	Identification of plants, bacteria and fungi proteins mainly involved in carbohydrate and energy metabolism, amino acids metabolism and stress/defense response	[12]
Rhizosphere of lettuce (*Lactuca sativa*)	Revealed higher amount of proteins related with virulence determinants, energy metabolism and stress/defense response in presence of pathogenic strain of *Fusarium oxysporum* that could be related to the interaction of the microbial consortium to this fungus	[69]
Rhizosphere of ratoon sugarcane and plant sugarcane	Demonstrated that ratoon sugarcane induced significant changes in soil enzyme activities and the catabolic diversity of microbial community, and that the expression level of soil proteins originated from plants, microbes and fauna. Also reported that 24.77 % of soil proteins are derived from bacteria and most of the microbial proteins were involved in membrane transport and signal transduction	[70]

5.9 Using Metaproteomics to Construct Metabolite Pathways and Environmental Signals

Soil represents a favorable environment for a wide range of microorganisms including algae, bacteria, and fungi. Almost all of the chemical changes that take place within the soil involve the active contribution of soil microflora. They mainly participate in carbon and nitrogen cycling, nutrient acquisition and soil formation processes, which are necessary for plant growth and survival. In addition, plants have profound effects on soil microbial communities, especially those colonizing the rhizosphere. This is because of the great carbon input to soils by plant root exudates. Furthermore, plants are immobile organisms that are often confronted with unfavorable conditions (e.g., salinity, drought, pathogen attacks). In order to evade abiotic and biotic stresses, one plant strategy is to establish associations with beneficial microbial organisms. One of the most complex tasks for a plant is to distinguish between mutualistic partners and parasites, especially in view of the fact that symbiotic and parasitic interactions share many common signaling pathways.

Recent studies have increased our understanding of phytobiome profiles and implications of the plant-associated microbial communities within the soil [4]. Metaproteomics data combined with data obtained from plant physiology, microbial biology, genomics, and computational biology are used to construct metabolic pathways. Therefore, we are able to address metabolic networks and their regulation on whole-plant microbial communities [52]. The metabolic and regulatory networks and biosynthetic capabilities of plant-associated microbes in real environments can be employed for biotechnological applications.

Newly initiated lines of research seek to investigate the phytobiome in various environments [53]. Understanding the influence of environmental conditions on the dynamics and activities of the phytobiome help in designing strategies which control emerging plant pathogens. The discovery of new antibiotic compounds and the development of new disease control strategies can be used to reduce pesticide use or to control diseases for which no other effective control approach exists [54].

5.10 Engineering Plant Microbial Communities

As we have already explained, the phytobiome in its natural state consists of a community of microbes that exist exogenously and endogenously in various forms within a plant. These associations appear symptomless at first glance, possibly representing a series of mutualistic or symbiotic relationships. Consistently, experiments performed under laboratory conditions revealed that the structure and components of the phytobiome provide strategic advantages to the plant, such as enhanced mineral acquisition and indirect pathogen protection. However, the role and function of the phytobiome is still largely unknown. The functional diversity and dynamics of microbial communities have recently been studied. Plant volatiles play an important role in microbe-microbe communications and can be studied by the metaproteomics approach. Engineering microbial communities can be part of the solution for plant growth and improved tolerance to harsh environmental conditions. Plant-associated microbial communities improve plants' responses to unfavorable conditions. Therefore, understanding the molecular basis of plant adaptations to stress conditions induced by microbial communities has received increased attention. However, the understanding of molecular and physiological mechanisms of plant adaptation to stress conditions gained through the association with microbial communities is still rudimentary. Through data obtained from investigating metabolic and signaling networks in controlling microbial communities within natural habitats, we advance our knowledge toward inducing tolerance to environmental stressors. Metaproteomics data holds the promise of enabling us to engineer metabolic junctions between plants and microbes to cope with environmental stress and increase plant yield and quality.

5.11 A Plant's Beneficial Microbes Help Them to Cope with Environmental Stresses

Drought and salinity are the major limiting factors of crop production worldwide. In order to establish sustainable agriculture, the application of symbionts can be a good alternative as symbionts not only enhance crop yields, but they can also increase plant tolerance against various stress conditions through adapting the plants' response to their environment. Research on the symbiont helps scientists to understand the stress adaptation mechanisms in plants and can lead us in designing better strategies to reduce plant loss due to environmental stress. Several proteomic-based investigations have provided new insight into host adaptation to environment conditions through the symbiotic interaction with beneficial microorganisms. In order to unveil the molecular mechanisms that enhance the plant's tolerance of salt and drought, we used proteomics to identify responsive proteins involved in plant stress tolerance.

Piriformospora indica is a root-interacting fungus, capable of enhancing plant growth by increasing plant resistance to a wide variety of pathogens and improving plant stress tolerance to extreme environmental conditions. The broad-spectrum root-colonized endophytic *P. indica* confers various beneficial effects to host plants, such as growth promotion, seed yield increase, abiotic stress tolerance and biotic stress resistance. Detailed understanding of how the symbiotic relationships are capable of enhancing the plant's adaption can lead to the development of sustainable agriculture.

We established proteomic approaches in order to unravel the molecular basis of enhanced salt tolerance in barley (*Hordeum vulgare* L.) conferred by *P. indica*. We identified 51 proteins involved in different functional categories including photosynthesis, cell antioxidant defense, protein translation and degradation, energy production, signal transduction and cell wall arrangement, and taken together, show that *P. indica* altered the host physiology to cope with salt stress [55]. We discovered the link between the expression of several novel proteins that lead to host adaptation against environmental stress [55, 56]. According to our results, it is likely that *P. indica* induced systemic response to salt stress by changing the physiology and proteome of *P. indica*-colonized barley plants, despite the fact that the fungus does not colonize plant leaves. Our research provides a deep understanding of how symbionts protect their plant hosts against the detrimental effects of soil salinity. We proposed that *P. indica* mediated stress tolerance through photosynthesis stimulation, improved energy release, and an enhancement in antioxidative capacity in the colonized plants [55, 56].

5.12 Current Limitations and Future Challenges of Metaproteomics Applications

In the last few years, much effort has been applied to understanding the multifaceted mechanisms of symbiosis through comparative proteomics of the whole plant. The analysis of cellular and subcellular proteomes clearly extends the depth of the proteome to be investigated [57]. It assists in targeting specific, often low-abundance, proteins by eliminating most other proteins from the whole plant proteome [58]. In contrast, only a few proteomic studies have been undertaken to study the symbiotic relationships at either the host specific tissue or cellular levels [59, 60]. The cellular proteome of membrane-bound proteins is important in understanding the initial cellular events that occur during symbiotic interactions [61–63]. It provides further insight into cell signaling events for specific tissue, cell type or organelles and thus increases the resolution of proteome profiling [64].

Applying new strategies to reduce the damage of environmental extremes through the application of microbial communities holds great potential. However, the analysis and monitoring of the real impacts require new approaches, which currently pose challenges to researchers. Because plant metaproteomics is a novel field of research, a standardized protocol has not yet been established [65]. The main goal should be to obtain novel insights into the molecular interactions existing between plant genomes and their associated phytobiome. The ultimate goal will be to apply microbial communities to improve environmental quality, crop production and ecosystem sustainability.

References

1. Delmont TO, Robe P, Cecillon S, Clark IM, Constancias F, Simonet P et al (2011) Accessing the soil metagenome for studies of microbial diversity. Appl Environ Microbiol 77:1315–1324
2. Guttman DS, Mchardy AC, Schulze-Lefert P (2014) Microbial genome-enabled insights into plant-microorganism interactions. Nat Rev Genet 15:797–813
3. Carvalhais LC, Dennis PG, Tyson GW, Schenk PM (2012) Application of metatranscriptomics to soil environments. J Microbiol Methods 91:246–251
4. Hettich RL, Pan C, Chourey K, Giannone RJ (2013) Metaproteomics: harnessing the power of high performance mass spectrometry to identify the suite of proteins that control metabolic activities in microbial communities. Anal Chem 85:4203–4214
5. Vandenkoornhuyse P, Quaiser A, Duhamel M, Le Van A, Dufresne A (2015) The importance of the microbiome of the plant holobiont. New Phytol 206:1196–1206

6. Wilmes P, Bond PL (2004) The application of two-dimensional polyacrylamide gel electrophoresis and downstream analyses to a mixed community of prokaryotic microorganisms. Environ Microbiol 6:911–920
7. Schlaeppi K, Bulgarelli D (2015) The plant microbiome at work. Mol Plant-Microbe Interact MPMI 28:212–217
8. De-La-Peña C, Badri DV, Lei Z, Watson BS, Brandão MM, Silva-Filho MC et al (2010) Root secretion of defense-related proteins is development-dependent and correlated with flowering time. J Biol Chem 285:30654–30665
9. Carvalhais LC, Dennis PG, Schenk MP (2014) Defence inducers rapidly influence the diversity of bacterial communities in a potting mix. Appl Soil Ecol 84:1–5
10. Carvalhais LC, Muzzi F, Tan C-H, Hsien-Choo J, Schenk PM (2013) Plant growth in Arabidopsis is assisted by compost soil-derived microbial communities. Frontiers Plant Sci 4:235
11. De-La-Peña C, Lei Z, Watson BS, Sumner LW, Vivanco JM (2008) Root-microbe communication through protein secretion. J Biol Chem 283:25247–25255
12. Wu L, Wang H, Zhang Z, Lin R, Zhang Z, Lin W (2011) Comparative metaproteomic analysis on consecutively Rehmannia glutinosa-monocultured rhizosphere soil. PLoS ONE 6:e20611
13. Hirsch PR, Mauchline TH (2012) Who's who in the plant root microbiome? Nat Biotechnol 30:961–962
14. Mendes R, Garbeva P, Raaijmakers JM (2013) The rhizosphere microbiome: significance of plant beneficial, plant pathogenic, and human pathogenic microorganisms. FEMS Microbiol Rev 37:634–663
15. Nannipieri P (2006) Role of stabilised enzymes in microbial ecology and enzyme extraction from soil with potential applications in soil proteomics. In: Nucleic acids and proteins in soil. Springer, pp 75–94
16. Ogunseitan OA (2006) Soil proteomics: extraction and analysis of proteins from soils. In: Nucleic acids and proteins in soil. Springer, pp 95–115
17. Hofman-Bang J, Zheng D, Westermann P, Ahring BK, Raskin L (2003) Molecular ecology of anaerobic reactor systems. In: Biomethanation I. Springer, pp 151–203
18. Fouts DE, Pieper R, Szpakowski S, Pohl H, Knoblach S, Suh M-J et al (2012) Integrated next-generation sequencing of 16S rDNA and metaproteomics differentiate the healthy urine microbiome from asymptomatic bacteriuria in neuropathic bladder associated with spinal cord injury. J Transl Med 10:174
19. Chourey K, Jansson J, Verberkmoes N, Shah M, Chavarria KL, Tom LM et al (2010) Direct cellular lysis/protein extraction protocol for soil metaproteomics. J Proteome Res 9:6615–6622
20. Jinjun K, Thomas H, Joy G, Kui W, Feng C (2005) Metaproteomic analysis of Chesapeake Bay microbial communities
21. Keiblinger KM, Wilhartitz IC, Schneider T, Roschitzki B, Schmid E, Eberl L et al (2012) Soil metaproteomics—comparative evaluation of protein extraction protocols. Soil Biol Biochem 54:14–24
22. Leary DH, Hervey Iv WJ, Li RW, Deschamps JR, Kusterbeck AW, Vora GJ (2012) Method development for metaproteomic analyses of marine biofilms. Anal Chem 84:4006–4013
23. Singleton I, Merrington G, Colvan S, Delahunty J (2003) The potential of soil protein-based methods to indicate metal contamination. Appl Soil Ecol 23:25–32
24. Ogunseitan O (1997) Direct extraction of catalytic proteins from natural microbial communities. J Microbiol Methods 28:55–63
25. Klaassens ES, De Vos WM, Vaughan EE (2007) Metaproteomics approach to study the functionality of the microbiota in the human infant gastrointestinal tract. Appl Environ Microbiol 73:1388–1392
26. Kolmeder CA, De Been M, Nikkilä J, Ritamo I, Mättö J, Valmu L et al (2012) Comparative metaproteomics and diversity analysis of human intestinal microbiota testifies for its temporal stability and expression of core functions. PLoS ONE 7:e29913

27. Abram F, Gunnigle E, O'flaherty V (2009) Optimisation of protein extraction and 2-DE for metaproteomics of microbial communities from anaerobic wastewater treatment biofilms. Electrophoresis 30:4149–4151
28. Sharma R, Dill BD, Chourey K, Shah M, Verberkmoes NC, Hettich RL (2012) Coupling a detergent lysis/cleanup methodology with intact protein fractionation for enhanced proteome characterization. J Proteome Res 11:6008–6018
29. Wisniewski JR, Zougman A, Nagaraj N, Mann M (2009) Universal sample preparation method for proteome analysis. Nat Methods 6:359
30. Tang Y, Underwood A, Gielbert A, Woodward MJ, Petrovska L (2014) Metaproteomics analysis reveals the adaptation process for the chicken gut microbiota. Appl Environ Microbiol 80:478–485
31. Pérez-Cobas AE, Gosalbes MJ, Friedrichs A, Knecht H, Artacho A, Eismann K et al (2013) Gut microbiota disturbance during antibiotic therapy: a multi-omic approach. Gut 62: 1591–1601
32. Verberkmoes NC, Russell AL, Shah M, Godzik A, Rosenquist M, Halfvarson J et al (2009) Shotgun metaproteomics of the human distal gut microbiota. ISME J 3:179–189
33. Schneider T, Keiblinger KM, Schmid E, Sterflinger-Gleixner K, Ellersdorfer G, Roschitzki B et al (2012) Who is who in litter decomposition? metaproteomics reveals major microbial players and their biogeochemical functions. ISME J 6:1749–1762
34. Köcher T, Pichler P, Swart R, Mechtler K (2012) Analysis of protein mixtures from whole-cell extracts by single-run nanoLC-MS/MS using ultralong gradients. Nat Protoc 7:882–890
35. Thakur SS, Geiger T, Chatterjee B, Bandilla P, Fröhlich F, Cox J et al (2011) Deep and highly sensitive proteome coverage by LC-MS/MS without prefractionation. Mol Cell Proteomics 10: M110.003699
36. Pirmoradian M, Budamgunta H, Chingin K, Zhang B, Astorga-Wells J, Zubarev RA (2013) Rapid and deep human proteome analysis by single-dimension shotgun proteomics. Mol Cell Proteomics 12:3330–3338
37. Yu Y, Suh M-J, Sikorski P, Kwon K, Nelson KE, Pieper R (2014) Urine sample preparation in 96-well filter plates for quantitative clinical proteomics. Anal Chem 86:5470–5477
38. Craig R, Beavis RC (2004) TANDEM: matching proteins with tandem mass spectra. Bioinformatics 20:1466–1467
39. Cottrell JS, London U (1999) Probability-based protein identification by searching sequence databases using mass spectrometry data. Electrophoresis 20:3551–3567
40. Coordinators NR, Acland A, Agarwala R, Barrett T, Beck J, Benson DA et al (2014) Database resources of the national center for biotechnology information. Nucleic Acids Res 42:D7
41. Consortium U (2011) Reorganizing the protein space at the Universal Protein Resource (UniProt). Nucleic Acids Res gkr981
42. Zakrzewski M, Goesmann A, Jaenicke S, Jünemann S, Eikmeyer F, Szczepanowski R et al (2012) Profiling of the metabolically active community from a production-scale biogas plant by means of high-throughput metatranscriptome sequencing. J Biotechnol 158:248–258
43. Jagtap P, Goslinga J, Kooren JA, Mcgowan T, Wroblewski MS, Seymour SL et al (2013) A two-step database search method improves sensitivity in peptide sequence matches for metaproteomics and proteogenomics studies. Proteomics 13:1352–1357
44. Bradshaw RA, Burlingame AL, Carr S, Aebersold R (2006) Reporting protein identification data the next generation of guidelines. Mol Cell Proteomics 5:787–788
45. Gupta N, Pevzner PA (2009) False discovery rates of protein identifications: a strike against the two-peptide rule. J Proteome Res 8:4173–4181
46. Lü F, Bize A, Guillot A, Monnet V, Madigou C, Chapleur O et al (2014) Metaproteomics of cellulose methanisation under thermophilic conditions reveals a surprisingly high proteolytic activity. ISME J 8:88–102
47. Ondov BD, Bergman NH, Phillippy AM (2011) Interactive metagenomic visualization in a Web browser. BMC Bioinf 12:1

48. Bernhardt J, Michalik S, Wollscheid B, Völker U, Schmidt F (2013) Proteomics approaches for the analysis of enriched microbial subpopulations and visualization of complex functional information. Curr Opin Biotechnol 24:112–119
49. Schulze WX, Gleixner G, Kaiser K, Guggenberger G, Mann M, Schulze E-D (2005) A proteomic fingerprint of dissolved organic carbon and of soil particles. Oecologia 142:335–343
50. Becher D, Bernhardt J, Fuchs S, Riedel K (2013) Metaproteomics to unravel major microbial players in leaf litter and soil environments: challenges and perspectives. Proteomics 13:2895–2909
51. Heyer R, Kohrs F, Benndorf D, Rapp E, Kausmann R, Heiermann M et al (2013) Metaproteome analysis of the microbial communities in agricultural biogas plants. New Biotechnol 30:614–622
52. Watrous J, Roach P, Alexandrov T, Heath BS, Yang JY, Kersten RD et al (2012) Mass spectral molecular networking of living microbial colonies. Proc Natl Acad Sci USA 109: E1743–1752
53. Bastida F, Hernández T, García C (2014) Metaproteomics of soils from semiarid environment: functional and phylogenetic information obtained with different protein extraction methods. J Proteomics 101:31–42
54. Raaijmakers JM, Mazzola M (2012) Diversity and natural functions of antibiotics produced by beneficial and plant pathogenic bacteria. Annu Rev Phytopathol 50:403–424
55. Alikhani M, Khatabi B, Sepehri M, Nekouei MK, Mardi M, Salekdeh GH (2013) A proteomics approach to study the molecular basis of enhanced salt tolerance in barley (Hordeum vulgare L.) conferred by the root mutualistic fungus Piriformospora indica. Mol BioSyst 9:1498–1510
56. Ghabooli M, Khatabi B, Ahmadi FS, Sepehri M, Mirzaei M, Amirkhani A et al (2013) Proteomics study reveals the molecular mechanisms underlying water stress tolerance induced by Piriformospora indica in barley. J Proteomics 94:289–301
57. Valot B, Gianinazzi S, Eliane DG (2004) Sub-cellular proteomic analysis of a Medicago truncatula root microsomal fraction. Phytochemistry 65:1721–1732
58. Lee S-J, Saravanan RS, Damasceno CM, Yamane H, Kim B-D, Rose JK (2004) Digging deeper into the plant cell wall proteome. Plant Physiol Biochem 42:979–988
59. Wan J, Torres M, Ganapathy A, Thelen J, Dague BB, Mooney B et al (2005) Proteomic analysis of soybean root hairs after infection by Bradyrhizobium japonicum. Mol Plant Microbe Interact 18:458–467
60. Muneer S, Ahmad J, Bashir H, Qureshi MI (2012) Proteomics of nitrogen fixing nodules under various environmental stresses. Plant Omics 5:167
61. Shahollari B, Peskan-Berghöfer T, Oelmüller R (2004) Receptor kinases with leucine-rich repeats are enriched in Triton X-100 insoluble plasma membrane microdomains from plants. Physiol Plant 122:397–403
62. Valot B, Dieu M, Recorbet G, Raes M, Gianinazzi S, Dumas-Gaudot E (2005) Identification of membrane-associated proteins regulated by the arbuscular mycorrhizal symbiosis. Plant Mol Biol 59:565–580
63. Valot B, Negroni L, Zivy M, Gianinazzi S, Dumas-Gaudot E (2006) A mass spectrometric approach to identify arbuscular mycorrhiza-related proteins in root plasma membrane fractions. Proteomics 6:S145–S155
64. Dubinin J, Braun H-P, Schmitz U, Colditz F (2011) The mitochondrial proteome of the model legume Medicago truncatula. Biochim Biophys Acta (BBA)-Proteins Proteomics 1814:1658–1668
65. Tanca A, Palomba A, Pisanu S, Deligios M, Fraumene C, Manghina V et al (2014) A straightforward and efficient analytical pipeline for metaproteome characterization. Microbiome 2:49
66. Delmotte N, Knief C, Chaffron S, Innerebner G, Roschitzki B, Schlapbach R et al (2009) Community proteogenomics reveals insights into the physiology of phyllosphere bacteria. Proc Natl Acad Sci 106:16428–16433

67. Wang H-B, Zhang Z-X, Li H, He H-B, Fang C-X, Zhang A-J et al (2010) Characterization of metaproteomics in crop rhizospheric soil. J Proteome Res 10:932–940
68. Knief C, Delmotte N, Chaffron S, Stark M, Innerebner G, Wassmann R et al (2012) Metaproteogenomic analysis of microbial communities in the phyllosphere and rhizosphere of rice. ISME J 6:1378–1390
69. Moretti M, Minerdi D, Gehrig P, Garibaldi A, Gullino ML, Riedel K (2012) A bacterial–fungal metaproteomic analysis enlightens an intriguing multicomponent interaction in the rhizosphere of Lactuca sativa. J Proteome Res 11:2061–2077
70. Lin W, Wu L, Lin S, Zhang A, Zhou M, Lin R et al (2013) Metaproteomic analysis of ratoon sugarcane rhizospheric soil. BMC Microbiol 13:1

Chapter 6
Proteomics in Energy Crops

Shiva Bakhtiari, Meisam Tabatabaei and Yusuf Chisti

Abstract The increasing global demand for energy, diminishing fossil fuel reserves, and the rise in greenhouse gas emissions created by their use, have fuelled efforts to identify renewable energy sources that are more sustainable and environment friendly. One alternative is to develop plants that are more efficient in utilizing solar energy and converting it into biomass which can be used as feedstock for biofuel production. The development of such improved bioenergy crops requires the use of more advanced biotechnological applications such as proteomics. This chapter will examine the use of proteomics in bioenergy crops in order to find quicker and more effective pathways of modifying them for optimum fuel production.

Keywords Biofuel · Energy crops · Biofuel crops · Biomass

The original version of this chapter was revised: The chapter was updated with the revised text. The erratum to this chapter is available at 10.1007/978-3-319-43275-5_13

S. Bakhtiari
Biology Department, Concordia University, 7141 Sherbrooke W.,
Montreal H4B 1R6, Canada

M. Tabatabaei (✉)
Microbial Biotechnology Department, Agricultural Biotechnology Research
Institute of Iran (ABRII), Agricultural Research, Education, and Extension
Organization (AREEO), Karaj, Iran
e-mail: meisam_tabatabaei@abrii.ac.ir; meisam_tab@yahoo.com

M. Tabatabaei
Biofuel Research Team (BRTeam), Karaj, Iran

Y. Chisti
School of Engineering, Massey University,
Private Bag 11 222, Palmerston North, New Zealand

© Springer International Publishing Switzerland 2016
G.H. Salekdeh (ed.), *Agricultural Proteomics Volume 1*,
DOI 10.1007/978-3-319-43275-5_6

6.1 Introduction

Bioenergy is a renewable form of energy. Plants and certain photosynthesizing microorganisms use sunlight to convert inorganic carbon and water to biochemical energy, the ultimate source of all bioenergy. Unlike other crops, bioenergy crops are grown specifically for energy. At present, bioenergy crops are exclusively higher plants.

Plant biomass can be used directly as fuel, or converted to biochar, to generate steam, heat and electrical power. Biomass and other plant chemicals (starch, sugars, lignoceullulose, oils) can be converted to various type of fuels including fuel alcohols (bioethanol, biobutanol), biodiesel and biogas. Production of some of these biofuels is well-established. Technology for making bioethanol from lignocellulosics is developing [1–3]. Plant biomass and other biochemcials can be used to make fuels that are currently obtained from petroleum feedstock. For example, diesel, gasoline, and kerosene can be made from biomass and biochemicals using chemical processes that are being developed.

Microbial biofuels are in development [4, 5], but are not likely to be commercialized in the near term [6]. This notwithstanding, photosynthesizing microorganisms can produce large quantities of biomass. Production of microbial biomass and chemicals can be independent of the weather and availability of arable land. Depending on the culture conditions, this biomass may be rich in starch, oils or other useful chemicals. Production of some plant-derived biofuels also depends on microbial action. For example, bioethanol and biobutanol produced by microbial action on plant-derived sugars.

The established and emerging biofuel crops are of three types. (1) Sugar and starch crops such as sugarcane, sugarbeet, corn (maize), wheat and cassava. Sugars and starch from these can be readily converted to ethanol and other fuel alcohols via microbial fermentation processes. (2) Oil crops such as oil palm, soybean, canola, sunflower, jatropha and camelina. Oils from these can be readily converted to biodiesel. (3) Lignocellulosic biomass crops including some grasses (e.g. switchgrass) and trees (e.g. poplar) that can generate relatively large quantities of biomass rapidly with a low input of resources. This biomass may be used to raise steam, or converted to fermentable sugars through methods now being developed.

Some of the above mentioned energy crops (e.g. corn, wheat, soybean) are also important food and feed crops. With limited arable land, water and fertilizers, diverting these crops to biofuels will certainly affect accessibility of food. Other crops such as Jatropha and switchgrass may be grown on marginal land not suited to food production. Biofuels derived from non-food crops are increasingly favored, but a full assessment of their long-term sustainability is required. Increasing the cropping area through deforestation and displacement of other natural ecosystems is not an option.

Improving energy yield and productivity of the bioenergy crops requires an indepth understanding of their proteomics: what proteins are being produced and how they impact metabolic function. This information is necessary for crop improvement and developing traits that affect survival under adverse conditions (e.g. drought, frost, temporary flooding, pest infestations). Although the proteomics of the major bioenergy crops are barely understood [7], progress is being made as reviewed in this chapter. Proteome maps are typically prepared using gel electrophoresis, especially two-dimensional polyacrylamide gel electrophoresis (2D-PAGE). Mass spectroscopic methods are generally used for identifying the proteins.

6.2 Starch and Sugar Crops

6.2.1 Corn (Zea mays)

Corn, after rice and wheat, is the most widely used food grain. Nearly a third of the global population subsists on corn [8]. Corn is a source of corn oil, a widely used food oil. Also, corn is a source of other industrial chemicals [9] and biofuels derived from corn starch [10]. Corn can be used as a bioenergy crop in two ways: the grains can be used to produce starch-based ethanol through processes known as dry or wet milling, and the corn stover has the potential of being converted to lignocellulosic biofuels through biochemical or thermochemical processes (Fig. 6.1a).

Biology and genetics of corn are well-known as it has long served as a model for studies of monocots [11]. Corn yield and biomass composition in relation to production of fuels have been discussed [12]. Genetic engineering approaches to improve bioethanol production from corn have been discussed [13].

Significant information exists on proteomics of corn (Table 6.1). Proteome profiles are available for several tissues including endosperm [14], leaf [15], root hair [16], primary root [17, 18], primary root pericycle [19], rachis [9], and egg cell [20]. Proteome associated with starch granules and seed flour [21, 22] is known.

Fig. 6.1 a Biofuel production pathways from, a corn and b sorghum

Table 6.1 Proteomic studies in corn[a]

Tissue	Treatment/stress	References
Leaf, bundle sheath	–	[12]
Seedling/chloroplast	–	[26]
Leaf base, tip/chloroplast	–	[27]
Root, hairs	–	[28]
Pollen, pistil	–	[29]
Basal region of seedling leaf/nuclear	–	[30]
Endosperm, embryo	–	[31]
Embryos	–	[32]
15-day-old plants	Drought	[33]
Embryos	Desiccation	[34]
Anther	Cold pretreatment	[35]
Pollen coat	–	[36]
Seedling/chloroplast	Salt	[37]
Root	Salt	[38]
Seedling	Nitric oxide/salt	[39]
Embryos	Salt	[40]
Seedling	Sugarcane mosaic virus	[41]
Leaf	Flooding	[18]
Seed flour	–	[29]
Leaf	Salicylic acid, abscisic acid	[42]
Leaf	–	[43]
Seed	–	[44]

[a]Chakraborty et al. [11], Copyright (2016), with permission from Elsevier

Subcellular proteome studies have been reported on mitochondria [23], root cell wall [18], chloroplast [24, 25], and stroma, membranes and nucleoids [9]. Developmental proteomics and stress proteomics have been discussed.

6.2.2 Sugarcane (Saccharum officinarum)

Sugarcane is a perineal grass. It is widely cultivated in tropical and subtropical regions mainly for the production of sugar [45] and to a lesser extent bioethanol. Sugarcane is probably the most efficient bioenergy crop for tropical and subtropical regions [46] and bioethanol from sugarcane is one of the best established biofuels [47]. The use of genetically modified sugarcane with improved traits for bioenergy production has been suggested [48, 49], but limited genomic information [7, 50] hinders progress. A lack of genome data has affected progress in proteomics [51]. The few studies have mostly focused on the proteins extracted from the leaf and root tissue [50]. Proteomics of sugarcane stalk, the major source of sugar, have received barely any attention. A proteome map of sugarcane stalk has been published [2]. Identification of the proteins involved in sugar metabolism and elucidation of their functions could be the basis of improving sugar yield and therefore the yield of bioethanol from sugarcane. The sugarcane proteomics have been reviewed by [6].

6.2.3 Sugar Beet (Beta vulgaris)

Sugar beet is a source of sucrose for conversion to bioethanol by fermentation. A number of proteomic studies are available on sugar beet [52–56]. A comprehensive proteome-wide characterization of sugar beet seed and seedlings, including the root, stem, cotyledons, and perisperm was reported byCatusse et al. [53]. Other studies focused on abiotic stresses such as drought [54], nutritional deficiencies [52] and salt tolerance [53, 55]. Proteomics of sugarbeet response to pathogens have been discussed [57, 58].

6.2.4 Wheat (Triticum spp.)

Wheat is a major food crop. It is a candidate crop for production of bioenergy, particularly in Europe. Wheat proteomics have received considerable attention (Table 6.2). Proteome reference maps have been made for wheat leaves [59], roots [60], endosperm [61], and amyloplasts [62]. Effects of various stress factors, radiation and toxic ions on wheat proteome have been reported [11].

6.2.5 Barley (Hordeum vulgare)

Barley is a major cereal crop grown mainly for feed and malting. Barley proteomics have spanned diverse areas (Table 6.3). Studies have addressed: (1) changes in protein abundance in grains of different barley varieties during development; (2) changes in leaf and shoot proteins in response to heat stress; (3) protein signatures associated with malting quality; (4) changes in root and shoot proteomes in response to nitrogen deficiency; (5) changes associated with radicle elongation in germinating seeds; (6) proteomics of cadmium accumulation in varieties with differing levels of cadmium tolerance; (7) effects of drought stress on leaf proteomes; (8) proteomics of leaf rust infection; (9) salinity stress responses; and (10) enhanced salt tolerance through mutualistic interactions with the root fungus *Piriformospora indica*. Functional proteomics of barley chloroplasts [107] and the proteomics of stress response in barely [108] have been reported.

Table 6.2 Proteomic studies in wheat[a]

Tissue	Treatment/stress	References
Spike development	Cold	[63]
Leaf	Drought, salinity	[64]
Peduncle	Drought, oxidative	[65]
Root	Flooding	[66]
Developing grain	–	[67]
Seed	–	[68]
Leaf	Salt	[68]
Leaf	Salt	[69]
Seed	Salt	[70]
Mitochondria	Salt	[71]
Root	Salt	[72]
Seed	Salt	[73]
Seed	Salt	[74]
Leaf	Drought	[75]
Grain	Drought	[76]
Leaf	Drought	[77]
Grain	Drought	[78]
Anthesis period	Drought	[79]
Seed	Drought	[80]
Seed	High temperature	[81]
Grain	High temperature	[82]
Leaf	Drought	[83]
Grain	Drought	[84]
Flour	–	[85]
Gluten proteins	–	[86]
Grain	–	[87]
Root	Abscisic acid	[88]
Leaf	–	[89]
Leafs and roots	Copper	[90]
Root	Aluminum	[91]
Seed	Cadmium	[92]
Seed	Enhanced UV-B radiation	[93]
Seed endosperm	–	[94]
Flour	–	[95]
Crown	Low temperature	[96]
Leaf	Low temperature	[97]
Grain	–	[98]
Grain	–	[99]
Grain	–	[100]
Leaf	Drought	[101]

(continued)

Table 6.2 (continued)

Tissue	Treatment/stress	References
Mature embryos	Late embryogenesis	[102]
Leaf	Low temperature	[103]
Leaf	–	[104]
Seed	Low temperature	[105]
Crown	Low temperature	[106]

[a]Chakraborty et al. [11], Copyright (2016), with permission from Elsevier

Table 6.3 Proteomic studies in barley[a]

Tissue	Treatment/stress	References
Roots and shoots/33 day seedlings	Nitrogen deficiency	[109]
Four extended leaf seedlings, leaf	Drought	[110]
Leaf sheath/14 day seedlings	Salt/fungal infection	[111]
Seed (review)	Grain filling/maturation	[112]
3 cm seedling roots	Nuclei isolation	[113]
Leaf/24 day hydroponic seedlings	Protein turnover	[114]
Leaf/first leaf stage seedlings	Leaf rust infection	[115]
Leaf	Drought/fungal infection	[116]
Malts	Malt filterability	[117]
Leaf/7 day seedlings	UV-B radiation	[118]
Shoots/3 day old seedlings	Drought	[119]
Leaf from seedlings	Salinity	[120]
Grains harvested/mature field-grown plants	Cadmium	[121]
Leaf, roots/seven day hydroponic seedlings	Salt	[122]
Grains harvested/mature plants	Salt	[123]

[a]Chakraborty et al. [11], Copyright (2016), with permission from Elsevier

6.2.6 Sorghum (Sorghum bicolor)

Sorghum bicolor is a C4 grass. It is among the top five of the economically important cereal crops [124]. Sorghum grain is a source of starch and its stalks contain sugar-rich juice. The soluble sugars in its stalk can easily be converted to ethanol using currently available, conventional fermentation technology (Fig. 6.1b). Sorghum is receiving much attention as a feedstock for bioethanol [125, 126]. It has the potential to provide more ethanol per hectare than corn. Sorghum genome has been fully sequenced, but studies of its proteome are limited [51]. Proteins secreted into the medium by sorghum cells grown in suspension culture have been studied [51]. Proteomics of salt stress responses [127] and response to draught stress [128] have been reported. Grain proteomics of certain sorghum variants have been discussed [129]. Functional distribution of proteins in sorghum leaf extracts is shown in Fig. 6.2. Leaf proteomics mostly relate to carbohydrate metabolism as leaf is where carbohydrate is produced via photosynthesis.

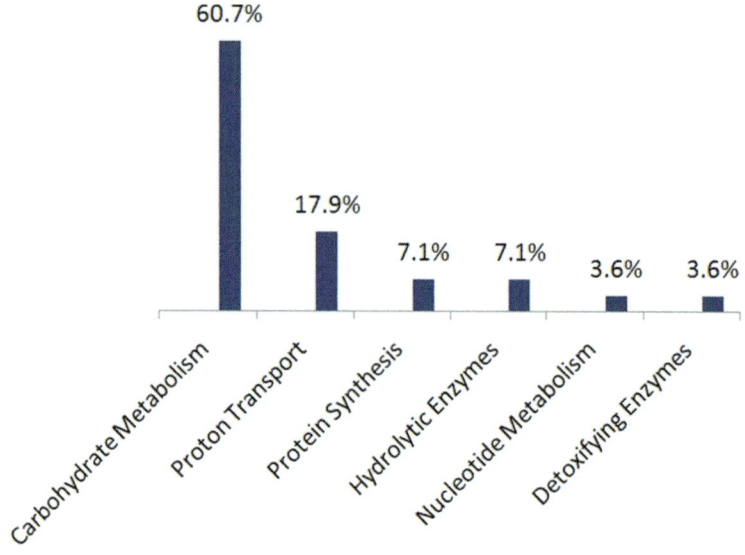

Fig. 6.2 Functional distribution of sorghum proteins in leaf extract. Based on [51]

6.3 Edible Oilseed Crops

6.3.1 Oil Palm (Elaeis guineensis)

Palm oil accounts for nearly half of the edible oil used worldwide [130]. It is also a main feedstock for the production of biodiesel. Apparently the first proteome analysis of the oil palm mesocarp was done only recently [131]. The aim was to identify the proteome changes associated with fruit maturation and oil content of palm fruits. Proteins from several important metabolic processes including starch and sucrose metabolism, glycolysis, pentose phosphate pathway, fatty acid biosynthesis, and oxidative phosphorylation were found to be differentially expressed during the different stages of fruit development after pollination. Other work has focused on the proteome of the oil palm leaf infected with *Ganoderma boninense*, a fungus that devastates oil palm plantings in Southeast Asia [130].

6.3.2 Rapeseed (Brassica napus)

Rapeseed (*Brassica napus*) is a source of a normally unpalatable (bitter) vegetable oil. Canola, a cultivar of rapeseed, produces edible oil that is low in glucosinolates (the bitter component of rapeseed oil) and erucic acid, a toxic substance. Canola oil is believed to be one of the healthiest of the vegetable oils. Canola oil is being commercially used to produce biodiesel. The entire genome of *B. napus* has been sequenced and this has helped with its proteomics studies.

Proteome-level changes in two different *B. napus* lines in response to the fungal pathogen *Alternaria brassicae* have been reported [132]. The dynamics of protein expression during pollen germination in Canola have been published [133]. Other studies have looked at the effects of different stress factors on the proteome of Canola. Effects of high temperature stress on proteome of early seedlings of Canola have been discussed [134]. Effects of salinity stress on proteome of Canola leaves [135, 136] and seedlings [136] have been reported. Proteomic analysis of Canola roots inoculated with bacteria and subjected to salt stress has been reported [137].

6.3.3 Sunflower *(Helianthus annuus)*

Sunflower is a source of a widely-consumed and healthy vegetable oil. The average annual sunflower oil yield is about 673 kg per hectare. Sunflower oil can be converted to biodiesel, but it is a more expensive feedstock compared to soybean and Canola oils. Proteomics of sunflower have received some attention [138]. Effects of drought stress [139], metal ion stress [140, 141], and salt stress [142] on proteomics of mostly the leaf tissue have been reported.

6.4 Non-edible Oil Crops

6.4.1 Jatropha *(Jatropha curcas)*

Jatropha curcas L. is a shrub that grows in arid and semi-arid areas. Jatropha seeds contain up to 40 % oil [143]. Production of Jatropha [144] and the biodiesel made from its oil [144, 145] have been reviewed in the literature.

Jatropha oil is inedible. The oil in Jatropha seeds is mostly stored in organelles known as oil bodies. Consequently, these oil bodies have been the focus of proteomic studies [146]. The oil body proteome of three varieties of Jatropha seeds has been discussed [146]. Oleosin and caleosin have been found to be the most abundant structural proteins in oil bodies [146].

The differential proteome of endosperm and embryo of mature Jatropha seeds has been reported [147]. The proteome profiles were found to be quite similar. The major groups of differential proteins were those associated with metabolism (25 %) and disease/defense (18 %) [147]. A detailed analysis of the Jatropha endosperm proteome of five different developmental stages has been published [148]. Proteome associated with stress responsive of Jatropha roots has been reported [51]. Information is available on the enzymes involved in fatty acid biosynthesis and their respective genes [51].

6.4.2 *Camelina* (Camelina sativa)

Camelina sativa, from the Brassicaceae family, is an oilseed plant [149] suited to relatively cold climates. Camelina oil is a potential biodiesel feedstock. Camelina seeds have an oil content of about 36–47 % [150]. Under optimal condtions, the oil yield may be as high as 420 kg/ha, but data are limited. Camelina oil contains more than 90 % unsaturated fatty acids [150].

Although the genome and transcriptome maps for Camelina are available [151], not much is known about its proteomics. Nearly 1500 proteins have been reported in Camelina seeds [152]. Studies have been published of the effects of overexpression of the Arabidopsis G-protein γ subunit 3 (AGG3) gene in Camelina on its proteome [152]. AGG3 is believed to regulate a range of fundamental growth and developmental pathways and has the potential to increase seed yields. Overexpression of AGG3 increased production of the proteins involved in primary and secondary metabolism and those involved in abscisic acid related responses [152]. The AGG3 protein was found to have a role in the regulation of oxidative stress and heavy metal stress tolerance [152].

6.5 Grasses

6.5.1 *Switchgrass* (Panicum virgatum)

Switchgrass (*Panicum virgatum*) is a perennial grass. It is a promising source of lignocellulosic biomass [153, 154]. Proteome analysis of the endomembrane of developing switchgrass coleoptiles (the protective sheaths that cover the growing shoot in grasses) identified 1750 unique proteins [155]. The relevant data are available from the ProteomeXchange (identifier PXD001351; [155]). This appears to be the only proteomic study relating to switchgrass.

6.5.2 *Miscanthus* (Miscanthus *spp.*)

Miscanthus is a perennial grass. It is a potential source of lignocellulosic biomass for combustion and conversion to fuels such as bioethanol [156]. Miscanthus x giganteus hybrid has been claimed to be more productive than the other Miscanthus species [157].

Not much data exist on nucleotide and protein sequences of Miscanthus [158]. A comprehensive proteomic study of *Miscanthus sinensis* leaves has provided a reference map [158]. The effects of heat stress on this reference map have been reported [158]. Proteome of *Miscanthus sinensis* leaves and roots exposed to antimony stress has been published [159]. Proteome changes in roots under chromium stress have been reported [158]. Chromium toxicity is linked to heavy metal tolerance and senescence pathways [160]. Chromium is known to affect vacuole sequestration, nitrogen metabolism and lipid peroxidation [160].

6.6 Tree Crops

6.6.1 Poplar (Populus spp.)

Trees of the genus *Populus* grow rapidly and, therefore, are among most attractive of the woody plants for the production of biofuels. Poplar wood is high in cellulose and low in lignin. Poplar was the first tree to have its genome sequenced [161, 162] and many proteomic studies relate to it.

A comprehensive proteomic analysis provided reference maps of eight poplar organs and tissues and a functional characterization of some proteins [162]. The effects of drought stress on proteome of poplar roots and leaves were reported [162]. Proteomes of poplar seeds with different vigor have been reported [163]. Sixty-five proteins relating to seed vigor were identified [163]. Most of these related to metabolism including protein synthesis and destination, energy generation, cell defense and rescue, and protein storage. These proteins accounted for 95 % of all the identified proteins [163]. The abundance of all these proteins was found to decrease as the seeds aged. This led to the conclusion that seed vigor (aging) was an energy-dependent process which required protein synthesis and degradation as well as cellular defense and rescue [163]. Proteomic analyses of poplar response to cadmium stress [164] and drought stress [165] have been reported. Protein profiles of *Populus* vascular tissue [166, 167] and chloroplast have been reported [168].

6.6.2 Willow (Salix spp.)

Willows comprise some 400 species of deciduous trees and shrubs. Rapid growing willow varieties are a good source of biomass for making fuels. Only a few proteomic studies relate to willow species.

Proteomics of salt stress responses in *Salix matsudana* Koidz, or Chinese Willow, have been discussed [169]. Proteins extracted from roots, stem and leaves that had been subjected to salinity stress were identified. Most (~ 54 %) of the differential protein spots were from roots. Around 24 % of the differential protein spots were from leaves and the rest were from the stem. A majority of these salt stress-responsive proteins were organ specific. The identified proteins were found to be involved in 12 metabolic pathways and processes [169].

Proteomic responses of different willow clones (*Salix fragilis* x *alba*) to sediments contaminated with heavy metals have been reported [170]. The low biomass producing clones subjected to heavy metal stress were as able to maintain the cellular activity as the high biomass producing clones, but had a less pronounced oxidative stress response [170].

6.7 Photosynthetic Microorganisms: Cyanobacteria and Microalgae

6.7.1 Cyanobacteria

Cyanobacteria are the only prokaryotes capable of oxygenic photosynthesis [4]. Some cyanobacteria produce large amounts of oils that can be converted to various fuels. Proteomic research on this group has mostly revolved around the model organism *Synechocystis* sp. PCC 6803. Organelle composition and stress responses have been the focus of this research. Proteomes of the outer membrane [171], the plasma membrane [172] and thylakoid membranes [173] have been characterized. Proteomics of high salt stress [174] and heat stress [175] have been the most commonly studied.

A comprehensive proteomic analysis *Synechocystis* sp. PCC 6803 characterized responses to 33 environmental factors including nutrient (nitrogen, iron, sulfate, phosphate) deprivation, varied trophic growth modes, cold stress and salt stress [176]. Proteomics of hexane resistance in *Synechocystis* sp. PCC 6803 have been reported [177] with the objective of investigating the low tolerance of this microorganism to alkanes.

6.7.2 Microalgae

The availability of genomic and transcriptomic data for several microalgae has helped with their proteomics analysis. Most studies have focused on the model green alga *Chlamydomonas reinhardtii*. Some of the key findings are reviewed here.

Proteomics of many organelles of *C. reinhardtii* have been characterized. This includes proteomics of mitochondria, chloroplast ribosomes [178, 179], thylakoid membranes [180], the light harvesting proteins [181–183], the eye spot [184] and cytoskeletal organelles such as the centrioles [185] and flagella [186].

Proteomics of lipid accumulation and storage in microalgae have attracted attention in view of the interest in fuels derived from algal lipids. Proteins associated with the oil droplets in *C. reinhardtii* have been examined by high throughput proteomics methods [187]. Proteomics have been used to identify the proteins occurring in *C. reinhardtii* under anaerobic conditions as this alga produces hydrogen (a potential biofuel) gas in an anaerobic environment [188].

Proteomics of the carbon concentrating mechanisms of *C. reinhardtii* have been investigated [189] and proteome changes linked to various environmental stress factors have received attention.

Proteomic studies have been reported on the model marine diatom *Phaeodactylum tricornutum*. These studies related to mechanisms involved in lipid accumulation in *P. tricornutum* under nutrient replete conditions [190] and under nitrogen deprivation [191]. Genomes of relatively few microalgae have been sequenced. Proteomics studies of such poorly characterized algae are few, but attempts are being made to bypass this lack of genomic data.

Acknowledgments The authors would like to thank Eng. Mohammad Ali Rajaeifar for his assistance in manuscript formatting.

References

1. Mood SH, Golfeshan AH, Tabatabaei M, Jouzani GS, Najafi GH, Gholami M et al (2013) Lignocellulosic biomass to bioethanol, a comprehensive review with a focus on pretreatment. Renew Sustain Energ Rev 27:77–93
2. Rehman MSU, Kim I, Chisti Y, Han J-I (2013) Use of ultrasound in the production of bioethanol from lignocellulosic biomass. EEST Part A Energ Sci Res 30:1391–1410
3. Sulaiman AZ, Ajit A, Chisti Y (2013) Ultrasound mediated enzymatic hydrolysis of cellulose and carboxymethyl cellulose. Biotechnol Prog 29:1448–1457
4. Tabatabaei M, Karimi K, Sárvári Horváth I, Kumar R (2015) Recent trends in biodiesel production. Biofuel Res J 2:258–267
5. Xu J, Du W, Zhao X, Zhang G, Liu D (2013) Microbial oil production from various carbon sources and its use for biodiesel preparation. Biofuels Bioprod Biorefin 7:65–77
6. Chisti Y (2013) Constraints to commercialization of algal fuels. J Biotechnol 167:201–214
7. Boaretto L, Mazzafera P (2013) The proteomes of feedstocks used for the production of second-generation ethanol: a lacuna in the biofuel era. Ann Appl Biol 163:12–22
8. Nuss ET, Tanumihardjo SA (2010) Maize: a paramount staple crop in the context of global nutrition. Compr Rev Food Sci Food Saf 9:417–436
9. Pechanova O, Takáč T, Šamaj J, Pechan T (2013) Maize proteomics: an insight into the biology of an important cereal crop. Proteomics 13:637–662
10. Bothast R, Schlicher M (2005) Biotechnological processes for conversion of corn into ethanol. Appl Microbiol Biotechnol 67:19–25
11. Chakraborty S, Salekdeh GH, Yang P, Woo SH, Chin CF, Gehring C et al (2015) Proteomics of important food crops in the Asia Oceania Region: current status and future perspectives. J Proteome Res 14:2723–2744
12. Dhugga KS (2007) Maize biomass yield and composition for biofuels. Crop Sci 47:2211–2227
13. Torney F, Moeller L, Scarpa A, Wang K (2007) Genetic engineering approaches to improve bioethanol production from maize. Curr Opin Biotechnol 18:193–199
14. Méchin V, Balliau T, Château-Joubert S, Davanture M, Langella O, Négroni L et al (2004) A two-dimensional proteome map of maize endosperm. Phytochemistry 65:1609–1618
15. Majeran W, Friso G, Ponnala L, Connolly B, Huang M, Reidel E et al (2010) Structural and metabolic transitions of C4 leaf development and differentiation defined by microscopy and quantitative proteomics in maize. Plant Cell 22:3509–3542

16. Li Z, Phillip D, Neuhäuser B, Schulze WX, Ludewig U (2015) Protein dynamics in young maize root hairs in response to macro-and micronutrient deprivation. J Proteome Res 14:3362–3371
17. Marcon C, Malik WA, Walley JW, Shen Z, Paschold A, Smith LG et al. (2015) A high resolution tissue-specific proteome and phosphoproteome atlas of maize primary roots reveals functional gradients along the root axis. Plant Physiol 00138.02015
18. Zhu J, Alvarez S, Marsh EL, LeNoble ME, Cho I-J, Sivaguru M et al (2007) Cell wall proteome in the maize primary root elongation zone. II. Region-specific changes in water soluble and lightly ionically bound proteins under water deficit. Plant Physiol 145:1533–1548
19. Dembinsky D, Woll K, Saleem M, Liu Y, Fu Y, Borsuk LA et al (2007) Transcriptomic and proteomic analyses of pericycle cells of the maize primary root. Plant Physiol 145:575–588
20. Okamoto T, Higuchi K, Shinkawa T, Isobe T, Lörz H, Koshiba T et al (2004) Identification of major proteins in maize egg cells. Plant Cell Physiol 45:1406–1412
21. Grimaud F, Rogniaux H, James MG, Myers AM, Planchot V (2008) Proteome and phosphoproteome analysis of starch granule-associated proteins from normal maize and mutants affected in starch biosynthesis. J Exp Bot 59:3395–3406
22. Pinheiro C, Sergeant K, CtM Machado, Renaut J, CnP Ricardo (2013) Two traditional maize inbred lines of contrasting technological abilities are discriminated by the seed flour proteome. J Proteome Res 12:3152–3165
23. Dahal D, Mooney BP, Newton KJ (2012) Specific changes in total and mitochondrial proteomes are associated with higher levels of heterosis in maize hybrids. Plant J 72:70–83
24. Liu X, Wu Y, Shen Z, Shen Z, Li H, Yu X et al (2010) Shotgun proteomics analysis on maize chloroplast thylakoid membrane. Front Biosci (Elite Ed) 3:250–255
25. Lonosky PM, Zhang X, Honavar VG, Dobbs DL, Fu A, Rodermel SR (2004) A proteomic analysis of maize chloroplast biogenesis. Plant Physiol 134:560–574
26. Friso G, Majeran W, Huang M, Sun Q, Van Wijk KJ (2010) Reconstruction of metabolic pathways, protein expression, and homeostasis machineries across maize bundle sheath and mesophyll chloroplasts: large-scale quantitative proteomics using the first maize genome assembly. Plant Physiol 152:1219–1250
27. Majeran W, Friso G, Asakura Y, Qu X, Huang M, Ponnala L et al (2012) Nucleoid-enriched proteomes in developing plastids and chloroplasts from maize leaves: a new conceptual framework for nucleoid functions. Plant Physiol 158:156–189
28. Nestler J, Schütz W, Hochholdinger F (2011) Conserved and unique features of the maize (Zea mays L.) root hair proteome. J Proteome Res 10:2525–2537
29. Yu J, Roy SK, Kamal AHM, Cho K, Kwon S-J, Cho S-W et al (2014) Protein profiling reveals novel proteins in pollen and pistil of W22 (ga1; Ga1) in maize. Proteomes 2:258–271
30. Guo B, Chen Y, Li C, Wang T, Wang R, Wang B et al (2014) Maize (Zea mays L.) seedling leaf nuclear proteome and differentially expressed proteins between a hybrid and its parental lines. Proteomics 14:1071–1087
31. Jin X, Fu Z, Ding D, Li W, Liu Z, Tang J (2013) Proteomic identification of genes associated with maize grain-filling rate. PLoS ONE 8:e59353
32. Guo B, Chen Y, Zhang G, Xing J, Hu Z, Feng W et al (2013) Comparative proteomic analysis of embryos between a maize hybrid and its parental lines during early stages of seed germination. PLoS ONE 8:e65867
33. Alvarez S, Marsh EL, Schroeder SG, Schachtman DP (2008) Metabolomic and proteomic changes in the xylem sap of maize under drought. Plant Cell Environ 31:325–340
34. Huang H, Møller IM, Song S-Q (2012) Proteomics of desiccation tolerance during development and germination of maize embryos. J Proteomics 75:1247–1262
35. Uváčková Ľ, Takáč T, Boehm N, Obert B, Šamaj J (2012) Proteomic and biochemical analysis of maize anthers after cold pretreatment and induction of androgenesis reveals an important role of anti-oxidative enzymes. J Proteomics 75:1886–1894
36. Wu X, Cai G, Gong F, An S, Cresti M, Wang W (2015) Proteome profiling of maize pollen coats reveals novel protein components. Plant Mol Biol Reporter 33:975–986

37. Zörb C, Herbst R, Forreiter C, Schubert S (2009) Short-term effects of salt exposure on the maize chloroplast protein pattern. Proteomics 9:4209–4220
38. Zörb C, Schmitt S, Mühling KH (2010) Proteomic changes in maize roots after short-term adjustment to saline growth conditions. Proteomics 10:4441–4449
39. Bai X, Yang L, Yang Y, Ahmad P, Yang Y, Hu X (2011) Deciphering the protective role of nitric oxide against salt stress at the physiological and proteomic levels in maize. J Proteome Res 10:4349–4364
40. Meng L-B, Chen Y-B, Lu T-C, Wang Y-F, Qian C-R, Yu Y et al (2014) A systematic proteomic analysis of NaCl-stressed germinating maize seeds. Mol Biol Rep 41:3431–3443
41. Wu L, Wang S, Chen X, Wang X, Wu L, Zu X et al (2013) Proteomic and phytohormone analysis of the response of maize (*Zea mays* L.) seedlings to sugarcane mosaic virus. PLoS ONE 8:e70295
42. Wu L, Zu X, Wang X, Sun A, Zhang J, Wang S et al (2013) Comparative proteomic analysis of the effects of salicylic acid and abscisic acid on maize (*Zea mays* L.) leaves. Plant Mol Biol Reporter 31:507–516
43. Facette MR, Shen Z, Björnsdóttir FR, Briggs SP, Smith LG (2013) Parallel proteomic and phosphoproteomic analyses of successive stages of maize leaf development. Plant Cell 25:2798–2812
44. Walley JW, Shen Z, Sartor R, Wu KJ, Osborn J, Smith LG et al (2013) Reconstruction of protein networks from an atlas of maize seed proteotypes. Proc Natl Acad Sci 110:E4808–E4817
45. Amalraj RS, Selvaraj N, Veluswamy GK, Ramanujan RP, Muthurajan R, Palaniyandi M et al (2010) Sugarcane proteomics: Establishment of a protein extraction method for 2-DE in stalk tissues and initiation of sugarcane proteome reference map. Electrophoresis 31:1959–1974
46. Waclawovsky AJ, Sato PM, Lembke CG, Moore PH, Souza GM (2010) Sugarcane for bioenergy production: an assessment of yield and regulation of sucrose content. Plant Biotechnol J 8:263–276
47. Arruda SM-JF-P (2009) The Brazilian experience of sugarcane ethanol industry. In Vitro Cell Dev Biol 1:372–381
48. Arruda P (2012) Genetically modified sugarcane for bioenergy generation. Curr Opin Biotechnol 23:315–322
49. Dal-Bianco M, Carneiro MS, Hotta CT, Chapola RG, Hoffmann HP, Garcia AAF et al (2012) Sugarcane improvement: how far can we go? Curr Opin Biotechnol 23:265–270
50. Barnabas L, Ramadass A, Amalraj RS, Palaniyandi M, Rasappa V (2015) Sugarcane proteomics: an update on current status, challenges, and future prospects. Proteomics 15:1658–1670
51. Ndimba BK, Ndimba RJ, Johnson TS, Waditee-Sirisattha R, Baba M, Sirisattha S et al (2013) Biofuels as a sustainable energy source: an update of the applications of proteomics in bioenergy crops and algae. J Proteomics 93:234–244
52. Rellán-Álvarez R, Andaluz S, Rodríguez-Celma J, Wohlgemuth G, Zocchi G, Álvarez-Fernández A et al (2010) Changes in the proteomic and metabolic profiles of Beta vulgaris root tips in response to iron deficiency and resupply. BMC Plant Biol 10:1
53. Catusse J, Strub J-M, Job C, Van Dorsselaer A, Job D (2008) Proteome-wide characterization of sugarbeet seed vigor and its tissue specific expression. Proc Natl Acad Sci 105:10262–10267
54. Hajheidari M, Abdollahian-Noghabi M, Askari H, Heidari M, Sadeghian SY, Ober ES et al (2005) Proteome analysis of sugar beet leaves under drought stress. Proteomics 5:950–960
55. Wakeel A, Asif AR, Pitann B, Schubert S (2011) Proteome analysis of sugar beet (*Beta vulgaris* L.) elucidates constitutive adaptation during the first phase of salt stress. J Plant Physiol 168:519–526
56. Yang L, Zhang Y, Zhu N, Koh J, Ma C, Pan Y et al (2013) Proteomic analysis of salt tolerance in sugar beet monosomic addition line M14. J Proteome Res 12:4931–4950

57. Larson RL, Hill AL, Nuñez A (2007) Characterization of protein changes associated with sugar beet (Beta vulgaris) resistance and susceptibility to Fusarium oxysporum. J Agr Food Chem 55:7905–7915
58. Webb KM, Broccardo CJ, Prenni JE, Wintermantel WM (2014) Proteomic profiling of sugar beet (Beta vulgaris) leaves during rhizomania compatible interactions. Proteomes 2:208–223
59. Donnelly BE, Madden RD, Ayoubi P, Porter DR, Dillwith JW (2005) The wheat (*Triticum aestivum* L.) leaf proteome. Proteomics 5:1624–1633
60. Song X, Ni Z, Yao Y, Xie C, Li Z, Wu H et al (2007) Wheat (*Triticum aestivum* L.) root proteome and differentially expressed root proteins between hybrid and parents. Proteomics 7:3538–3557
61. Vensel WH, Tanaka CK, Cai N, Wong JH, Buchanan BB, Hurkman WJ (2005) Developmental changes in the metabolic protein profiles of wheat endosperm. Proteomics 5:1594–1611
62. Balmer Y, Vensel WH, DuPont FM, Buchanan BB, Hurkman WJ (2006) Proteome of amyloplasts isolated from developing wheat endosperm presents evidence of broad metabolic capability. J Exp Bot 57:1591–1602
63. Zheng YS, Guo JX, Zhang JP, Gao AN, Yang XM, Li XQ et al (2013) A proteomic study of spike development inhibition in bread wheat. Proteomics 13:2622–2637
64. Peng Z, Wang M, Li F, Lv H, Li C, Xia G (2009) A proteomic study of the response to salinity and drought stress in an introgression strain of bread wheat. Mol Cell Proteomics 8:2676–2686
65. Bazargani MM, Sarhadi E, Bushehri A-AS, Matros A, Mock H-P, Naghavi M-R et al (2011) A proteomics view on the role of drought-induced senescence and oxidative stress defense in enhanced stem reserves remobilization in wheat. J Proteomics 74:1959–1973
66. Kong F-J, Oyanagi A, Komatsu S (2010) Cell wall proteome of wheat roots under flooding stress using gel-based and LC MS/MS-based proteomics approaches. Biochim Biophys Acta (BBA)-Proteins Proteomics 1804:124–136
67. Guo G, Lv D, Yan X, Subburaj S, Ge P, Li X et al (2012) Proteome characterization of developing grains in bread wheat cultivars (*Triticum aestivum* L.). BMC Plant Biol 12:1
68. Guo H, Zhang H, Li Y, Ren J, Wang X, Niu H et al (2011) Identification of changes in wheat (*Triticum aestivum* L.) seeds proteome in response to anti–trx s gene. PLoS ONE 6:e22255
69. Capriotti AL, Borrelli GM, Colapicchioni V, Papa R, Piovesana S, Samperi R et al (2014) Proteomic study of a tolerant genotype of durum wheat under salt-stress conditions. Anal Bioanal Chem 406:1423–1435
70. Maleki M, Naghavi M, Alizadeh H, Poostini K, Mishani CA (2014) Comparison of protein changes in the leaves of two bread wheat cultivars with different sensitivity under salt stress. Annu Res Rev Biol 4:1784
71. Jacoby RP, Millar AH, Taylor NL (2010) Wheat mitochondrial proteomes provide new links between antioxidant defense and plant salinity tolerance. J Proteome Res 9:6595–6604
72. Guo G, Ge P, Ma C, Li X, Lv D, Wang S et al (2012) Comparative proteomic analysis of salt response proteins in seedling roots of two wheat varieties. J Proteomics 75:1867–1885
73. Fercha A, Capriotti AL, Caruso G, Cavaliere C, Gherroucha H, Samperi R et al (2013) Gel-free proteomics reveal potential biomarkers of priming-induced salt tolerance in durum wheat. J Proteomics 91:486–499
74. Caruso G, Cavaliere C, Guarino C, Gubbiotti R, Foglia P, Laganà A (2008) Identification of changes in Triticum durum L. leaf proteome in response to salt stress by two-dimensional electrophoresis and MALDI-TOF mass spectrometry. Anal Bioanal Chem 391:381–390
75. Ford KL, Cassin A, Bacic AF (2011) Quantitative proteomic analysis of wheat cultivars with differing drought stress tolerance. Front Plant Sci 2:44
76. Zhang S, Guoqi S, Yulian L, Jie G, Jiao W, Guiju C et al (2014) Comparative proteomic analysis of cold responsive proteins in two wheat cultivars with different tolerance to spring radiation frost. Front Agr Sci Eng 1:37–45

77. Budak H, Akpinar BA, Unver T, Turktas M (2013) Proteome changes in wild and modern wheat leaves upon drought stress by two-dimensional electrophoresis and nanoLC-ESI–MS/MS. Plant Mol Biol 83:89–103
78. Jiang S-S, Liang X-N, Li X, Wang S-L, Lv D-W, Ma C-Y et al (2012) Wheat drought-responsive grain proteome analysis by linear and nonlinear 2-DE and MALDI-TOF mass spectrometry. Int J Mol Sci 13:16065–16083
79. Peremarti A, Marè C, Aprile A, Roncaglia E, Cattivelli L, Villegas D et al (2014) Transcriptomic and proteomic analyses of a pale-green durum wheat mutant shows variations in photosystem components and metabolic deficiencies under drought stress. BMC Genom 15:1
80. Kang G, Li G, Xu W, Peng X, Han Q, Zhu Y et al (2012) Proteomics reveals the effects of salicylic acid on growth and tolerance to subsequent drought stress in wheat. J Proteome Res 11:6066–6079
81. Laino P, Shelton D, Finnie C, De Leonardis AM, Mastrangelo AM, Svensson B et al (2010) Comparative proteome analysis of metabolic proteins from seeds of durum wheat (cv. Svevo) subjected to heat stress. Proteomics 10:2359–2368
82. Yang F, Jørgensen AD, Li H, Søndergaard I, Finnie C, Svensson B et al (2011) Implications of high-temperature events and water deficits on protein profiles in wheat (*Triticum aestivum* L. cv. Vinjett) grain. Proteomics 11:1684–1695
83. Caruso G, Cavaliere C, Foglia P, Gubbiotti R, Samperi R, Laganà A (2009) Analysis of drought responsive proteins in wheat (*Triticum durum*) by 2D-PAGE and MALDI-TOF mass spectrometry. Plant Sci 177:570–576
84. Ge P, Ma C, Wang S, Gao L, Li X, Guo G et al (2012) Comparative proteomic analysis of grain development in two spring wheat varieties under drought stress. Anal Bioanal Chem 402:1297–1313
85. Mamone G, Caro SD, Luccia AD, Addeo F, Ferranti P (2009) Proteomic-based analytical approach for the characterization of glutenin subunits in durum wheat. J Mass Spectrom 44:1709–1723
86. Pompa M, Giuliani MM, Palermo C, Agriesti F, Centonze D, Flagella Z (2013) Comparative analysis of gluten proteins in three durum wheat cultivars by a proteomic approach. J Agric Food Chem 61:2606–2617
87. Liu W, Zhang Y, Gao X, Wang K, Wang S, Zhang Y et al (2012) Comparative proteome analysis of glutenin synthesis and accumulation in developing grains between superior and poor quality bread wheat cultivars. J Sci Food Agric 92:106–115
88. Alvarez S, Roy Choudhury S, Pandey S (2014) Comparative quantitative proteomics analysis of the ABA response of roots of drought-sensitive and drought-tolerant wheat varieties identifies proteomic signatures of drought adaptability. J Proteome Res 13:1688–1701
89. Pascovici D, Gardiner DM, Song X, Breen E, Solomon PS, Keighley T et al (2013) Coverage and consistency: bioinformatics aspects of the analysis of multirun iTRAQ experiments with wheat leaves. J Proteome Res 12:4870–4881
90. Li G, Peng X, Xuan H, Wei L, Yang Y, Guo T et al (2013) Proteomic analysis of leaves and roots of common wheat (*Triticum aestivum* L.) under copper-stress conditions. J Proteome Res 12:4846–4861
91. Oh MW, Roy SK, Kamal AHM, Cho K, Cho S-W, Park C-S et al (2014) Proteome analysis of roots of wheat seedlings under aluminum stress. Mol Biol Rep 41:671–681
92. Wang Y, Qian Y, Hu H, Xu Y, Zhang H (2011) Comparative proteomic analysis of Cd-responsive proteins in wheat roots. Acta Physiologiae Plant 33:349–357
93. Duan J, Tian X, Jia Z (2013) Proteomics uncovers a role for enhanced ultraviolet-B radiation on wheat leaves
94. Kamal AHM, Kim K-H, Shin D-H, Seo H-S, Shin K-H, Park C-S et al (2009) Proteomics profile of pre-harvest sprouting wheat by using MALDI-TOF mass spectrometry. Plant Omics 2:110

95. Vensel WH, Tanaka CK, Altenbach SB (2014) Protein composition of wheat gluten polymer fractions determined by quantitative two-dimensional gel electrophoresis and tandem mass spectrometry. Proteome Sci 12:1
96. Kosová K, Prášil IT, Vítámvás P, Dobrev P, Motyka V, Floková K et al (2012) Complex phytohormone responses during the cold acclimation of two wheat cultivars differing in cold tolerance, winter Samanta and spring Sandra. J Plant Physiol 169:567–576
97. Rinalducci S, Egidi MG, Karimzadeh G, Jazii FR, Zolla L (2011) Proteomic analysis of a spring wheat cultivar in response to prolonged cold stress. Electrophoresis 32:1807–1818
98. Gao L, Wang A, Li X, Dong K, Wang K, Appels R et al (2009) Wheat quality related differential expressions of albumins and globulins revealed by two-dimensional difference gel electrophoresis (2-D DIGE). J Proteomics 73:279–296
99. Nadaud I, Girousse C, Debiton C, Chambon C, Bouzidi MF, Martre P et al (2010) Proteomic and morphological analysis of early stages of wheat grain development. Proteomics 10:2901–2910
100. Tasleem-Tahir A, Nadaud I, Girousse C, Martre P, Marion D, Branlard G (2011) Proteomic analysis of peripheral layers during wheat (*Triticum aestivum* L.) grain development. Proteomics 11:371–379
101. Demirevska K, Zasheva D, Dimitrov R, Simova-Stoilova L, Stamenova M, Feller U (2009) Drought stress effects on Rubisco in wheat: changes in the Rubisco large subunit. Acta Physiologiae Plant 31:1129–1138
102. Irar S, Brini F, Goday A, Masmoudi K, Pagès M (2010) Proteomic analysis of wheat embryos with 2-DE and liquid-phase chromatography (ProteomeLab PF-2D)—a wider perspective of the proteome. J Proteomics 73:1707–1721
103. Han Q, Kang G, Guo T (2013) Proteomic analysis of spring freeze-stress responsive proteins in leaves of bread wheat (*Triticum aestivum* L.). Plant Physiol Biochem 63:236–244
104. Kang G, Li G, Ma H, Wang C, Guo T (2013) Proteomic analysis on the leaves of TaBTF3 gene virus-induced silenced wheat plants may reveal its regulatory mechanism. J Proteomics 83:130–143
105. Sarhadi E, Mahfoozi S, Hosseini SA, Salekdeh GH (2010) Cold acclimation proteome analysis reveals close link between the up-regulation of low-temperature associated proteins and vernalization fulfillment. J Proteome Res 9:5658–5667
106. Vítámvás P, Prášil IT, Kosova K, Planchon S, Renaut J (2012) Analysis of proteome and frost tolerance in chromosome 5A and 5B reciprocal substitution lines between two winter wheats during long-term cold acclimation. Proteomics 12:68–85
107. Petersen J, Rogowska-Wrzesinska A, Jensen ON (2015) Functional proteomics of barley and barley chloroplasts–strategies, methods and perspectives. Sub-cellular Proteomics 208
108. Kosová K, Vítámvás P, Prášil IT (2014) Proteomics of stress responses in wheat and barley—search for potential protein markers of stress tolerance. Front Plant Sci 5:711
109. Møller AL, Pedas P, Andersen B, Svensson B, Schjoerring JK, Finnie C (2011) Responses of barley root and shoot proteomes to long-term nitrogen deficiency, short-term nitrogen starvation and ammonium. Plant Cell Environ 34:2024–2037
110. Ashoub A, Beckhaus T, Berberich T, Karas M, Brüggemann W (2013) Comparative analysis of barley leaf proteome as affected by drought stress. Planta 237:771–781
111. Alikhani M, Khatabi B, Sepehri M, Nekouei MK, Mardi M, Salekdeh GH (2013) A proteomics approach to study the molecular basis of enhanced salt tolerance in barley (*Hordeum vulgare* L.) conferred by the root mutualistic fungus Piriformospora indica. Mol BioSyst 9:1498–1510
112. Finnie C, Svensson B (2009) Barley seed proteomics from spots to structures. J Proteomics 72:315–324
113. Petrovská B, Jeřábková H, Chamrád I, Vrána J, Lenobel R, Uřinovská J et al (2014) Proteomic analysis of barley cell nuclei purified by flow sorting. Cytogenet Genome Res 143:78–86

114. Nelson CJ, Alexova R, Jacoby RP, Millar AH (2014) Proteins with high turnover rate in barley leaves estimated by proteome analysis combined with in planta isotope labeling. Plant Physiol 166:91–108
115. Bernardo L, Prinsi B, Negri AS, Cattivelli L, Espen L, Valè G (2012) Proteomic characterization of the Rph15 barley resistance gene-mediated defence responses to leaf rust. BMC Genom 13:1
116. Ghabooli M, Khatabi B, Ahmadi FS, Sepehri M, Mirzaei M, Amirkhani A et al (2013) Proteomics study reveals the molecular mechanisms underlying water stress tolerance induced by Piriformospora indica in barley. J Proteomics 94:289–301
117. Jin Z, Li X-M, Gao F, Sun J-Y, Mu Y-W, Lu J (2013) Proteomic analysis of differences in barley (Hordeum vulgare) malts with distinct filterability by DIGE. J Proteomics 93:93–106
118. Kaspar S, Matros A, Mock H-P (2010) Proteome and flavonoid analysis reveals distinct responses of epidermal tissue and whole leaves upon UV-B Radiation of barley (*Hordeum vulgare* L.) seedlings. J Proteome Res 9:2402–2411
119. Kausar R, Arshad M, Shahzad A, Komatsu S (2013) Proteomics analysis of sensitive and tolerant barley genotypes under drought stress. Amino Acids 44:345–359
120. Rasoulnia A, Bihamta MR, Peyghambari SA, Alizadeh H, Rahnama A (2011) Proteomic response of barley leaves to salinity. Mol Biol Rep 38:5055–5063
121. Sun H, Cao F, Wang N, Zhang M, Ahmed IM, Zhang G et al (2013) Differences in grain ultrastructure, phytochemical and proteomic profiles between the two contrasting grain Cd-accumulation barley genotypes. PLoS ONE 8:e79158
122. Witzel K, Weidner A, Surabhi G-K, Börner A, Mock H-P (2009) Salt stress-induced alterations in the root proteome of barley genotypes with contrasting response towards salinity. J Exp Bot 60:3545–3557
123. Witzel K, Weidner A, SURABHI GK, Varshney RK, Kunze G, BUCK-SORLIN GH et al. (2010) Comparative analysis of the grain proteome fraction in barley genotypes with contrasting salinity tolerance during germination. Plant Cell Environ 33:211–222
124. Baskaran P, Jayabalan N (2005) In vitro plant regeneration and mass propagation system for Sorghum bicolor-a valuable major cereal crop. J Agric Technol 12:345–363
125. Almodares A, Hadi M (2009) Production of bioethanol from sweet sorghum: A review. Afr J Agric Res 4:772–780
126. Rooney WL, Blumenthal J, Bean B, Mullet JE (2007) Designing sorghum as a dedicated bioenergy feedstock. Biofuels Bioprod Biorefin 1:147–157
127. Swami AK, Alam SI, Sengupta N, Sarin R (2011) Differential proteomic analysis of salt stress response in Sorghum bicolor leaves. Environ Exp Bot 71:321–328
128. Jedmowski C, Ashoub A, Beckhaus T, Berberich T, Karas M, Brüggemann W (2014) Comparative analysis of Sorghum bicolor proteome in response to drought stress and following recovery. Int J Proteomics 2014
129. Cremer JE, Bean SR, Tilley MM, Ioerger BP, Ohm JB, Kaufman RC et al (2014) Grain sorghum proteomics: integrated approach toward characterization of endosperm storage proteins in kafirin allelic variants. J Agric Food Chem 62:9819–9831
130. Al-Obaidi JR, Mohd-Yusuf Y, Razali N, Jayapalan JJ, Tey C-C, Md-Noh N et al (2014) Identification of proteins of altered abundance in oil palm infected with Ganoderma boninense. Int J Mol Sci 15:5175–5192
131. Loei H, Lim J, Tan M, Lim TK, Lin QS, Chew FT et al (2013) Proteomic analysis of the oil palm fruit mesocarp reveals elevated oxidative phosphorylation activity is critical for increased storage oil production. J Proteome Res 12:5096–5109
132. Sharma N, Rahman MH, Strelkov S, Thiagarajah M, Bansal VK, Kav NN (2007) Proteome-level changes in two Brassica napus lines exhibiting differential responses to the fungal pathogen Alternaria brassicae. Plant Sci 172:95–110
133. Sheoran IS, Pedersen EJ, Ross AR, Sawhney VK (2009) Dynamics of protein expression during pollen germination in canola (*Brassica napus*). Planta 230:779–793

134. Ismaili A, Salavati A, Pour Mohammadi P (2015) A comparative proteomic analysis of responses to high temperature stress in hypocotyl of canola (*Brassica napus* L.). Protein Pept Lett 22:285–299
135. Bandehagh A, Salekdeh GH, Toorchi M, Mohammadi A, Komatsu S (2011) Comparative proteomic analysis of canola leaves under salinity stress. Proteomics 11:1965–1975
136. Yıldız M, Akçalı N, Terzi H (2015) Proteomic and biochemical responses of canola (*Brassica napus* L.) exposed to salinity stress and exogenous lipoic acid. J Plant Physiol 179:90–99
137. Banaei-Asl F, Bandehagh A, Uliaei ED, Farajzadeh D, Sakata K, Mustafa G et al (2015) Proteomic analysis of canola root inoculated with bacteria under salt stress. J Proteomics 124:88–111
138. Alireza S (2014) Differential Proteomics Analysis in Sunflower (*Helianthus annuus* L.). Biotechnology 13:245
139. Fulda S, Mikkat S, Stegmann H, Horn R (2011) Physiology and proteomics of drought stress acclimation in sunflower (*Helianthus annuus* L.). Plant Biology 13:632–642
140. Garcia JS, Souza GHMF, Eberlin MN, Arruda MAZ (2009) Evaluation of metal-ion stress in sunflower (*Helianthus annuus* L.) leaves through proteomic changes. Metallomics 1:107–113
141. Júnior CAL, de Sousa Barbosa H, Galazzi RM, Koolen HHF, Gozzo FC, Arruda MAZ (2015) Evaluation of proteome alterations induced by cadmium stress in sunflower (*Helianthus annuus* L.) cultures. Ecotoxicol Environ Saf 119:170–177
142. Messaitfa ZH, Shehata AI, El Quraini F, Al Hazzani AA, Rizwana H, El wahabi M (2014) Proteomics analysis of salt stressed Sunflower (*Helianthus annuus*). Int J Pure Appl Biosci 2:68–17
143. Jingura RM, Kamusoko R (2015) A multi-factor evaluation of Jatropha as a feedstock for biofuels: the case of sub-Saharan Africa. Biofuel Res J 2:254–257
144. Contran N, Chessa L, Lubino M, Bellavite D, Roggero PP, Enne G (2013) State-of-the-art of the Jatropha curcas productive chain: from sowing to biodiesel and by-products. Ind Crops Prod 42:202–215
145. Koh MY, Ghazi TIM (2011) A review of biodiesel production from Jatropha curcas L. oil. Renew Sustain Energy Rev 15:2240–2251
146. Liu H, Wang C, Chen F, Shen S (2015) Proteomic analysis of oil bodies in mature Jatropha curcas seeds with different lipid content. J Proteomics 113:403–414
147. Liu H, Yang Z, Yang M, Shen S (2011) The differential proteome of endosperm and embryo from mature seed of Jatropha curcas. Plant Sci 181:660–666
148. Shah M, Soares EL, Carvalho PC, Soares AA, Domont GB, FbC Nogueira et al (2015) Proteomic analysis of the endosperm ontogeny of Jatropha curcas L. seeds. J Proteome Res 14:2557–2568
149. Vollmann J, Eynck C (2015) Camelina as a sustainable oilseed crop: contributions of plant breeding and genetic engineering. Biotechnol J 10:525–535
150. Kagale S, Koh C, Nixon J, Bollina V, Clarke WE, Tuteja R et al. (2014) The emerging biofuel crop Camelina sativa retains a highly undifferentiated hexaploid genome structure. Nat Commun 5
151. Nguyen HT, Silva JE, Podicheti R, Macrander J, Yang W, Nazarenus TJ et al (2013) Camelina seed transcriptome: a tool for meal and oil improvement and translational research. Plant Biotechnol J 11:759–769
152. Alvarez S, Roy Choudhury S, Sivagnanam K, Hicks LM, Pandey S (2015) Quantitative proteomics analysis of Camelina sativa seeds overexpressing the AGG3 gene to identify the proteomic basis of increased yield and stress tolerance. J Proteome Res 14:2606–2616
153. David K, Ragauskas AJ (2010) Switchgrass as an energy crop for biofuel production: a review of its ligno-cellulosic chemical properties. Energy Environ Sci 3:1182–1190
154. Parrish DJ, Fike JH (2005) The biology and agronomy of switchgrass for biofuels. BPTS 24:423–459

155. Lao J, Sharma MK, Sharma R, Fernández-Niño SMG, Schmutz J, Ronald PC et al (2015) Proteome profile of the endomembrane of developing coleoptiles from switchgrass (Panicum virgatum). Proteomics 15:2286–2290
156. Widholm J (2010) Miscanthus: a promising biomass crop. Adv Bot Res 56:75137Hu
157. Laurent A, Pelzer E, Loyce C, Makowski D (2015) Ranking yields of energy crops: a meta-analysis using direct and indirect comparisons. Renew Sustain Energy Rev 46:41–50
158. Sharmin SA, Alam I, Rahman MA, Kim K-H, Kim Y-G, Lee B-H (2013) Mapping the leaf proteome of Miscanthus sinensis and its application to the identification of heat-responsive proteins. Planta 238:459–474
159. Xue L, Ren H, Li S, Gao M, Shi S, Chang E et al (2015) Comparative proteomic analysis in Miscanthus sinensis exposed to antimony stress. Environ Pollut 201:150–160
160. Sharmin SA, Alam I, Kim K-H, Kim Y-G, Kim PJ, Bahk JD et al (2012) Chromium-induced physiological and proteomic alterations in roots of Miscanthus sinensis. Plant Sci 187:113–126
161. Brunner AM, Busov VB, Strauss SH (2004) Poplar genome sequence: functional genomics in an ecologically dominant plant species. Trends Plant Sci 9:49–56
162. Plomion C, Lalanne C, Claverol S, Meddour H, Kohler A, Bogeat-Triboulot MB et al (2006) Mapping the proteome of poplar and application to the discovery of drought-stress responsive proteins. Proteomics 6:6509–6527
163. Zhang H, Wang W-Q, Liu S-J, Møller IM, Song S-Q (2015) Proteome analysis of poplar seed vigor. PLoS ONE 10:e0132509
164. Yang Y, Li X, Yang S, Zhou Y, Dong C, Ren J et al (2015) Comparative physiological and proteomic analysis reveals the leaf response to cadmium-induced stress in poplar (Populus yunnanensis). PLoS ONE 10:e0137396
165. Xiao X, Yang F, Zhang S, Korpelainen H, Li C (2009) Physiological and proteomic responses of two contrasting Populus cathayana populations to drought stress. Physiol Plant 136:150–168
166. Abraham P, Adams R, Giannone RJ, Kalluri U, Ranjan P, Erickson B et al (2011) Defining the boundaries and characterizing the landscape of functional genome expression in vascular tissues of Populus using shotgun proteomics. J Proteome Res 11:449–460
167. Kalluri UC, Hurst GB, Lankford PK, Ranjan P, Pelletier DA (2009) Shotgun proteome profile of Populus developing xylem. Proteomics 9:4871–4880
168. Yuan H-M, Li K-L, Ni R-J, Guo W-D, Shen Z, Yang C-P et al (2011) A systemic proteomic analysis of Populus chloroplast by using shotgun method. Mol Biol Rep 38:3045–3054
169. Qiao G, Zhang X, Jiang J, Liu M, Han X, Yang H et al (2014) Comparative proteomic analysis of responses to salt stress in Chinese willow (Salix matsudana Koidz). Plant Mol Biol Reporter 32:814–827
170. Evlard A, Sergeant K, Ferrandis S, Printz B, Renaut J, Guignard C et al (2014) Physiological and proteomic responses of different willow clones (Salix fragilis X alba) exposed to dredged sediment contaminated by heavy metals. Int J Phytorem 16:1148–1169
171. Huang K, Fingar DC (2014) Growing knowledge of the mTOR signaling network. In: Seminars in cell & developmental biology. Elsevier, pp 79–90
172. Pisareva T, Shumskaya M, Maddalo G, Ilag L, Norling B (2007) Proteomics of synechocystis sp. PCC 6803. FEBS J 274:791–804
173. Srivastava R, Pisareva T, Norling B (2005) Proteomic studies of the thylakoid membrane of Synechocystis sp. PCC 6803. Proteomics 5:4905–4916
174. Fulda S, Huang F, Nilsson F, Hagemann M, Norling B (2000) Proteomics of Synechocystis sp. strain PCC 6803. Eur J Biochem 267:5900–5907
175. Suzuki I, Simon WJ, Slabas AR (2006) The heat shock response of Synechocystis sp. PCC 6803 analysed by transcriptomics and proteomics. J Exp Bot 57:1573–1578
176. Wegener KM, Singh AK, Jacobs JM, Elvitigala T, Welsh EA, Keren N et al (2010) Global proteomics reveal an atypical strategy for carbon/nitrogen assimilation by a cyanobacterium under diverse environmental perturbations. Mol Cell Proteomics 9:2678–2689

177. Liu J, Chen L, Wang J, Qiao J, Zhang W (2012) Proteomic analysis reveals resistance mechanism against biofuel hexane in Synechocystis sp. PCC 6803. Biotechnol Biofuels 5:1
178. Yamaguchi K, Subramanian AR (2003) Proteomic identification of all plastid-specific ribosomal proteins in higher plant chloroplast 30S ribosomal subunit. Eur J Biochem 270:190–205
179. Yamaguchi K, Beligni MV, Prieto S, Haynes PA, McDonald WH, Yates JR et al (2003) Proteomic characterization of the Chlamydomonas reinhardtii chloroplast ribosome identification of proteins unique to the 70 S ribosome. J Biol Chem 278:33774–33785
180. Hippler M, Klein J, Fink A, Allinger T, Hoerth P (2001) Towards functional proteomics of membrane protein complexes: analysis of thylakoid membranes from Chlamydomonas reinhardtii. Plant J 28:595–606
181. Stauber EJ, Fink A, Markert C, Kruse O, Johanningmeier U, Hippler M (2003) Proteomics of Chlamydomonas reinhardtii light-harvesting proteins. Eukaryot Cell 2:978–994
182. Yamaguchi K, Prieto S, Beligni MV, Haynes PA, McDonald WH, Yates JR et al (2002) Proteomic characterization of the small subunit of chlamydomonas reinhardtii chloroplast ribosome identification of a novel S1 domain-containing protein and unusually large orthologs of bacterial S2, S3, and S5. Plant Cell 14:2957–2974
183. Stauber EJ, Hippler M (2004) Chlamydomonas reinhardtii proteomics. Plant Physiol Biochem 42:989–1001
184. Schmidt M, Geßner G, Luff M, Heiland I, Wagner V, Kaminski M et al (2006) Proteomic analysis of the eyespot of Chlamydomonas reinhardtii provides novel insights into its components and tactic movements. Plant Cell 18:1908–1930
185. Keller LC, Romijn EP, Zamora I, Yates JR, Marshall WF (2005) Proteomic analysis of isolated chlamydomonas centrioles reveals orthologs of ciliary-disease genes. Curr Biol 15:1090–1098
186. Pazour GJ, Agrin N, Leszyk J, Witman GB (2005) Proteomic analysis of a eukaryotic cilium. J Cell Biol 170:103–113
187. Moellering ER, Benning C (2010) RNA interference silencing of a major lipid droplet protein affects lipid droplet size in Chlamydomonas reinhardtii. Eukaryot Cell 9:97–106
188. Terashima M, Specht M, Naumann B, Hippler M (2010) Characterizing the anaerobic response of Chlamydomonas reinhardtii by quantitative proteomics. Mol Cell Proteomics 9:1514–1532
189. Baba M, Suzuki I, Shiraiwa Y (2011) Proteomic analysis of high-CO2-inducible extracellular proteins in the unicellular green alga, Chlamydomonas reinhardtii. Plant Cell Physiol 52:1302–1314
190. Ge F, Huang W, Chen Z, Zhang C, Xiong Q, Bowler C et al (2014) Methylcrotonyl-CoA carboxylase regulates triacylglycerol accumulation in the model diatom Phaeodactylum tricornutum. Plant Cell 26:1681–1697
191. Yang Z-K, Ma Y-H, Zheng J-W, Yang W-D, Liu J-S, Li H-Y (2014) Proteomics to reveal metabolic network shifts towards lipid accumulation following nitrogen deprivation in the diatom Phaeodactylum tricornutum. J Appl Phycol 26:73–82

Chapter 7
The Proteome of Orchids

Chiew Foan Chin

Abstract Orchids are the most diverse species in the family of flowering plants. Apart from its beautiful flowers that make it well known as an ornamental plant, some orchids are known to have medicinal values and some can produce useful products such as perfumes and flavouring essence. Many studies have been conducted to elucidate the biological and molecular mechanisms in orchids. This review takes a look at the use of proteomic studies conducted in various aspects of orchid research in order to provide a more in-depth understanding on the biology of orchids.

Keywords Orchids · Proteomics

7.1 Introduction

Orchids are highly priced ornamental plants due to the exquisite beauty of their flowers. In the global floriculture trade, orchids, both as cut flowers and potted plants, is the second most popular commodity after rose that contributes multi-billion dollars to the floral industry. Orchid cut flower is estimated to comprise around 10 % of international fresh cut flower trade [1].

Much research effort has been put into enhance the value of orchid plants. More recently, advance research tools available in plant science such as genetic engineering, functional genomics, proteomics and metabolomics [2, 3] have been applied to gain an in depth understanding of the orchid plants. This review will focus on the use of proteomic studies on orchids.

C.F. Chin (✉)
School of Biosciences, Faculty of Science, University of Nottingham Malaysia Campus, Jalan Broga, 43500 Semenyih, Selangor Darul Ehsan, Malaysia
e-mail: chiew-foan.chin@nottingham.edu.my

7.2 Uses of Orchids

Orchids are not only known as the queen of flowers, they have many other uses. The scent of orchid flower is used as potential fragrance chemical components to produce perfume while the pods of vanilla orchids are used for flavouring cakes, ice cream and soft drinks [4]. They are also important for use in producing medicines for treating different diseases such as asthma, arthritis, boils, blood dysentery, bone fractures and sores. For example, the roots of Cypripedium pubescence contain fixed oil, volatile oils, sugars, starch, resins and tannins, which are useful in treating nervous irritability, spasm, fits and hysteria. The compounds extracted from the leaves, flowers, roots or pseudobulbs of orchids have also been used as aphrodisiac, bronchodilator, contraceptive or cooling agents [5].

7.3 The Orchid Botany

The Orchidaceae or commonly referred to as the orchid family is one of the world's largest plant family with a rich diversity of species. The orchid family comprises of 880 genera with an estimated number of species that ranges between 22,075 and 26,567. Orchids made up of approximately 30 % of monocotyledons or 10 % of all the world's flowering plants [6, 7]. The most common orchid species are Vanda, Dendrobium, Cattleya, Cymbidium and Phalaenopsis. Over ecological time, orchids have evolved to adapt and modify their features to take advantage of adverse environmental conditions on earth. Therefore, to date, orchids can be found in most habitats except in the deserts and Antarctica [8]. The tropical regions such as Asia, Central and South America have the highest orchid varieties in the world.

7.4 Growth Habit of Orchids

Orchids have two distinct main types of growth habits namely, sympodial or monopodial [9]. Sympodial orchids have a horizontal stem called rhizome at the base of the plants (Fig. 7.1). As the rhizome grows, it bends upward and form small plants with leaves as well as flower scapes. These orchids form swollen shoots called pseudobulbs, which are used for water and nutrients storage. Cattleya, Cymbidium, Oncidium and Dendrobium are examples of sympodial orchids. While rhizome and pseudobulbs are observed in sympodial orchids, these features are lacking in monopodial orchids. Monopodial orchids have a single main stem where a series of leaves are produced at the apical bud (Fig. 7.1). Nodes above each leaf are where the roots and flower stems emerge. The stem may sometimes branch but this occurrence is rare in most types of orchids. Orchids that exhibit this type of growth habit include Phalaenopsis, Vandas and Vanilla. Besides distinct habits of

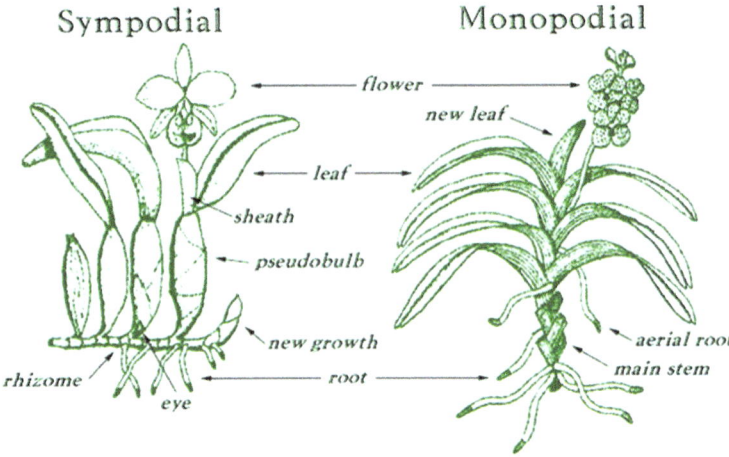

Fig. 7.1 The differences between sympodial and monopodial growth habit of orchids (http://www.aos.org/images/img_content/newsletter_issues/aug09.html)

growth, orchids can also be classified as epiphytes, terrestrial, subterranean, or lithophytes. Approximately 70 % of the orchid family are epiphytic while 25 % are terrestrial and the remaining 5 % are either subterranean or lithophytes [10]. Most of the tropical orchids are either epiphytic or lithophytic while the temperate orchids are terrestrial.

7.5 Cytogenetics of Orchids

Orchidaceae has reported to have a high diversity of chromosome number [9]. For example, the variations in chromosome number were observed within the subfamilies of Spiranthoideae (2n = 28, 36, 46, 48 and 92) and Orchidoideae (2n = 42, 44, 48, 80, 84, 168). The primary basic chromosome number of orchids has been confirmed to be ×1 = 7.

7.6 Genomic and Transcriptomic Studies in Orchids

Orchids are currently the angiosperm family with the most variable genome sizes, with values varying up to 168-fold, ranging from 1C = 0.33 pg in *Trichocentrum maduroi* to 55.4 pg in *Pogonia ophioglossoides* [11]. Recently, the whole genome of the orchid, *Phalaenopsis equestris* (*P. equestris*) has been completely sequenced [12]. *P. equestris* is known to be a popular ornamental plant with great commercial features for floral industry. This orchid plant species also serves as an important breeding parent due to the many colours of its flowers. The availability of whole genome sequence of *P. equestris* will open up a new vista for the research studies of orchids.

In addition, transcriptomic databases of orchids are also available. OrchidBase (http://orchidbase.itps.ncku.edu.tw/est/home2012.aspx) is a collection of EST sequences derived from cDNA libraries constructed from ten Phalaenopsis orchids and ten orchid species across five subfamilies of Orchidaceae namely, Apostasioideae, Vanilloideae, Cypripedioideae, Orchidoideae and Epidendroideae [13]. Another orchid database is OrchidStra (http://orchidstra.abrc.sinica.edu.tw/none/index.php), which is a database with transcriptomic sequences derived from the *Phalaenopsis aphrodite* subsp. *formosana* using Roche 454 and Illumina/Solexa platform [14]. OncidiumOrchidGenomeBase is an orchid database consists of cDNA libraries for six different organs from the orchid plant namely, leaves, pseudobulbs, young inflorescences, inflorescences, flower buds and mature flowers (http://predictor.nchu.edu.tw/oogb/) [15].

7.7 Proteomic Studies in Orchids

The study of proteins derived from the full complement of the genome through proteomic studies has provided an insight into the molecular and biochemical mechanisms taking place in the biological systems of a cell or organism. Nevertheless, the information on the proteomic studies in orchids is scarce and under-represented. Most of the proteomic studies on orchids are on flower development and tissue culture of orchids for mass production (Table 7.1). In addition, another important aspect of orchid study i.e., on mycorrhizal fungi interaction has been conducted. Other orchid studies involved proteomic technologies were on drought stress and cell cycle regulation.

7.8 Reproductive Biology

The flower of the orchids is the most remarkable organ part of the plant. Not only that the orchid flowers come in many colours, shapes and forms, it produces a diverse range of floral odours in the mixtures of alkanes and alkenes to attract pollinators [16]. Some orchids have this unique feature of displaying mimicry to fool its pollinators to facilitate fertilisation. The flower of the orchids can mimic the appearance of the female insect pollinator or produce the floral odour that mimics the sex pheromones of the female pollinator to attract the male pollinator to visit the flower. Such adaptations have contributed to the successful survival of the orchid families through generations. Thus, the flower structure and the evolutionary adaptation of orchids have been an intense area of study for the evolutionary biologists.

7 The Proteome of Orchids

Table 7.1 Proteomic studies on various species of orchids

Orchid species	Plant organ used	Proteomic technique	Treatment	Remarks	References
Ophrys exaltata subsp. *archipelagi, O. garganica, O. sphegodes*	Flower labellum	LC-MS/MS, LTQ (HPLC)	Pollinator attraction	Identification of candidate genes for pollinator attraction and reproductive isolation (for e.g., genes for hydrocarbon and anthocyanin biosynthesis and regulation, and the development of floral morphology)	[20]
Cymbidium ensifolium	Flower	2 DE MALDI ToF/ToF	Flower structure	Comparative study on labellum and inner lateral petals of the flower	[18]
Phalaenopsis equestris	Flower	Yeast 2 hybrid system	Flower development	Study the DNA binding properties and protein–protein interactions of the floral homeotic MADS-box protein complexes	[19]
Vanilla planifolia	Nodes, callus	2 DE, MS ToF/ToF	Tissue culture	Callus development in tissue culture	[4]
Vanilla planifolia	Organogenic calli	2 DE, MS-TOF-TOF	Tissue culture	Investigated the initial biochemical and molecular mechanisms that trigger shoot organogenesis	[24]
Dendrobium officinale	Roots of tissue cultured seedlings	2DE, MALDI-TOF-MS (DIGE)	Tissue culture	Investigate effects of *Mycena dendrobii* on the survival of *Dendrobium officinale* tissue cultured seedlings	[26]
Oncidium sphacelatum	Protocorm	2 DE LC MS/MS, iTRAQ	Germination	Investigate the effect of mycorrhizal fungus, *Ceratobasidium* sp. isolate on the growth of protocorm	[27]
Anoectochilus formosanus	Leaf	2 DE	Drought stress	Investigation on drought effects	[28]
Phalaenopsis aphrodite	Protocorm and meristem tissue	Yeast 2 hybrid system	Cell cycle regulation	Study the protein–protein interaction on the expression and regulation of cell cycle genes	[29]

The flower of orchids has a unique structure. Apart from the sepal and petal, there is a modified petal structure known as labellum. Being located close to the pollens and with other decorative structures such as glands and distinctive colorations as well as sites of hydrocarbons synthesis, labellum could acts as insect attractants for orchids [17]. Therefore, an in depth studies on labellum of orchids has attracted much interests. Li et al. [18] used proteomic approach to investigate the differential proteins between labellum and petals. A total of 30 differential proteins were found and identified using MALDI-TOF/TOF (Table 7.1). The high beta-glucosidase protein was found present in the labellum indicated its role in floral odour emission and thus act as pollinator attractant. On the other hand, Tsai et al. [19] used Yeast 2 hybrid system to study the protein-protein interactions of the floral homeotic MADS-box protein complexes that determined the flower structure of orchids.

The study of another evolutionary adaption of orchids on pollinator deception was undertaken by Sedeek et al. [20]. Three Ophrys species namely, *O. exaltata*, *O. sphegodes* and *O. garganica* were used in the study because the genus Ophrys were known to be sexually deceptive orchids. The proteins were extracted from the flower labellum tissue of Ophrys orchids. The proteins extracted from the flowers of the three Ophrys species were analysed through LC MS/MS (Table 7.1) and identified by comparing to the protein database created from the Ophrys floral transcriptomic data. The proteins were found to map to enzymes such as stearoyl-ACP desaturase (SAD) and b-ketoacyl-CoA synthase, which are involved in hydrocarbon biosynthesis. In addition, Sedeek et al. [20] were the first research group to have deposited orchid proteome onto PRIDE database [21] under accession numbers 27721–27914 and the ProteomeXchange Consortium (http://proteomecentral.proteomexchange.org) under accession number PXD000069 (doi: 10.6019/PXD000069).

7.9 Micropropagation

Micropropagation is an important tool for mass multiplications of ornamental plants. Since this technology is rather labour intensive and can be quite costly, it would be only economically viable to apply the technology to high market value plant such as orchids. In fact, orchid is the first plant that has successfully been introduced in vitro more than 100 years ago [22]. Today, mass propagation of orchids through tissue culture is established and well documented. However, due to the genotype dependant nature of tissue culture, different species of orchids were found to respond differently under different in vitro conditions [23]. Also, different orchid species will pose different challenges in the in vitro environments. Hence, investigations on different aspects of tissue culture such as browning of leaf explants [23], initiation of shoot organogenesis [24] and callus development [4] were carried out using proteomic technologies in order to provide a more in depth analysis to the challenges.

7.10 Mycorrhizal Interactions

Orchids are known to associate mutualistically with a heterogeneous group of fungi known as the orchid mycorrhiza. One such association with the mycorrhizal fungi is in the seed germination of orchids. Since the orchid seeds are minute and contain only a few nutrients storage, its association with mycorrhizal fungi is important for the seed germination and seedling development [25]. Using proteomic studies, Xu et al. [26] found that the mycorrhizal fungus, *Mycena dendrobii*, enhanced stress tolerance and promoted new root formation, which helped to improve the survival and growth of *Dendrobium officinale* tissue culture seedlings. Another mycorrhizal interaction work has been conducted by Valadres et al. [27]. In this study, 2DE LC MS/MS coupled with iTRAQ has been used to isolate and identify proteins associated with the effect of mycorrhizal fungi on the growth of *Oncidium sphacelatum* protocorms. The results suggest that several phytohormone and secondary metabolites, reactive oxygen species, and defense-related proteins may play a role in orchid mycorrhizal interactions.

7.11 Other Studies

Other proteomic studies in orchids include plant stress response and cell cycle.

Pandey et al. [28] used 2DE method to investigate the protein content of the leaf of the orchids, *Anoectochilus formosanus*, under drought stress condition. The study found that the protein content in the leaf was significantly reduced under drought condition.

Since cell proliferation is governed by cell cycle regulation, which in turn determines the growth and development of a plant, it is important to have a more in depth understanding on the mechanisms of cell cycle regulation. Lin et al. [29] used the yeast 2 hybrid protein-protein interaction system to study the major cell-cycle regulators in the moth orchids, *Phalaenopsis aphrodite*. The study found that the cell cycle regulator such as cyclin dependent kinase A is conserved. In addition, expression of the major cell cycle genes was co-ordinately regulated during pollination induced reproductive development.

7.12 Conclusion

Proteomics is a promising tool for the elucidation of gene products and provide the closest molecular link with the phenotypic expression. The analysis of proteomics will be enhanced with the availability of genomic and transcriptomic databases. With the recent complete genome sequence of Phalaenopsis orchid [12], it is therefore foreseeable that there will be more useful proteomic data of orchids being generated in the near future.

References

1. De L, Singh D (2015) Biodiversity, conservation and bio-piracy in orchids-an overview. J Glob Biosci 4:2030–2043
2. Hsiao YY, Pan ZJ, Hsu CC, Yang YP, Hsu YC, Chuang YC et al (2011) Research on orchid biology and biotechnology. Plant Cell Physiol 52:1467–1486
3. Hossain MM, Kant R, Van PT, Winarto B, Zeng S, Teixeira Da Silva JA (2013) The application of biotechnology to orchids. Crit Rev Plant Sci 32:69–139
4. Tan BC, Chin CF, Liddell S, Alderson P (2013) Proteomic analysis of callus development in Vanilla planifolia Andrews. Plant Mol Biol Report 31:1220–1229
5. Hossain MM (2011) Therapeutic orchids: traditional uses and recent advances—an overview. Fitoterapia 82:102–140
6. Dressler RL (1993) Phylogeny and classification of the orchid family. Cambridge University Press
7. Ong PT, O'byrne P, Yong SYW, Saw LG, Kiew R (2012) Wild Orchids of Peninsular Malaysia
8. Arditti J (2009) Micropropagation of orchids. Wiley
9. Pridgeon AM, Cribb PJ, Chase MW, Rasmussen FN (1999) Genera Orchidacearum: vol. 1. General Introduction, Apostasioideae, Cypripedioideae. Oxford University Press, Oxford (xiv, 197p.-illus., col. illus. ISBN)
10. Atwood JT (1986) The size of the Orchidaceae and the systematic distribution of epiphytic orchids. Selbyana 171–186
11. Leitch IJ, Kahandawala I, Suda J, Hanson L, Ingrouille MJ, Chase MW et al (2009) Genome size diversity in orchids: consequences and evolution. Ann Bot 104:469–481
12. Cai J, Liu X, Vanneste K, Proost S, Tsai W-C, Liu K-W et al (2015) The genome sequence of the orchid Phalaenopsis equestris. Nat Genet 47:65–72
13. Fu CH, Chen YW, Hsiao YY, Pan ZJ, Liu ZJ, Huang YM et al (2011) Orchidbase: a collection of sequences of the transcriptome derived from orchids. Plant Cell Physiol 52:238–243
14. Su C-L, Chao Y-T, Chang Y-CA, Chen W-C, Chen C-Y, Lee A-Y et al (2011) De novo assembly of expressed transcripts and global analysis of the Phalaenopsis aphrodite transcriptome. Plant Cell Physiol 52:1501–1514
15. Chang Y-Y, Chu Y-W, Chen C-W, Leu W-M, Hsu H-F, Yang C-H (2011) Characterization of Oncidium 'Gower Ramsey'transcriptomes using 454 GS-FLX pyrosequencing and their application to the identification of genes associated with flowering time. Plant Cell Physiol 52:1532–1545
16. Schluter PM, Schiestl FP (2008) Molecular mechanisms of floral mimicry in orchids. Trends Plant Sci 13:228–235
17. Rudall PJ, Bateman RM (2002) Roles of synorganisation, zygomorphy and heterotopy in floral evolution: the gynostemium and labellum of orchids and other lilioid monocots. Biol Rev Camb Philos Soc 77:403–441
18. Li X, Xu W, Chowdhury MR, Jin F (2014) Comparative proteomic analysis of labellum and inner lateral petals in Cymbidium ensifolium flowers. Int J Mol Sci 15:19877–19897
19. Tsai WC, Pan ZJ, Hsiao YY, Jeng MF, Wu TF, Chen WH et al (2008) Interactions of B-class complex proteins involved in tepal development in Phalaenopsis orchid. Plant Cell Physiol 49:814–824
20. Sedeek KE, Qi W, Schauer MA, Gupta AK, Poveda L, Xu S et al (2013) Transcriptome and proteome data reveal candidate genes for pollinator attraction in sexually deceptive orchids. PLoS ONE 8:e64621
21. Vizcaino JA, Cote R, Reisinger F, Barsnes H, Foster JM, Rameseder J et al (2010) The Proteomics Identifications database: 2010 update. Nucleic Acids Res 38:D736–742
22. Yam TW, Arditti J (2009) History of orchid propagation: a mirror of the history of biotechnology. Plant Biotechnol Report 3:1–56

23. Chen J-T, Chang W-C (2002) Effects of tissue culture conditions and explant characteristics on direct somatic embryogenesis in OncidiumGower Ramsey'. Plant Cell Tissue Organ Cult 69:41–44
24. Palama TL, Menard P, Fock I, Choi YH, Bourdon E, Govinden-Soulange J et al (2010) Shoot differentiation from protocorm callus cultures of Vanilla planifolia (Orchidaceae): proteomic and metabolic responses at early stage. BMC Plant Biol 10:82
25. Dearnaley JD (2007) Further advances in orchid mycorrhizal research. Mycorrhiza 17:475–486
26. Xu X, Ma X, Lei H, Song H, Ying Q, Xu M et al (2015) Proteomic analysis reveals the mechanisms of Mycena dendrobii promoting transplantation survival and growth of tissue culture seedlings of Dendrobium officinale. J Appl Microbiol 118:1444–1455
27. Valadares R, Perotto S, Santos E, Lambais M (2014) Proteome changes in Oncidium sphacelatum (Orchidaceae) at different trophic stages of symbiotic germination. Mycorrhiza 24:349–360
28. Pandey DM, Wu RZ, Hahn E-J, Paek K-Y (2006) Drought effect on electrophoretic protein pattern of Anoectochilusformosanus. Sci Hortic 107:205–209
29. Lin H-Y, Chen J-C, Wei M-J, Lien Y-C, Li H-H, Ko S-S et al (2014) Genome-wide annotation, expression profiling, and protein interaction studies of the core cell-cycle genes in Phalaenopsis aphrodite. Plant Mol Biol 84:203–226

Chapter 8
Proteomic Tools for the Investigation of Nodule Organogenesis

Nagib Ahsan and Arthur R. Salomon

Abstract Over the last decades, proteomics approaches are increasingly being utilized to develop a more comprehensive picture of nodulation in many models and/or economically important legume species. This chapter provides an overview of recent developments in the application of proteomic technologies including gel-based, gel-free, isotopic labeling for quantitation, and post translational modifications that target proteomic analysis of nodule organogenesis. These approaches provide a deeper understanding of protein regulation and interaction among the possible pathways that are associated with nodulation in legume plants. In addition, the challenges faced by proteomics in understanding nodulation are discussed, and some possible future strategies for meeting these challenges are proposed.

Keywords Nodulation · Subcellular and tissue specific proteomics · Single cell proteomics · Targeted proteomics

8.1 Introduction

Nodulation is a complex organogenesis process involving signal exchanges between the host plant roots and their rhizobial symbionts in certain members of the leguminosae family plants. This symbiotic interaction results the formation of nodule in roots wherein plants supplies reduced carbon derived from photosynthesis for the bacteroides in exchange for fixed nitrogen available as ammonium.

N. Ahsan (✉)
Division of Biology and Medicine, Brown University, Providence, RI 02903, USA
e-mail: nagib_ahsan@brown.edu; nagib.ahsan@lifespan.org

N. Ahsan · A.R. Salomon
Center for Cancer Research and Development, Proteomics Core Facility,
Rhode Island Hospital, Providence, RI 02903, USA

A.R. Salomon
Department of Molecular Biology, Cell Biology, and Biochemistry,
Brown University, Providence, RI 02903, USA

Key genes involved in the nodulation process have been discovered in several studies using molecular and genetic approaches for some legume species including *Lotus japonicas* and *Medicago truncatula* [1–5]. Gaining an understanding of the biological function of any novel gene and/or mutant surely provides valuable information of that particular gene and protein gene product. However, protein expression is regulated not only at the transcriptional level, but also at the translational and post-translational levels. In addition, protein networks and/or pathways that are common for all legumes and regulated during nodulation are poorly investigated. Obtaining information at the translational and post-translational levels, proteomics approaches are extremely efficient tools that offer deeper into complex biological protein networks in any developmental process.

Over the last decades, proteomics approaches are increasingly being utilized to develop a more comprehensive picture of nodulation in many models and/or economically important legume species including Medicago, soybean, lotus and pea (Table 8.1; Fig. 8.1). Recently Imin [6] described the applications of gel-based quantitative and comparative proteomics in analyzing root nodule samples. However, recent proteomics research revealed that gel-free proteomics approaches are quite capable of increasing the yield of protein identifications and high quality quantitation. In addition, several post-translational modifications regulated during nodulation process can also be determined and quantified by various gel-free methods (Fig. 8.1).

This chapter provides an overview of recent developments in the application of proteomic technologies including gel-based, gel-free, isotopic labeling for quantitation, and post translational modifications that target proteomic analysis of nodule organogenesis. These approaches provide a deeper understanding of protein regulation and interaction among the possible pathways that are associated with nodulation in legume plants. In addition, the challenges faced by proteomics in understanding nodulation are discussed, and some possible future strategies for meeting these challenges are proposed.

8.2 Proteomics Methodologies for Analyzing Nodulation

Optimization of protein extraction and purification from biological samples in a proteomic experiment is a fundamentally critical step that greatly influences the breadth of proteome characterized. Most plant tissues consist of a wide range of compounds including proteases, polysaccharides, lipids, phenolic compounds, and secondary metabolites which are the most common interfering components in gel electrophoresis. However, a significant effort has been made over the last two decades to optimize protein extraction methods from recalcitrant plant tissue samples [7, 8].

8 Proteomic Tools for the Investigation of Nodule Organogenesis

Table 8.1 A summary of nodule proteomics related papers published in the period of 2000 October, 2015

Plant species	Rhizobia	Proteomics methods	IP[a]	Key results	References
Glycine max cv. Stevens	Bradyrhizobium japonicum USDA110	2-DE, N-terminal sequencing	17	The identification of homologs of HSP70 and HSP60 associated with the peribacteroid membrane (PBM) is the first evidence that the molecular machinery for co- or post-translational import of cytoplasmic proteins is present in symbiosomes	[9]
Melilotus alba	Sinorhizobium meliloti strain 1021	2-DE, N-terminal sequencing MALDI-TOF	100	Bacteroid cells showed down-regulation of several proteins involved in nitrogen acquisition, including glutamine synthetase, urease, a urea-amide binding protein, and a PII isoform, indicating that the bacteroids were nitrogen proficient	[10]
Pisum sativum L. cv. Solara	Rhizobium leguminosarum bv. Viceae strain Riso 18a	2-DE, ESI-MS/MS	46	Identification of a number of endomembrane proteins including V-ATPase, BIP, and an integral membrane protein known from COPI-coated vesicles, supporting the role of the endomembrane system in PBM biogenesis	[57]
Medicago truncatula cv. Jemalong J5	Sinorhizobium meliloti strain RCR 2011	2-DE, MALDI-TOF, Q-TOF	~20	Nodulated roots and mycorrhizal roots were analyzed	[58]
Lotus japonicus GIFU (B-129)	Mesorhizobium loti strain R7A	BN-PAGE, gel free, LC-MS/MS	160	Root nodule peribacteroid membrane proteins were identified	[27]
Medicago truncatula	Sinorhizobium meliloti strain 1021	2-DE, MALDI-TOF	2224	Total of 27 putative nodule specific proteins and 35 nutrient-stress-specific proteins were identified in S. meliloti cells in nodules	[59]
Medicago truncatula A17	Sinorhizobium meliloti	2-DE, LC-MS/MS	51	Identification of a large number of symbiosome membrane proteins provides a basis to hypothesize mechanisms of symbiosome membrane formation and function	[60]
Glycine max L. Merr. cv. Akishirome	Bradyrhizobium japonicum USDA110	2-DE, MALDI-TOF, ESI-MS, N-terminal sequencing	34	Of the proteins that were detected only in nodule mitochondria, phosphoserine aminotransferase, flavanone 3-hydroxylase, coproporphyrinogen III oxidase, one ribonucleoprotein	[35]

(continued)

Table 8.1 (continued)

Plant species	Rhizobia	Proteomics methods	IP[a]	Key results	References
Medicago truncatula, Melilotus alba	Sinorhizobium meliloti strain 1021	2-DE, MALDI-TOF	420	A selective suite of ABC-type transporters was present in nodule bacteria that were biased towards the transport of amino acids and inorganic ions (P and Fe)	[61]
Medicago truncatula cv. Jemalong J5	N/A	2-DE, MALDI-TOF	96	Root microsomal proteome map has been developed	[62]
Glycine max cv. William 82	Bradyrhizobium japonicum USDA110	2-DE, MALDI-TOF, LC-MS/MS	27	First study of root hair proteomics. Together with known proteins respond to rhizobial inoculation some novel proteins including phospholipase D and phosphoglucomutase were also identified	[24]
Glycine max cv. William 82	Bradyrhizobium japonicum USDA110	2-DE, MALDI-TOF	182	A partial proteome map of B. japonicum was constructed	[63]
Glycine max cv. William 82	Bradyrhizobium japonicum USDA110	2-DE, MALDI-TOF	313	Proteins related to fatty acid, nucleic acid and cell surface synthesis were significantly higher in cultured cells whereas Nitrogen metabolism was more pronounced in bacteroids	[64]
Medicago truncatula cv. Jemalong A17, mutant	Sinorhizobium meliloti strain 1021	2-DE, MALDI-TOF, LC-MS/MS	11	A total of 11 proteins were differentially affected in wild-type and skl in response to both ACC and nodulation hypothesized that during early nodule development, the plant induces ethylene-mediated stress-responses to limit nodule numbers	[65]
Medicago truncatula cv. Jemalong A17	Sinorhizobium meliloti strain 1021	FPLC, Gel-free, LC-MS/MS	377	Sucrose synthase (SuSy) identified as a key enzyme involved in drought stress and also identified new marker enzymes such as plant Met synthase and bacteroid Ser hydroxy methyltransferase	[17]
Medicago truncatula cv. Jemalong A17, mutant	Sinorhizobium meliloti strain 1021	2-DIGE, MALDI-TOF/TOF MS	170	Results indicates that proteins differentially accumulated between untreated wild-type and supernodulation mutant (sunn) roots also showed changes in auxin response, consistent with altered auxin levels in sunn	[13]

(continued)

Table 8.1 (continued)

Plant species	Rhizobia	Proteomics methods	IP[a]	Key results	References
Glycine max cv. William 82	Bradyrhizobium japonicum USDA110	2-DE, MALDI-TOF	69	The partial soybean nodule cytosol proteome map was developed and proteins were categorized by function	[30]
Medicago truncatula cv. Jemalong A17		Gel-free, AQUA MRM	15	A selective LC-MS/MS-based method using synthetic isotope-labelled peptides has been applied for accurate quantification of SuSy proteins and other nodule metabolic enzymes	[66]
Medicago truncatula cv. Jemalong A17	Sinorhizobium meliloti strain 2011	Gel-free, LC-MS/MS, AQUA MRM	310	Drought induced a reduction of SNF rates and major changes in the metabolic profile of nodules, mostly an accumulation of amino acids and. This accumulation was coincidental with a decline in the levels of bacteroid proteins involved in SNF and C metabolism, along with a partial reduction of the levels of plant sucrose synthase 1	[16]
Glycine max cv. Snimaldalkong 2, mutant SS2-2	Bradyrhizobium japonicum USDA110	2-DE, MALDI-TOF, ESI-MS/MS	103	Comparative proteome analysis indicated lower expression of malate dehydrogenase, leghemoglobins and nitrogenase in the supernodulation mutant (SS2-2), as compared to the wild type, indicating SS2-2 forms functionally immature nodules in higher numbers with the lower activity of nitrogen fixation	[67]
Medicago truncatula cv. Jemalong A17	NA	2-DE, SCX, IMAC, LC-MS/MS	829	The first large-scale plant phosphoproteomic study to utilize ETD. Analysis of the identified phosphorylation sites revealed phosphorylation motifs not previously observed in plants	[47]
Medicago truncatula cv. Jemalong A17	Sinorhizobium meliloti strain 2011	1D SDS-PAGE, LC-MS/MS	117	Identification of many sulfenylated proteins indicating sulfenylation may regulate the activity of proteins playing major roles in the development and functioning of the symbiotic interaction	[41]

(continued)

Table 8.1 (continued)

Plant species	Rhizobia	Proteomics methods	IP[a]	Key results	References
Glycine max L. cv. Eb-b0-1, En1282, Enrei	Bradyrhizobium japonicum MAFF 211342	2-DE, LC-MS/MS	56	The supernodulating and non-nodulating varieties responded oppositely to bacterial inoculation with respect to the expression of 11 proteins. The suppression of the autoregulatory mechanism in the supernodulating variety might be due to negative regulation of defense and signal transduction-related processes	[68]
Glycine max cv. William 82	Bradyrhizobium japonicum USDA110	iTRAQ labeling, phospho enrichment, LC-MS/MS	1126	Identification of a large number (1659) of non-redundant phosphorylation sites, in root hair and stripped roots suggesting a complex network of kinase-substrate and phosphatase-substrate interactions in response to rhizobial inoculation	[18]
Medicago truncatula cv. Jemalong A17, C31 (nfp-1), TRV25 (dmi3-1)	Sinorhizobium meliloti strain Rm1021	TMT, iTRAQ labeling, phospho enrichment, LC-MS/MS	7739	Identification of the largest dataset for nod factor-induced changes in the phosphorylation status of 13,506 phosphosites in 7739 proteins from the model legume Medicago truncatula	[48]
Phaseolus vulgaris L. cv. Contender	Rhizobium leguminosarum bv. phaseoli strain 3622	Gel-free, LC-MS/MS	224	Nodule mitochondria are the early target of oxidative modifications and a likely source of redox signals. Alternative oxidase and MnSOD may play important roles in controlling ROS concentrations and the redox state of mitochondria	[19]
Lotus japonicus Gifu B-129 and snf1	Mesorhizobium loti strain MAFF303099	2-DE, MALDI TOF/TOF	1100	Development of the most extensive root and nodule proteome map in two developmental stages and compared with the spontaneous nodule formation mutant (snf1)	[11]
Pisum sativum cv. Sugar-lace	Rhizobium leguminosarum	2-DE, MALDI TOF/TOF	18	Among the identified proteins, 3 proteins related to flavonoid metabolism, 2 to sulfur metabolism and 3 RNA-binding proteins could be molecular targets for future studies focused on the improvement of legumes tolerance to drought	[69]

(continued)

8 Proteomic Tools for the Investigation of Nodule Organogenesis 143

Table 8.1 (continued)

Plant species	Rhizobia	Proteomics methods	IP[a]	Key results	References
Medicago truncatula cv. Jemalong A17	Sinorhizobium meliloti strain 2011	AQUA MRM	6	Both MetS and SAMS show a decline in protein content in drought-stressed roots and nodules and at least the expression of one of the ACS homologs is rapidly down-regulated. These results suggest that ethylene production is reduced at early stress stages, with possible implications in nitrogen fixation signaling	[70]
Aeschynomene indica	Bradyrhizobium sp. strain ORS278	1D SDS-PAGE, LC-MS/MS	1429	Photosynthetic Bradyrhizobium proteome during the symbiotic process with A. indica that forms root and stem nodules were analyzed. Mutant analysis suggested that in addition to the EtfAB system, the fixA locus is required for symbiotic efficiency	[71]
Medicago truncatula cv. Jemalong A17	Sinorhizobium meliloti strain WSM419	Gel-free, LC-MS/MS	138	The presence of early nodule-specific cysteine-rich (NCR) peptides in nitrogen-fixing bacteroids indicates their high stability, and their long-term maintenance suggests persisting biological roles in the bacteroids	[20]
Glycine max cv. Stephens	Bradyrhizobium japonicum	Gel-free, LC-MS/MS	212	Among proteins, a number of putative transporters for compounds such as sulfate, calcium, hydrogen ions, peptide/dicarboxylate, and nitrate, as well as transporters were identified for which the substrate is not easy to predict	[21]
Lotus japonicus MG-20	Mesorhizobium loti strain MAFF303099	Gel-free, LC-MS/MS	1024	Results suggested that M. loti enters a nitrogen-deficient condition during the early stages of nodule development, and then a nitrogen-rich condition during the intermediate stages of nodule development. In addition, M. loti assimilated ammonia during the intermediate stages of nodule development	[22]
Glycine max	Bradyrhizobium liaoningense CCBAU05525	2-DE, MALDI TOF/TOF	27	Results showed that water-soluble humic materials stimulated cell metabolism and nutrient transport, which resulted in increased cell density of CCBAU05525 and nod gene and nitrogen fixation related proteins expression thus prepared the bacteria for better bacteroid development	[72]

[a]Identified proteins

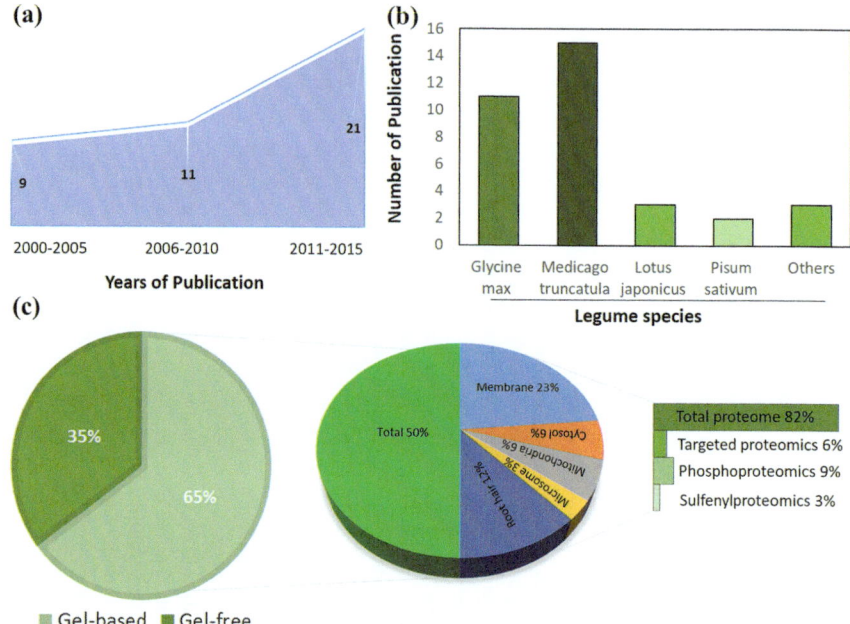

Fig. 8.1 Current status of nodule proteomic studies in legume plants. **a** The growth in the numbers of papers published between 2000 and 2015. The numbers of publications were compiled by searches in PubMed (http://www.ncbi.nlm.nih.gov/pubmed/) with the key words "legume nodule proteomic". **b**, **c** Represents the comparative number of studies carried on various types of legume species and proteomics methodologies, respectively

8.2.1 Gel-Based Proteomics

In classical gel-based proteomics, protein separation by one dimensional (1-DE) and/or 2-dimensional polyacrylamide gel electrophoresis (2-DE) coupled with Edman sequencing, MALDI-TOF MS, or ESI-MS/MS analysis are the most frequent methodologies used to investigate protein regulation in the nodulation process (Table 8.1). In conventional 1-DE and 2-DE based analyses, gels are mostly stained with coomassie brilliant blue and/or silver nitrate. For the first time, Panter et al. [9] used the 2-DE approach to demonstrate protein regulation during nodulation in soybean roots. Using N-terminal sequencing they were able to identify a total of 17 proteins of which nine did not show any significant results. This outcome might be related to the lack of complete genome sequences of many of the legume species including soybean. Later, the same research group (Dr. Udvardi) developed the proteome maps of control *M. alba* roots, wild-type nodules, cultured *S. meliloti*, and *S. meliloti* bacteroides [10]. Among the 600 differentially regulated proteins 250 were up-regulated in the nodule, compared with the root, and over 350 proteins were down regulated in the bacteroid form of the rhizobia, compared with cultured cells. Among these a total of 100 proteins were identified with peptide mass fingerprinting. To date, the most extensive 2-DE gel-based experiment was conducted to develop a root and nodule proteome map of Lotus by Dam et al. [11] where over 1000 proteins have been successfully identified.

The conventional 2-DE gel-based approach has been extensively used as recently as 2010 to demonstrate protein regulation and to develop proteome map of many legumes and rhizobia during nodulation. However, this approach has a number of limitations including gel to gel variation, low resolution in terms of protein separation and quantitation. To address some of these weaknesses, researchers have employed 2-D differentially gel electrophoresis (2D-DIGE), another gel-based approach that utilizes various fluorescent reagents (CyDye) to label protein samples before 2-DE. This method allows several samples to be separated simultaneously and visualized in one gel, considerably reducing gel-to-gel variation, and improving the precision of the quantitative analysis [12]. Thus far, only one proteomic experiment has employed this technology to investigate the relation of auxin in nodule formation in Medicago [13].

8.2.2 Gel-Free Proteomics

2-DE gel methodologies remain the most widely used method for analyzing root nodule protein regulation in legume plants and bacteroides (Fig. 8.1). However, the technique is not satisfactory and in some respects impractical for separating and identifying soluble/insoluble and/or highly hydrophobic core components of multi-subunit complexes in organelle proteomes such as membrane proteomes. To overcome this limitation, gel-free proteomics approaches have been used in many nodule proteomics studies (Table 8.1; Fig. 8.1). In this method, proteins are digested with trypsin in solution followed by loading of desalted peptides onto a C18 reversed-phase column which are in line with the MS/MS. A gel free shotgun proteomic analysis can yield confident identification of 13,000 proteins [14] whereas only 2000 protein spots can be clearly visualized and processed prior to identification by mass spectrometry using 2-DE approach. Most importantly the 2-DE approach is low throughput. Characterization of the protein identity within a 2-DE spot requires a separate LC-MS/MS experiment, making wide-scale discovery impractical.

Accurate quantitation of relative peptide abundance is critical to gain biological understanding from shotgun proteomic experiments. In intensity-based, label-free quantitative analysis, relative quantitation is provided by monitoring individual peptide abundance in the MS spectrum through different cellular states. Alternatively, label-free quantitative analysis may be provided by spectral counting of peptides. Samples may also be quantified through stable isotope labeling either in cell culture (SILAC method) or by adding the isotopic labels after protein digestion. Isotope tagging for relative and absolute protein quantitation (iTRAQ) based on isobaric mass labels at the N-termini and lysine side chains of peptides in a digest mixture [15], is one of the most reliable and promising gel-free proteomics methods. This approach offers confident protein identification with excellent reproducible quantitation. This quantitation is essential to fully characterize protein dynamics, turnover, and interaction partners.

However, quantitative shotgun proteomic analysis has only been applied to the analysis of nodule proteome in a limited number of cases thus far such as identification and regulation of nodule proteins during development or in response to stress conditions [16–22]. Recently, label-free quantitative proteomic analysis was applied to the *M. loti* bacteroide proteome during the course of nodule maturation. This analysis revealed that *M. loti* enters a nitrogen-deficient condition during the early stages of nodule development, and then a nitrogen-rich condition during the intermediate stages of nodule development [22].

Earlier studies showed that, gel-free shotgun proteomic approaches can identify almost all types of proteins, including highly acidic, basic, hydrophobic and low abundant proteins, as well as protein complexes (Table 8.1). These techniques are considered to be among the most promising techniques available for large-scale quantitative proteome analysis in plant biology [23].

8.3 Subcellular and Tissue Specific Proteomics

During nodule formation, rhizobial infection of root hairs exhibits fascinating cell biology. Rhizobia enter the root hair via endocytosis and are enclosed in a new subcellular structure called the infection thread, which is later formed from the invagination of the root-hair cell plasma membrane [24]. In this complex organogenesis process a number genes and/or proteins from different subcellular organs including plasma membrane, microsome, mitochondria, and cytosol are directly involved. Subcellular proteomics can be defined as the analysis of the expressed proteins of purified individual cell compartments. This approach has emerged as an interesting tool to complement total cell lysate proteomic data [25]. The current status of plant subcellular proteomics has been extensively described by Agrawal et al. [26]. Subcellular and tissue specific proteomics is therefore considered to be an essential approach for the nodule biologist to investigate the subcellular and tissue specific protein networks associated with nodulation.

8.3.1 Membrane Proteomics

Rhizobia enter root cells by endocytosis and form a unique compartment in the plant cell known as symbiosome which is surrounded by a peribacteroid membrane (PBM) formed from the plant plasma membrane during endocytosis of the bacteria. The symbiosome plays a central role in the exchange of compounds between the organisms [27].

As the root plasma membrane is the first organelle that physically interacts with and responds to the rhizobia, it is considered one of the key cellular structures for proteomic analysis. For the first time, Panter et al. [9] used 2-DE coupled with N-terminal sequencing to identify a handful of PBM proteins such as HSP70 and HSP60. This study provided the first indication of the molecular machinery for co- or post-translational import of cytoplasmic proteins in symbiosomes of the soybean nodule. However, the identification of a very low number of PBM proteins was probably due to the low resolution of separation of the PBM proteins by 2-DE and the lack of complete genome sequence. This limitation was further resolved by using gel-free proteomics technology which led to identification of over 100 PBM proteins including a number of membrane proteins like transporters for sugars, sulfate, and receptor kinases of *L. japonicus* [27]. Recently, soybean symbiosome membrane (SM) and the PBM proteome was further analyzed by Clarke et al. [21]. In this study, the investigators used bicarbonate stripping and chloroform-methanol extraction of isolated SM to reduce the complexity of the samples and enrich for hydrophobic integral membrane proteins. Shotgun proteomic analysis led to the confident identification of over 200 proteins from SM and PBM samples which included a number of putative transporters for compounds such as sulfate, calcium, hydrogen ions, peptide/dicarboxylate, and nitrate, as well as unknown transporters and most of the previously identified SM and PBM proteins. Taken together these earlier studies provided very clear messages that gel-free proteomics approaches are the most efficient way of analyzing the nodule membrane proteome.

8.3.2 Cytosolic Proteomics

The plant cytosol is a complex intracellular fluid containing all the cellular components that allow interactions between partitioned metabolic processes [28]. It has been predicted that in eukaryotes over 50 % of total cellular protein reside in the cytosol [29]. It is now well known that the cytosol is involved in a number of prominent biochemical processes in the eukaryotic cell, however to the best of our knowledge there are only two studies that have been conducted to analyze the cytosolic proteome of nodules [19, 30]. Using conventional 2-DE based proteomic approach Oehrle et al. [30] identified a total of 69 cytosolic proteins from soybean nodule wherein the largest categories of proteins were involved in carbon metabolism (~ 28 %) followed by nitrogen metabolism (~ 12 %), oxygen protection (~ 12 %), and protein trafficking (11 %).

In the second study, Matamoros et al. [19] demonstrated the relative contribution of cell organelles of host cells in nodule aging by analyzing the pea mitochondrial and cytosolic proteome. Using a gel-free shotgun proteomic approach, a total of 81 cytosolic proteins were identified. Their results showed consistent reductions in the protein concentrations of carbon metabolism enzymes, inhibition of protein synthesis and increase in serine proteinase activity, disorganization of cytoskeleton, and a sharp reduction of cytosolic proteins during the senescence process.

The sparsity of nodule cytosolic proteomic data sets is a major limitation for the better understanding of the complex response and involvement of the cytosolic proteome in nodule growth and development. Therefore, future comparative analysis of nodule cytosolic proteomes of different legume species is essential. Isolation of pure nodule cytosolic fractions from a diverse set of legume species for LC–MS/MS analysis will be critically essential to achieve this aim.

8.3.3 Mitochondrial Proteomics

Among the various subcellular organelles, mitochondria display a very noticeable morphological, physiological and biochemical transformation during nodule formation. For example, the shape of mitochondria was found to be larger, elongated and cristae-rich by intensive folding of the inner membranes in the soybean root rhizobia infected zone [31]. It has also been reported that mitochondria isolated from soybean nodules exhibited higher oxidative and phosphorylative enzyme activities than those from the roots [32]. Additionally, it has been proposed that mitochondria might have functions for establishing symbiotic signal exchange because mitochondria are believed to be evolved from ancestral bacteria [33] and play a key role in apoptosis [34]. These earlier studies provide valuable clues that motivate more in-depth analysis of the mitochondrial proteome.

Thus far, only two studies have been conducted to characterize the nodule mitochondrial proteome. For the first time, soybean nodule and root mitochondrial proteome have been analyzed by 2-DE and a total of 34 proteins were successfully identified using MALDI-TOF, LC-MS/MS and N-terminal amino acid sequencing [35]. Together with previously established mitochondrial proteins, this study newly revealed a number of mitochondrial proteins such as 27, 22.5 kDa subunits of NADH ubiquinone reductase, ferrochelatase, hypothetical protein 11 (coxI 5′ region), NADPH quinone oxidoreductase and pseudo-atpA. These proteins were not previously cataloged in the plant mitochondrial proteome of Arabidopsis, pea and rice [35]. More recently, the nodule mitochondrial proteome of common bean has been analyzed in order to demonstrate the relative contribution of the mitochondria and cytosol of host cells to nodule aging [19]. Comparative proteome analysis of different developmental stages revealed that mitochondria are the early target of oxidative modifications and a likely source of redox signals during the nodule senescence process.

8.4 Single Cell Proteomics

Root hair, a single cell structural extension of root epidermal cells in the elongation zone are the primary sites for rhizobial infection which lead to the formation of a new organ, the nodule in legumes [36]. Due to several advantages, root hair is considered one of the most important cellular targets for analysis using system biology [37]. The Stacey group demonstrated for the first time the protein changes in soybean root hairs upon rhizobial inoculation [24]. Among the 37 proteins identified by a conventional 2-DE method coupled with MALDI-TOF, a number of root hair and rhizobial inoculation specific proteins associated with root hair deformation, infection and legume nodulation in response to bacterial infection were also identified. Using gel-based approach Brechenmacher et al. [38] developed a soybean root hair proteome map encompassing a catalogue of nearly 1500 proteins. Later, a very comprehensive proteome reference map of the soybean root hair cell was developed by the same research group wherein over 5700 proteins were identified using the Accurate Mass and Time (AMT) tag approach coupled with LC-MS/MS [39]. To date, very limited information is available for the root hair proteomic composition of other legumes. Therefore, this well-established soybean root hair proteome protocol offers an opportunity to demonstrate other legumes root hair cell proteome which will ultimately build an integrated predictive model for nodule biology.

8.5 Post Translational Modification

Protein post-translational modifications (PTMs) can affect protein function, interactions with other proteins, subcellular targeting, and stability. PTMs play essential roles in protein signaling, localization, function, degradation and other important biological processes. A number of recent studies have shown that a large number of nodule proteins become targets of posttranslational modification including, phosphorylation [40], sulfenylation [41], and ubiquitination [42]. Results of these earlier studies indicate that nodule formation is a tightly regulated process that integrates a variety of signaling events by various post translational modifications. Recent advanced technologies in mass spectrometry are capable of identification of a variety of peptide modifications such as phosphorylation, sulfenylation, nitrosylation and ubiquitination. Quantitative analysis of PTMs by mass spectrometry offers a promising way to understand the role of these modifications and the target proteins in signaling networks involved either in nodule organogenesis or in the infection process.

8.5.1 Phosphoproteomics

It has become evident that many kinases such as LysM-receptor kinase (LYK3), receptor-like kinase (DMI2), and calcium/calmodulin-dependent protein kinases (DMI3) are activated as an early event of nod factor perception in response to rhizobial infection of legume roots [43, 44]. In addition, a number of other protein kinases have been identified by genetic approaches to be associated with nodule organogenesis [45, 46] indicating that protein phosphorylation plays a central role in symbiotic signaling networks. Certainly, these studies emphasize the importance of quantitative proteomic analysis of proteins that are rapidly phosphorylated after rhizobial inoculation.

Recent advances in MS-based methods for phosphoproteomic analysis offer the prospect of a wide-scale view of cellular phosphorylation across the proteome. However, few large scale phosphoproteomic studies have been conducted so far for legume nodules and/or roots [18, 47, 48]. For the first time, the Medicago root phosphoproteome was analyzed by Grimsrud et al. [47]. In this study, proteins were isolated from a root whole cell lysate and membrane-enriched fractions and phosphopeptides were enriched by immobilized metal affinity chromatography (IMAC) and identified by electron transfer dissociation (ETD)-enabled LTQ Orbitrap mass spectrometer. With these technologies a total of 3457 unique phosphopeptides corresponding to 829 unique proteins containing a total of 3404 non-redundant phosphorylation sites were identified. Until now, the most comprehensive phosphoproteome analysis of Medicago roots were performed by Rose et al. [48] wherein early response of nod factor treatment in wildtype versus mutant nfp and the dmi3 Medicago root proteome were compared. For quantitative analysis, proteins were labeled with TMT and iTRAQ followed by strong cation exchange and nano-LC-MS/MS analysis. This study revealed 7739 proteins containing 13,506 non-redundant phosphosites in response to nod factor of the model legume *Medicago truncatula*. This large scale quantitative phosphoprotemic dataset revealed enrichment in proteins implicated in the NF signaling cascade with or without nod factor.

Nguyen et al. [18] performed a comparative phosphoproteomic analysis of the soybean root hair and stripped roots with or without inoculation of the soybean-specific rhizobium *B. japonicum*. Proteins were labeled with the isobaric tag eight-plex iTRAQ and phosphopeptides were enriched by Ni-NTA magnetic beads followed by nanoLC-MS/MS coupled with HCD and decision tree guided CID/ETD for identification. A total of 1625 unique phosphopeptides with 1659 non-redundant phosphorylation sites corresponding to 1126 non-redundant phosphoproteins were identified from both root hairs and stripped roots wherein 273 phosphopeptides were significantly differentially regulated in response to *B. japonicum* infection. This study indicated a complex kinase-phosphatase-substrate network during rhizobial infection [18].

8.5.2 Sulfenyl Proteomics

Sulfenylation is another type of protein modification wherein the protein cysteine residues are oxidized by hydrogen peroxide to form sulfenic acid (-SOH) [49]. Some recent studies showed that sulfenylated proteins are involved in metabolic processes in plants [50]. Reactive oxygen species such as hydrogen peroxide appear to be essential for optimal nodule development [51], underscoring the importance of identification of protein targets of legume nodules.

Recently, Oger et al. [41] identified the sulfenylated proteins of *Medicago truncatula* and the *Sinorhizobium meliloti* during different developmental stages. Two different methods: chemical Bio-DCP1 probe and YAP1-cCRD genetic probe were used to trap the S-OH oxidized proteins followed by affinity chromatography for purification. The trapped purified proteins were further separated by SDS-PAGE followed by trypsin digestion and identification by LC-MS/MS analysis. These two methods led to the identification of a total of 91 *M. truncatula* and 20 *S. meliloti* sulfenylated proteins. These results indicate that sulfenylation may regulate the activity of proteins playing important roles in nodule development and function [41]. In summary, this proteomic analysis opens a new avenue of study of how other post-translational modifications can regulate the nodulation process.

8.6 Targeted Proteomics

In the *M. truncatula* genome, more than 500 nodule-specific cysteine-rich (NCR) peptide corresponding genes have been expressed at the mRNA level but only eight of them have been validated at the protein level [52, 53]. The mature NCR peptides are mostly 30–50 residues long including a signal peptide. The secreted mature peptides feature conserved cysteine patterns [20]. Recently, Farkas et al. [53] showed that NCR247 penetrates the bacteria and forms complexes with many bacterial proteins leading to arrested bacterial cell division and initiation of cell elongation. Therefore, the comprehensive identification of all the NCR peptides is critically important. Thus far only a single proteomic analysis has been performed to catalog the NCR proteins of legume species [20]. This study successfully identified a total of 138 NCRs and provided the first evidence of translation of the NCR genes and high level accumulation of the NCR peptides in the bacteroides.

However, recent advances in mass spectrometry such as the availability of the newest generation of Orbitrap instruments such as Q Exactive hybrid quadrupole-Orbitrap and other triple quadrupole instruments such as TSQ Quantiva or Vantage offers an efficient and user-friendly alternative way for further targeted analysis of any specific peptide. Targeted proteomics using these new instruments is able to provide absolute quantification at a wide scale.

8.7 Future Perspectives

Certainly, comparative and quantitative proteomic technologies have greatly enhanced our understating at the proteome level of how legume roots respond to rhizobia during the nodulation process. In addition, wide-scale analysis of post-translationally modified proteins revealed phosphorylation of thousands of proteins during the nodulation process suggesting a complex network of kinase-substrate and phosphatase-substrate interactions in response to rhizobial infections. In the future, it will be crucial to identify the direct interactions between kinases and phosphatases and the individual phosphorylation sites identified in these phosphoproteomic studies. Recent advances in the mass spectrometry technologies could potentially identify the targets and the phosphorylation sites of many of kinases in vitro and/or in vivo [54–56]. Although phosphorylation is one of the most prevalent types of post-translational regulations thought to be happening during the nodulation process, identification of large number of sulfenylated proteins also suggested that nodulation process could also be regulated by other type protein modifications. Other post translational modifications including phosphorylation, sulfenylation, nitrosylation, N-glycosylation, ubiquitination, methionine oxidation, S-nitrosylation, and acetylation have been detected in plants. Wide-scale characterization of these other PTMs could be imperative for understanding their role during the nodulation process.

References

1. Radutoiu S, Madsen LH, Madsen EB, Felle HH, Umehara Y, Grønlund M et al (2003) Plant recognition of symbiotic bacteria requires two LysM receptor-like kinases. Nature 425: 585–592
2. Madsen EB, Madsen LH, Radutoiu S, Olbryt M, Rakwalska M, Szczyglowski K et al (2003) A receptor kinase gene of the LysM type is involved in legumeperception of rhizobial signals. Nature 425:637–640
3. Limpens E, Franken C, Smit P, Willemse J, Bisseling T, Geurts R (2003) LysM domain receptor kinases regulating rhizobial Nod factor-induced infection. Science 302:630–633
4. Madsen LH, Tirichine L, Jurkiewicz A, Sullivan JT, Heckmann AB, Bek AS et al (2010) The molecular network governing nodule organogenesis and infection in the model legume *Lotus japonicus*. Nat Commun 1:10
5. Domonkos A, Horvath B, Marsh JF, Halasz G, Ayaydin F, Oldroyd GE et al (2013) The identification of novel loci required for appropriate nodule development in *Medicago truncatula*. BMC Plant Biol 13:157
6. Imin N (2013) Proteomics and the analysis of nodulation. Methods Mol Biol 1069:259–269
7. Saravanan RS, Rose JK (2004) A critical evaluation of sample extraction techniques for enhanced proteomic analysis of recalcitrant plant tissues. Proteomics 4:2522–2532
8. Carpentier SC, Panis B, Vertommen A, Swennen R, Sergeant K, Renaut J et al (2008) Proteome analysis of non-model plants: a challenging but powerful approach. Mass Spectrom Rev 27:354–377
9. Panter S, Thomson R, De Bruxelles G, Laver D, Trevaskis B, Udvardi M (2000) Identification with proteomics of novel proteins associated with the peribacteroid membrane of soybean root nodules. Mol Plant Microbe Interact 13:325–333
10. Natera SH, Guerreiro N, Djordjevic MA (2000) Proteome analysis of differentially displayed proteins as a tool for the investigation of symbiosis. Mol Plant Microbe Interact 13:995–1009
11. Dam S, Dyrlund TF, Ussatjuk A, Jochimsen B, Nielsen K, Goffard N et al (2014) Proteome reference maps of the *Lotus japonicus* nodule and root. Proteomics 14:230–240
12. Tonge R, Shaw J, Middleton B, Rowlinson R, Rayner S, Young J et al (2001) Validation and development of fluorescence two-dimensional differential gel electrophoresis proteomics technology. Proteomics 1:377–396

13. Van Noorden GE, Kerim T, Goffard N, Wiblin R, Pellerone FI, Rolfe BG et al (2007) Overlap of proteome changes in *Medicago truncatula* in response to auxin and *Sinorhizobium meliloti*. Plant Physiol 144:1115–1131
14. Baerenfaller K, Grossmann J, Grobei MA, Hull R, Hirsch-Hoffmann M, Yalovsky S et al (2008) Genome-scale proteomics reveals *Arabidopsis thaliana* gene models and proteome dynamics. Science 320:938–941
15. Dunkley TP, Hester S, Shadforth IP, Runions J, Weimar T, Hanton SL et al (2006) Mapping the Arabidopsis organelle proteome. Proc Natl Acad Sci U S A 103:6518–6523
16. Larrainzar E, Wienkoop S, Scherling C, Kempa S, Ladrera R, Arrese-Igor C et al (2009) Carbon metabolism and bacteroid functioning are involved in the regulation of nitrogen fixation in *Medicago truncatula* under drought and recovery. Mol Plant Microbe Interact 22:1565–1576
17. Larrainzar E, Wienkoop S, Weckwerth W, Ladrera R, Arrese-Igor C, González EM (2007) *Medicago truncatula* root nodule proteome analysis reveals differential plant and bacteroid responses to drought stress. Plant Physiol 144:1495–1507
18. Nguyen THN, Brechenmacher L, Aldrich JT, Clauss TR, Gritsenko MA, Hixson KK et al (2012) Quantitative phosphoproteomic analysis of soybean root hairs inoculated with *Bradyrhizobium japonicum*. Mol Cell Proteomics 11:1140–1155
19. Matamoros MA, Fernández-García N, Wienkoop S, Loscos J, Saiz A, Becana M (2013) Mitochondria are an early target of oxidative modifications in senescing legume nodules. New Phytol 197:873–885
20. Durgo H, Klement E, Hunyadi-Gulyas E, Szucs A, Kereszt A, Medzihradszky KF et al (2015) Identification of nodule-specific cysteine-rich plant peptides in endosymbiotic bacteria. Proteomics 15:2291–2295
21. Clarke VC, Loughlin PC, Gavrin A, Chen C, Brear EM, Day DA et al (2015) Proteomic analysis of the soybean symbiosome identifies new symbiotic proteins. Mol Cell Proteomics 14:1301–1322
22. Nambu M, Tatsukami Y, Morisaka H, Kuroda K, Ueda M (2015) Quantitative time-course proteome analysis of *Mesorhizobium loti* during nodule maturation. J proteomics 125:112–120
23. Thelen JJ, Peck SC (2007) Quantitative proteomics in plants: choices in abundance. Plant Cell 19:3339–3346
24. Wan J, Torres M, Ganapathy A, Thelen J, Dague BB, Mooney B et al (2005) Proteomic analysis of soybean root hairs after infection by *Bradyrhizobium japonicum*. Mol Plant Microbe Interact 18:458–467
25. Jung E, Heller M, Sanchez JC, Hochstrasser DF (2000) Proteomics meets cell biology: the establishment of subcellular proteomes. Electrophoresis 21:3369–3377
26. Agrawal GK, Bourguignon J, Rolland N, Ephritikhine G, Ferro M, Jaquinod M et al (2011) Plant organelle proteomics: collaborating for optimal cell function. Mass Spectrom Rev 30:772–853
27. Wienkoop S, Saalbach G (2003) Proteome analysis. Novel proteins identified at the peribacteroid membrane from *Lotus japonicus* root nodules. Plant Physiol 131:1080–1090
28. Ito J, Batth TS, Petzold CJ, Redding-Johanson AM, Mukhopadhyay A, Verboom R et al (2011) Analysis of the Arabidopsis cytosolic proteome highlights subcellular partitioning of central plant metabolism. J Proteome Res 10:1571–1582
29. Lahav M, Schoenfeld N, Epstein O, Atsmon A (1982) A method for obtaining high recovery of purified subcellular fractions of rat liver homogenate. Anal Biochem 121:114–122
30. Oehrle NW, Sarma AD, Waters JK, Emerich DW (2008) Proteomic analysis of soybean nodule cytosol. Phytochemistry 69:2426–2438
31. Werner D, Mörschel E (1978) Differentiation of nodules of *Glycine max*. Planta 141:169–177
32. Suganuma N, Yamamoto Y (1987) Respiratory metabolism of mitochondria in soybean root nodules. Soil Sci Plant Nutr 33:93–101
33. Karlin S, Campbell AM (1994) Which bacterium is the ancestor of the animal mitochondrial genome? Proc Natl Acad Sci 91:12842–12846
34. Green DR, Reed JC (1998) Mitochondria and apoptosis. Science 281:1309

35. Hoa LT-P, Nomura M, Kajiwara H, Day DA, Tajima S (2004) Proteomic analysis on symbiotic differentiation of mitochondria in soybean nodules. Plant Cell Physiol 45:300–308
36. Oldroyd GE, Dixon R (2014) Biotechnological solutions to the nitrogen problem. Curr Opin Biotechnol 26:19–24
37. Hossain MS, Joshi T, Stacey G (2015) System approaches to study root hairs as a single cell plant model: current status and future perspectives. Front Plant Sci 6
38. Brechenmacher L, Lee J, Sachdev S, Song Z, Nguyen TH, Joshi T et al (2009) Establishment of a protein reference map for soybean root hair cells. Plant Physiol 149:670–682
39. Brechenmacher L, Nguyen TH, Hixson K, Libault M, Aldrich J, Pasa-Tolic L et al (2012) Identification of soybean proteins from a single cell type: the root hair. Proteomics 12: 3365–3373
40. Lima L, Seabra A, Melo P, Cullimore J, Carvalho H (2006) Post-translational regulation of cytosolic glutamine synthetase of *Medicago truncatula*. J Exp Bot 57:2751–2761
41. Oger E, Marino D, Guigonis J-M, Pauly N, Puppo A (2012) Sulfenylated proteins in the *Medicago truncatula–Sinorhizobium meliloti* symbiosis. J Proteomics 75:4102–4113
42. Małolepszy A, Urbański DF, James EK, Sandal N, Isono E, Stougaard J et al (2015) The deubiquitinating enzyme AMSH1 is required for rhizobial infection and nodule organogenesis in *Lotus japonicus*. Plant J 83:719–731
43. Smit P, Limpens E, Geurts R, Fedorova E, Dolgikh E, Gough C et al (2007) Medicago LYK3, an entry receptor in rhizobial nodulation factor signaling. Plant Physiol 145:183–191
44. Shimoda Y, Han L, Yamazaki T, Suzuki R, Hayashi M, Imaizumi-Anraku H (2012) Rhizobial and fungal symbioses show different requirements for calmodulin binding to calcium calmodulin–dependent protein kinase in *Lotus japonicus*. Plant Cell 24:304–321
45. Lefebvre B, Timmers T, Mbengue M, Moreau S, Hervé C, Tóth K et al (2010) A remorin protein interacts with symbiotic receptors and regulates bacterial infection. Proc Natl Acad Sci 107:2343–2348
46. Chen T, Zhu H, Ke D, Cai K, Wang C, Gou H et al (2012) A MAP kinase kinase interacts with SymRK and regulates nodule organogenesis in *Lotus japonicus*. Plant Cell 24:823–838
47. Grimsrud PA, Den Os D, Wenger CD, Swaney DL, Schwartz D, Sussman MR et al (2010) Large-scale phosphoprotein analysis in *Medicago truncatula* roots provides insight into in vivo kinase activity in legumes. Plant Physiol 152:19–28
48. Rose CM, Venkateshwaran M, Volkening JD, Grimsrud PA, Maeda J, Bailey DJ et al (2012) Rapid phosphoproteomic and transcriptomic changes in the rhizobia-legume symbiosis. Mol Cell Proteomics 11:724–744
49. Leonard SE, Carroll KS (2011) Chemical 'omics' approaches for understanding protein cysteine oxidation in biology. Curr Opin Chem Biol 15:88–102
50. Hancock JT, Henson D, Nyirenda M, Desikan R, Harrison J, Lewis M et al (2005) Proteomic identification of glyceraldehyde 3-phosphate dehydrogenase as an inhibitory target of hydrogen peroxide in Arabidopsis. Plant Physiol Biochem 43:828–835
51. Pauly N, Pucciariello C, Mandon K, Innocenti G, Jamet A, Baudouin E et al (2006) Reactive oxygen and nitrogen species and glutathione: key players in the legume–Rhizobium symbiosis. J Exp Bot 57:1769–1776
52. Van De Velde W, Zehirov G, Szatmari A, Debreczeny M, Ishihara H, Kevei Z et al (2010) Plant peptides govern terminal differentiation of bacteria in symbiosis. Science 327:1122–1126
53. Farkas A, Maróti G, Dürgő H, Györgypál Z, Lima RM, Medzihradszky KF et al (2014) *Medicago truncatula* symbiotic peptide NCR247 contributes to bacteroid differentiation through multiple mechanisms. Proc Natl Acad Sci 111:5183–5188
54. Feilner T, Hultschig C, Lee J, Meyer S, Immink RG, Koenig A et al (2005) High throughput identification of potential Arabidopsis mitogen-activated protein kinases substrates. Mol Cell Proteomics 4:1558–1568
55. Dephoure N, Howson RW, Blethrow JD, Shokat KM, O'shea EK (2005) Combining chemical genetics and proteomics to identify protein kinase substrates. Proc Natl Acad Sci U S A 102:17940–17945

56. Ahsan N, Huang Y, Tovar-Mendez A, Swatek KN, Zhang J, Miernyk JA et al (2013) A versatile mass spectrometry-based method to both identify kinase client-relationships and characterize signaling network topology. J Proteome Res 12:937–948
57. Saalbach G, Erik P, Wienkoop S (2002) Characterisation by proteomics of peribacteroid space and peribacteroid membrane preparations from pea (*Pisum sativum*) symbiosomes. Proteomics 2:325–337
58. Bestel-Corre G, Dumas-Gaudot E, Poinsot V, Dieu M, Dierick JF, Van TD et al (2002) Proteome analysis and identification of symbiosis-related proteins from *Medicago truncatula* Gaertn. by two-dimensional electrophoresis and mass spectrometry. Electrophoresis 23:122–137
59. Djordjevic MA, Chen HC, Natera S, Van Noorden G, Menzel C, Taylor S et al (2003) A global analysis of protein expression profiles in *Sinorhizobium meliloti*: discovery of new genes for nodule occupancy and stress adaptation. Mol Plant Microbe Interact 16:508–524
60. Catalano CM, Lane WS, Sherrier DJ (2004) Biochemical characterization of symbiosome membrane proteins from *Medicago truncatula* root nodules. Electrophoresis 25:519–531
61. Djordjevic MA (2004) *Sinorhizobium meliloti* metabolism in the root nodule: a proteomic perspective. Proteomics 4:1859–1872
62. Valot B, Gianinazzi S, Eliane DG (2004) Sub-cellular proteomic analysis of a *Medicago truncatula* root microsomal fraction. Phytochemistry 65:1721–1732
63. Sarma AD, Emerich DW (2005) Global protein expression pattern of *Bradyrhizobium japonicum* bacteroids: a prelude to functional proteomics. Proteomics 5:4170–4184
64. Sarma AD, Emerich DW (2006) A comparative proteomic evaluation of culture grown vs nodule isolated *Bradyrhizobium japonicum*. Proteomics 6:3008–3028
65. Prayitno J, Imin N, Rolfe BG, Mathesius U (2006) Identification of ethylene-mediated protein changes during nodulation in *Medicago truncatula* using proteome analysis. J Proteome Res 5:3084–3095
66. Wienkoop S, Larrainzar E, Glinski M, Gonzalez EM, Arrese-Igor C, Weckwerth W (2008) Absolute quantification of *Medicago truncatula* sucrose synthase isoforms and N-metabolism enzymes in symbiotic root nodules and the detection of novel nodule phosphoproteins by mass spectrometry. J Exp Bot 59:3307–3315
67. Lim CW, Park JY, Lee SH, Hwang CH (2010) Comparative proteomic analysis of soybean nodulation using a supernodulation mutant, SS2-2. Biosci Biotechnol Biochem 74:2396–2404
68. Salavati A, Bushehri AA, Taleei A, Hiraga S, Komatsu S (2012) A comparative proteomic analysis of the early response to compatible symbiotic bacteria in the roots of a supernodulating soybean variety. J Proteomics 75:819–832
69. Irar S, Gonzalez EM, Arrese-Igor C, Marino D (2014) A proteomic approach reveals new actors of nodule response to drought in split-root grown pea plants. Physiol Plant 152:634–645
70. Larrainzar E, Molenaar JA, Wienkoop S, Gil-Quintana E, Alibert B, Limami AM et al (2014) Drought stress provokes the down-regulation of methionine and ethylene biosynthesis pathways in *Medicago truncatula* roots and nodules. Plant Cell Environ 37:2051–2063
71. Delmotte N, Mondy S, Alunni B, Fardoux J, Chaintreuil C, Vorholt JA et al (2014) A proteomic approach of bradyrhizobium/aeschynomene root and stem symbioses reveals the importance of the fixA locus for symbiosis. Int J Mol Sci 15:3660–3670
72. Gao TG, Xu YY, Jiang F, Li BZ, Yang JS, Wang ET et al (2015) Nodulation characterization and proteomic profiling of *Bradyrhizobium liaoningense* CCBAU05525 in response to water-soluble humic materials. Sci Rep 5:10836

Chapter 9
Proteomic Applications for Farm Animal Management

Ehsan Oskoueian, William Mullen and Amaya Albalat

Abstract The implementation of proteomics is an important step towards a better understanding of the complex biological systems that define animal health and production. The role that proteomics can play in the context of farm animal production is increasingly recognized and to date proteomics has been applied to characterize the physiology behind animal growth and development, reproduction, welfare and animal products. Furthermore, recent advances in mass spectrometry technologies have led to the development of novel strategies aimed at the identification of biomarkers present in different tissues and body fluids. Identification of valid biomarkers in animal tissue or body fluids such as serum, urine, milk, saliva, cerebrospinal fluid and semen to enable bio-monitoring on animal health and provide valuable information, on production, feeding status, and animal-environment interaction is a priority in this field. Therefore, analysis of the proteome linked with biomarker discovery is emerging into a field of high interest, with the aim of improving farm animal productivity and welfare. The present book chapter addresses the recent specific advances of interest in farm animal proteomics and introduces biomarker approaches that are relevant in animal health, production and quality.

Keywords Proteomics · Biomarker research · Farm animals · Bio-monitoring · Ruminant · Poultry · Fish · Growth

---

E. Oskoueian (✉)
Agricultural Biotechnology Research Institute of Iran (ABRII),
East and North-East Branch, Agricultural Research, Education,
and Extension Organization, P.O.B. 91735/844, Mashhad, Iran
e-mail: e.oskoueian@abrii.ac.ir

W. Mullen
Institute of Cardiovascular and Medical Sciences,
University of Glasgow, Glasgow, UK
e-mail: William.Mullen@glasgow.ac.uk

A. Albalat
School of Natural Sciences, University of Stirling, Stirling, UK
e-mail: amaya.albalat@stir.ac.uk

9.1 Introduction

Animal production has transformed to a largely industrialized sector over the past half-century. Increased effectiveness in animal production systems is a pressing issue given current estimates in global population growth of around 9 billion people by 2050 Godfray et al. [1]. In this scenario food producers in general and animal producers in particular are expected to experience increased competition of available resources. Farm animals are mainly raised to obtain animal products that are destined for use (i.e. wool and leather) or direct consumption (i.e. milk, meat and eggs). The main animal species being commercially farmed at large scale include terrestrial (cattle, swine, poultry and sheep) and aquatic species (several species of fish and prawns). The end products of these intensive animal farming systems have traditionally been mainly meat and milk from terrestrial farming and fish and shellfish products from the aquaculture sector, which are gaining importance in terms of volume and nutritional properties [2]. The end products of both terrestrial and aquatic farmed animal species will be highly depended on the farming conditions, level of domestication (higher in terrestrial than aquatic species) and selective breeding strategies. The role that proteomics or the study of all proteins in a given tissue or biofluid can play in animal production and food sciences is increasingly being recognized [3]. The study of the proteome adds a layer of functional information to the available genomic resources. For this reason, in the last decade, different proteomics approaches have been applied to increase our understanding at production stage and post-production stages (Fig. 9.1). Proteomic applications include studies to unravel the physiology behind animal growth and development, reproduction, animal welfare and health/disease conditions at production stage and there are also studies to characterize the properties of the products produced and study their alterations during different storage and/or post-harvest treatments at post-production stage [4–8]. In many cases, fundamental proteomic research has also been focused on the discovery of protein biomarkers for different diagnostic purposes and establishing quality markers [3, 9–12] (Fig. 9.1).

A biomarker in a clinical context can be defined as a molecule that is objectively measured and evaluated as an indicator of a normal biological processes, pathogenic processes, or pharmacologic responses to therapeutic intervention [13]. This is one of the main applications of proteomics in the human biomedical field and it is relevant also in a veterinary medicine context [14, 15]. Moreover, in a food science context proteomics has also been used to identify individual proteins or protein patterns that can be used as quality markers of the final products. This approach, which is particularly important in food production, requires taking into account not only the animal biology production conditions per se but also the slaughtering, processing and storage conditions [7]. This interaction between animal physiology, biochemistry and technology will for instance determine changes in the proteome that will take place in the conversion from muscle to meat and finally dictate key quality properties such as water holding capacity, tenderness, and color [7].

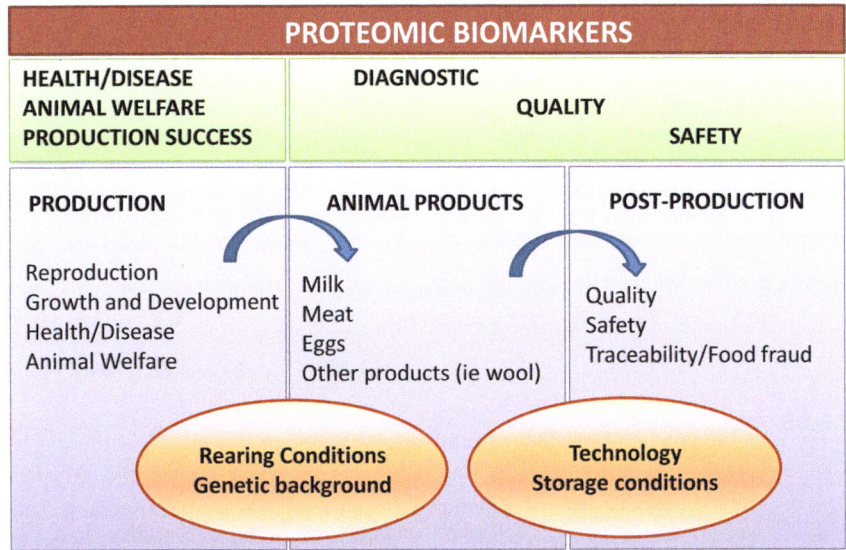

Fig. 9.1 Schematic illustration of proteomics approaches to improve farm animal production and management

In this chapter we will introduce the main analytical workflows that have been used in farm animal/food science and also present proteomic studies that aimed at a better understanding of the biology of the animals being farmed or the discovery of biomarkers for farm animal management in general. This will include studies in the areas of animal nutrition and growth, gastrointestinal health, some key animal products (meat, milk and eggs) and animal welfare.

9.2 Analytical Workflows Used in Farm Animal Proteomics

The regulation of gene product expression can take place at gene sequence, transcription, translation and post-translation levels. Arguably the expression levels of all proteins would provide the most relevant single data set to characterize a biological system as proteins are responsible for the actual cellular work [16]. Thus from mid 1970s the analysis of the proteome has been attempted using different analytical workflows. This is also true when examining studies dealing with proteomics in farm animal, aquaculture and food sciences. In contrast to transcriptomics where very low expressed genes can be detected the complexity and high dynamic range observed in the proteome means that proteins and/or peptides need to be separated using different strategies, gel-based or gel-free methods respectively before they are quantified and identified by mass spectrometry (MS) (Fig. 9.2).

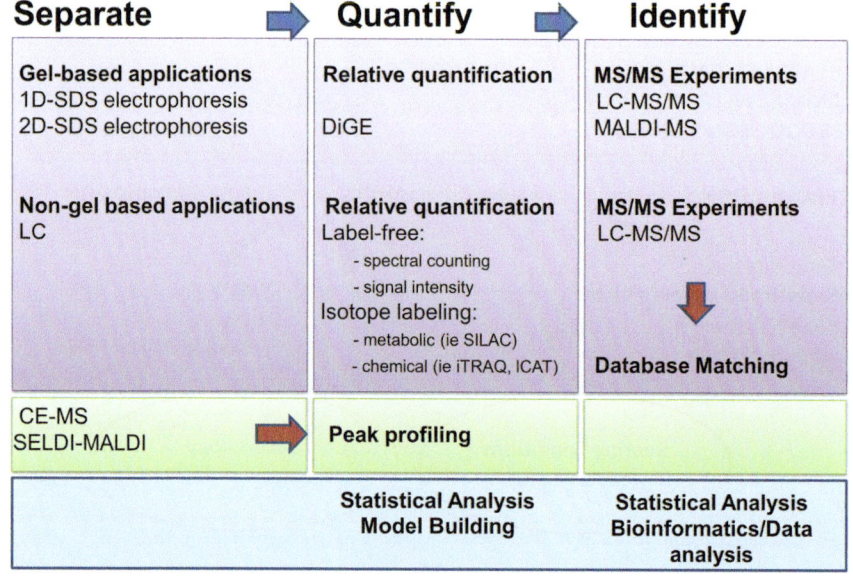

Fig. 9.2 The workflow showing the proteins and peptides separation, quantification and identification techniques using non-targeted approaches

Generally, gel-based proteomics approach consists of several stages including sample preparation to gel imaging, image to spot digestion, digest to MS analysis, and search of results into a database, data analysis and archiving. This workflow has been successfully applied in many studies, especially since the development of two-dimensional protein gel electrophoresis (2DE) [3]. In 2DE, proteins are separated according to their isoelectric point (first dimension) and molecular weight (MW) (second dimension, sodium dodecyl sulphate electrophoresis). Detection of proteins is achieved by using visible stains such as colloidal Coomassie or silver stain. Due to gel reproducibility issues if relative quantification is required differential gel electrophoresis (2D-DiGE) is normally the method of choice (Fig. 9.2). In 2D-DiGE the use of amino-selective fluorescence dyes with different excitation wavelengths is used before the 2DE step so that two different samples plus and internal standard can be run at the same time, allowing better comparisons and gel-to-gel variation. Spots can be then digested typically using trypsin into peptides and analysed using MALDI-MS or LC-MS/MS to gather information on their amino acid sequence. This approach has, so far, been the most widely used for farmed animal research. While 2DE allows for the detection of post-translational modifications (PTMs), protein isoforms and visualization of protein complexes this platform is less effective in detecting low abundance, very large or very hydrophobic proteins [17]. However, possibly one of the main limitations of this approach is that 2DE cannot be coupled on-line with an MS-detector and therefore it is relatively labour intensive, especially if compared with on-line fractionation systems such as liquid chromatography (LC-MS) and capillary electrophoresis (CE-MS) [18]. These systems have been reported to be rather complementary, although for identification purposes LC-MS/MS has been the method of choice [19, 20] while CE-MS and direct peptide fingerprints using matrix assisted laser desorption (MALDI-MS) have used for peptide profiling [21–23].

In general terms, LC is more sensitive than gel electrophoresis as there is no need to recover proteins or peptides from a gel matrix. However, sample complexity increases with the higher the number of analytes making consistent, reproducible data acquisition a challenge and also the pipeline for data analysis more complex, especially is a label-free method is used. In shotgun LC-MS/MS proteins that have been previously digested into peptides can be quantified using label-free or label-based methods (Fig. 9.2).

In reality, in a proteomics context absolute protein quantification is difficult and often not needed per se as in many cases the identification of differential proteins can be achieved using relative quantification. Relative quantification of intact proteins can be achieved in 2DE using labeling strategies such as the popular difference gel electrophoresis (DiGE), a strategy commonly used in farm animal research [24, 25]. Relative quantification at peptide level can be achieved using chemically labeled methods such as iTRAQ or by metabolically labeling proteins using heavy/light amino acids (stable isotope label with amino acids in culture; SILAC). Examples of studies using iTRAQ in farm animal context are varied and increasing indicating the relevance of this approach [26–28]. However, there is evidence that iTRAQ has a lower coverage of the proteome, as well as other issues, than label free approaches [29, 30].

9.2.1 Proteomics in Nutrition and Growth

The growth and development of farm animals requires a balanced diet with the right mixture of nutrients. The proteome analysis of tissues or body fluids can elucidate growth mechanisms and reveals how the nutrients are contributing to this process. Among nutritional biomarkers, proteins related to amino acids metabolism such as hepatic alpha enolase, elongation factor 2, calreticulin, cytochrome b5, apolipoprotein A-I and catalase proteins have been reported to change in lactating dairy goats fed with a high-grain diet [31]. In another study by Romao et al. [32] who studied the proteomic changes in subcutaneous adipose tissue from cattle fed a high-grain diet, in an effort to understand the molecular mechanisms involved in fat development. The low abundance of lipogenic proteins in the subcutaneous fat implied that the nutrients consumed by cattle were channeled into muscle development instead of fat deposition and as consequence studies in this area could help develop new strategies to manipulate adiposity in beef cattle improving meat quality and animal productivity [32].

One of the important economic traits in the poultry industry is fast growth and development. The proteome analysis of breast meat by Phongpa-Ngan et al. [33]

and Doherty et al. [34] indicated that a high abundance of proteins, including pyruvate kinase, creatine kinase, triosephosphate isomerase, ubiquitin, heat shock protein, myosin heavy chain and actin are associated with a fast growth rate of chickens. In addition, the study conducted by Teltathum and Mekchay [35] unveiled the important role of other proteins classified as metabolic and stress-related such like phosphoglycerate mutase 1, triosephosphate isomerase 1, apolipoprotein a1 and fatty acid binding protein 3 which changed significantly with chicken growth and development.

The study of the proteome in saliva has increased in recent years. Saliva is a watery fluid secreted by the salivary glands containing electrolyte, hormones, and serum-derived proteins. It is easily obtainable from large numbers of animals in a noninvasive manner. One of the interesting properties of saliva is the presence of proteins that encompass a wide range of activities including control of feeding behavior [36, 37], feed conversion efficiency [38] and detoxification of anti-nutritional factors [39]. For instance, the function of proteins such as carbonic anhydrase II and VI are to maintain the constant bicarbonate concentration in the digestive tract and adaptations to new diets are important [40]. The decline in the abundance of these proteins could be an indicative marker of acidosis in ruminants [41].

In the aquaculture sector, optimization of fish diets to meet animal nutritional requirements, while reducing the inclusion of fishmeal and fish oil has been a priority for many years. Changes in dietary protein and oil sources mainly due to the substitution of fish meal and fish oil by vegetable protein and oil sources has been studies in different species. In all cases, proteomics data indicated how dietary manipulation affects a number of biological pathways. For instance, using 2DE proteome analysis identified increased protein catabolism and protein turnover in rainbow trout fed with a diet with a partial substitution of fish meal for soybean meal for 12 weeks [42]. When rainbow trout were fed a diet containing no fishmeal a significant reduction in growth rate was observed concomitant with an increase in proteins involved in primary energy metabolism and two proteasome subunits, indicative of protein degradation suggesting overall a higher energy demand in fish fed diets with high content of plant protein [43]. On the other hand, the relevance of phospholipid content in diets has been accessed by analysing the proteome in the liver of pikeperch larvae. In this case, high phospholipid levels in the diet were associated with a down-regulation of the glycolytic pathway and an increase in sarcosine dehydrogenase (involved in methionine metabolism) while proteins involved in cellular stress were up-regulated in larvae fed lowest phospholipid diet (increase in glutathione S-transferase M and glucose regulated protein 75). However, at this point it is important to emphasise that the impact that changes in diet will have on fish metabolism will depend not only on the species and stage of development but also on the genetic background of the species [6]. Another area of interest has been the effect of probiotics included in fish diets [44, 45]. Effects reported in early Atlantic cod larvae indicated up-regulation of some growth-related proteins upon probacteria administration while up-regulation of proteins with immunity function was not always observed [45].

9.2.2 Proteomics in Gastrointestinal Health

The gastrointestinal tract (GIT) is responsible for feed digestion and nutrient absorption. The site of digestion is different in the monogasteric and polygasteric terrestrial farm animals. Monogasteric animals such as poultry, pigs and horses possess single-chambered stomach while ruminants like cow, sheep or goats have four-chambered complex stomach. The complex stomach in ruminants adapted to the wide range of feed resources and the presence of rumen microbes enables the utilization of cellulosic materials [46].

According to Yang et al. [47] metabolic processes taking place in the GIT could be monitored through proteome analysis. For instance, the proteome analysis of rumen epithelial cells in goats fed high energy diet using 2DE indicated a significant decrease in chaperone proteins with cellular protective function such as heat shock cognate 71 kDa protein, peroxiredoxin-6, serpin HQ, protein disulfide-isomerase and selenium-binding protein [48]. Downregulation of these proteins together with acidosis could result in the impairment of rumen barrier function. Similarly, a proteomics study using DiGE on intestine tissue of the chickens supplemented with *Enterococcus faecium* as a probiotic, indicated that although there was no effect on daily weight gain an enhanced expression of proteins related to intestinal structure, which may increase the absorptive surface area was observed. Furthermore, other proteins related to substance metabolism, immune and antioxidant systems were altered increasing our understanding on the probiotic mechanisms involved on broiler intestine biology [49].

Indeed, the rumen acts as fermentation chamber hosting various microbes including bacteria, protozoa, fungi and archaea. The microbial enzymes digest and convert the fibrous material to the valuable metabolites to be absorbed by the host. The animal gut microbiota plays an important role in feed digestion, protection from pathogens and provision of key metabolites. Recently, metaproteomics has been introduced as a technique to study the gut microbiota in terrestrial farm animals. The identification of bacteria proteins recovered directly from fecal samples provides valuable information on microbial population present in the gut. The metaproteomics is still developing in the animal science and to date only a limited number of studies are available on ruminants and poultry [49–51].

9.3 Proteomics in Some Key Animal Products

9.3.1 Meat Proteomics

The conversion of muscle to meat is a complex process and proteomics has been applied in many studies to unravel the biochemical reactions that are involved in this process as well as the molecular processes that determine meat quality [7, 52, 53]. From a biochemical perspective, briefly, upon animal slaughter, with no circulating blood supplying oxygen, muscle cells activate anaerobic metabolism eventually depleting glycogen reserves to generate adenosine triphosphate (ATP). As there is no removal of metabolic end products lactic acid is accumulated producing muscle acidification. This process causes not only loss in water holding capacity (WHC) but also calcium release and eventually formation of cross-bridges between myosin and actin filaments that cannot be resolved beginning the onset of *rigor mortis*. A second step involves the action of different enzymes mainly involved in proteolysis and oxidation. Factors such as genotype, sex, age, nutritional status and management strategies such as stress during transportation and slaughter protocols have all been shown to affect these initial post-mortem metabolic processes ultimately affecting meat quality. Proteomic studies have showed the importance of these metabolic processes such as early post-mortem glycolysis [54]. After 1 h of the slaughtering process key proteins from the glycolysis and the tricarboxylic acid cycle increased significantly together with chaperone proteins often associated with cellular responses to environmental stress such as crystalline and heat shock proteins HSP27 and HSP60. The role of small, sHSPs in meat quality was initially presented by several authors around a decade ago [55, 56]. With muscle post-mortem acidification and nutrient depletion apoptosis is induced and in response sHSPs are recruited to impede the onset of apoptosis and to chaperone the unfolding of muscle proteins. Thus heat shock proteins such as HSP27 have been proposed as biomarkers of tenderness [57]. Other makers for meat tenderness have been recently reviewed by Ouali et al. [55] and Gobert et al. [9] and include proteins from glycolytic and oxidative energy production, cell detoxification, protease inhibitors and HSPs.

Peroxiredoxin 1 and 6, protein DJ (PARK7) [58], troponin T, alpha actin, or heavy (MYH1) and light (MYL1, MYL2 and MYL6B) myosin II chains have been shown to change significantly [59] and are identified as biomarkers for the meat tenderness. In addition, another key trait in defining meat quality is the color of the meat as it is perceived by consumers as a spoilage indicator. The desirable color of the meat varies according to different products but in general terms for the beef, poultry and pork is cherry red, pale pink and pink, respectively. In accordance with the results of Joseph et al. [60] color stability of *Longissimus* could be attributed to higher abundance of chaperones and antioxidant proteins including peroxiredoxin-2, HSP27 and peptide methionine sulfoxide reductase. In general terms, it appears that darker meat is more oxidative oriented as shown by more abundant mitochondrial enzymes of the respiratory chain, hemoglobin, and chaperone or regulatory proteins while in muscle leading to lighter meat is more abundant on proteins involved in glycolysis and glutathione S-transferase [61].

Apart from tenderness and color, the flavor of the meat in pork and beef has also shown to affect meat palatability. Meat flavor is determined among other factors by the deposition of fat within the skeletal muscle known as marbling fat. Using 2DE Kim et al. [62] reported that the triosephosphate isomerase and succinate dehydrogenase proteins correlated with intra muscular fat deposition (marbling fat) in Korean cattle steers. Other studies also using 2DE have studied the contribution of different proteins and peptides in meat flavor. Although further studies are needed in order to identify biomarkers of meat flavor [63].

Proteomics-based techniques have been very useful to characterize different breeds [64]. Research in this area has also been applied to determine food adulteration. For instance, the study by Sentandreu et al. [65] confirmed that the proteomics is capable of detecting contaminating chicken in pork at the concentration as low as 0.5 %. In addition, detecting animals treated with the illegal growth promoting agents such as prednisolone, dexamethasone and oestradiol has also been possible with the aid of proteomics [66].

9.3.2 Milk Proteomics

Milk is a biological fluid with unique quality and complexity. It contains macro- and micronutrients essential for growth and development [67]. To date, proteomics approaches have been applied to characterise the milk proteome [68], study mechanisms underlying milk production [69, 70], evaluation of milk quality [71] and to identify the type of bacterial infection [23].

As reviewed by Bendixen et al. [72] while the major components in milk were biochemically characterized more than two decades ago, the analysis of less abundant but interesting proteins using proteomic techniques is more recent. Milk composition has been shown to vary very significantly according to genetic, epigenetic and environmental factors [73, 74]. Negative energy balance (NEB) in cows produced an increase in the abundance of stomatin and galactose-1-phosphate proteins in the milk which could be used as indicators of NEB [75].

As already mentioned milk has also been used in several studies as a diagnostic fluid. Traditionally, the most common strategy for the detection of bacteria pathologies in bovine milk is by the use of 16S rRNA sequencing [76]. However, using MALDI-MS it is possible to identify food-borne pathogens such as *Listeria monocytogenes* and *Staphylococcus aureus* in the milk [77–80]. Besides, the identification of microorganisms using mass spectrometry, milk has been analysed using different proteomic platforms with the aim to find early biomarkers of mastitis. Bovine mastitis is a disease that is responsible for major economic losses to the dairy industry [81, 82]. Significant milk changes described include the presence of clots in milk, milk discoloration and high levels of leukocyte numbers, which leads to a rise in somatic cell count. Mastitis is usually caused by bacterial infection with major species being *Streptococcus agalactiae*, *Staphylococcus aureus* and *Mycoplasma bovis* as well as environmental pathogens (i.e. *Streptococcus uberis*), environmental coliforms gram negative (i.e. *Escherichia coli*) and other Gram negative bacteria (i.e. *Serratia*, *Pseudomonas* and *Proteus*). Clinical mastitis is detected by electrical conductivity, somatic cell count and lactate dehydrogenase activity [83–85]. At proteome level, Ibeagha-Awemu et al. [86] found 73 proteins significantly different using GeLC-MS/MS between normal whey and whey from quarters infected with either *E. coli* and *S. aureus*. Several authors have looked in detail the proteolytic events that take place under a mastitis infection as indeed a potential cause of milk deterioration is proteolysis of milk proteins by bacterial or endogenous proteases. Larsen et al. [87] identified approximately 20 different peptides in milk samples from mammary glands infused lipoteichoic acid (toxin of *S. aureus*). Peptides detected mainly through the action of endogenous proteases originated mainly from α_{S1}- and β-caseins. Using CE-MS Mansor et al. [23] developed a biomarker panel based on the naturally occurring peptides present in milk from healthy versus milk from mastitis cows infected either with *S. aureus* or *E. coli*. Differences in peptide profiling were also detected between the bacteria species, which could help inform on antibiotic treatment.

9.3.3 Egg Proteomics

Eggs are considered as an excellent source of inexpensive high quality protein for human consumption and food production, proteins of pharmaceutical interest and proteins that have been widely used in the biomedical research field and protein chemistry [88–90]. Overall, the egg industry is interested in optimization of egg production, strengthen the physical egg properties against bacterial entry, increase egg shelf life and identification of protein biomarkers for egg quality.

There are two distinctly different part of an egg, the egg yolk is the part which has the function to act as the food source for the developing embryo inside and it is suspended in the egg white. The egg yolk represents a major reservoir of vitamins and minerals and it contains the entire egg lipid content and about 50 % of the protein. Mann and Mann [91] identified 119 proteins in chicken egg yolk plasma using GeLC-MS/MS. The most abundant proteins were serum albumin, vitellogenin, apovitellenins, immunoglobulin Y, ovalbumin and a 12 kDa serum protein with cross-reactivity to β2-microglobulin. This initial list was expanded to 255 unique proteins by Farinazzo et al. [92] using combinatorial peptide ligand libraries.

The egg white is the other part of the egg that has been described using proteomics. Guérin-Dubiard et al. [93] identified sixteen proteins in the egg white using 2-DE. Application of GeLC-MS/MS by Mann and Mann [94] increased the identified proteins to 158. Proteomics approaches have also been applied to study egg quality. Egg white thinning or the loss of egg white viscosity was studied with storage time using 2DE [95]. Proteins that significantly changed over time at ambient temperature (22 °C) were identified as ovalbumin, clusterin, ovoinhibitor, ovotransferrin and prostaglandin D2 synthase. Furthermore, when looking at the effect of storage temperature an accelerated degradation of ovalbumin was observed at higher temperature. In this study, the decrease in clusterin was suggested as an effective biomarker for egg quality evaluation [96].

One of the major concerns in the egg industry is bacterial contamination. The eggshell, associated cuticle and shell membranes are the egg's first line of defense against contamination [97]. Although the eggshell is mainly composed of calcium carbonate several proteins involved in defense to resist contamination has been identified. Egg cuticle proteome analysis using GeLC-MS/MS by Rose-Martel et al. [98] showed the presence of a number of proteins known to have antimicrobial activity (lysozyme C, ovotransferrin, ovocalyxin-32, cystatin and ovoinhibitor).

9.4 Animal Welfare and Proteomics

Generally farm animals are reared under an intensive system and the presence of stressors or lack of healthy condition may increase the animal susceptibility to the diseases. The role of that stress and welfare play in animal production is important not only from a quality and safety perspective but also because of public perception and product acceptance [99]. In a farming context, stressors may arise due to obvious suboptimal rearing and environmental conditions such as excessive crowding/high stocking densities but also due to less obvious suboptimal conditions such not allowing animals to express their 'natural' behavior or personality [100, 101]. At production stage, the consequences or impact of stress on farm animal production has been well studied and has been linked not only to impaired growth but also immune dysfunction [102]. Furthermore, at post-production stage stress and welfare conditions can affect the quality of the final products [103, 104]. In this area, proteomic studies have mainly been applied to evaluate stress conditions and their impact on animal health and welfare. The identification of objective laboratory-based biomarkers of stress and welfare is also a current priority as many current stress markers such as cortisol have limitations [105]. Stress-related studies so far have mostly focused on the effects of density [106], handling [107], environmental conditions [108] and pre-slaughter stresses [5].

In pigs, 2D-DiGE was applied to examine the proteome changes of skeletal muscle in response to acute heat stress [108]. In predominantly white fiber portions, heat stress decreased the abundance of tubulins and soluble actin and increased phosphorylated cofilin 2 indicating a loss of microtubule structure. Overall, proteomic data indicated significant changes in carbohydrate metabolism, structure and an antioxidant response in skeletal muscle. The effect of high stocking density has also been studied in pigs, as this is a determinant parameter in pig welfare. Serum proteins were separated and quantified using 2D-DiGE and a significant increase in β-actin was found and validated by western blot. This increase in serum β-actin could be related to tissue damage associated with high stocking density [106]. In cows, the physiological adaptations to different management systems were studied in order to reveal new stress/welfare biomarkers. Redox system was also shown to be significantly affected (glutathione peroxidase and paraoxonase) in cows living in challenging environmental conditions [109]. Proteomics has also been applied to assess fish welfare and their response to stress such as cold stress [110] and repeated handling/crowding [111]. Alves et al. [111] compared the liver proteome of gilthead sea bream (*Sparus aurata*) grown under low-stressful and stressful conditions (repeated handling and crowding). Proteins affected included heat shock proteins, fatty acid binding protein, mitochondrial porine, calmodulin, cofilin, glutamine synthetase, hemoglobin, beta-tubulin and proteins involved in carbohydrate metabolism were differentially expressed in the fish grown under stressful condition.

9.5 Conclusions

The literature survey indicated that applications of proteomics approaches in the farm animals have been expand significantly during the last decade and its implementation is an important step toward understanding the complex biological systems that control animal growth and development, reproduction, welfare and animal products (milk, meat, egg, wool). Apart from that, the recent advances in the area of biomarker discovery resulted in identification of specific, sensitive, quick and cost effective biomarkers in body fluids such as serum, urine, milk, saliva and semen which enabled the bio-monitoring and provide valuable information on animal health, production, feeding status, and animal-environment interaction. Therefore, the proteomics approaches are going to emerge into a field of high interest with purpose of improving farm animal productivity and welfare.

References

1. Godfray HCJ, Beddington JR, Crute IR, Haddad L, Lawrence D, Muir JF et al (2010) Food security: the challenge of feeding 9 billion people. Science 327:812–818
2. Jones AC, Mead A, Kaiser MJ, Austen MC, Adrian AW, Auchterlonie NA et al (2015) Prioritization of knowledge needs for sustainable aquaculture: a national and global perspective. Fish Fish 16:668–683
3. Almeida A, Bassols A, Bendixen E, Bhide M, Ceciliani F, Cristobal S et al (2015) Animal board invited review: advances in proteomics for animal and food sciences. Animal 9:1–17
4. Morzel M, Chambon C, Hamelin M, Santé-Lhoutellier V, Sayd T, Monin G (2004) Proteome changes during pork meat ageing following use of two different pre-slaughter handling procedures. Meat Sci 67:689–696
5. Morzel M, Chambon C, Lefèvre F, Paboeuf G, Laville E (2006) Modifications of trout (*Oncorhynchus mykiss*) muscle proteins by preslaughter activity. J Agric Food Chem 54:2997–3001
6. Morais S, Silva T, Cordeiro O, Rodrigues P, Guy DR, Bron JE et al (2012) Effects of genotype and dietary fish oil replacement with vegetable oil on the intestinal transcriptome and proteome of Atlantic salmon (*Salmo salar*). BMC Genom 13:448
7. Paredi G, Raboni S, Bendixen E, De Almeida AM, Mozzarelli A (2012) "Muscle to meat" molecular events and technological transformations: the proteomics insight. J Proteomics 75:4275–4289
8. Montowska M, Pospiech E (2013) Species-specific expression of various proteins in meat tissue: proteomic analysis of raw and cooked meat and meat products made from beef, pork and selected poultry species. Food Chem 136:1461–1469
9. Gobert M, Sayd T, Gatellier P, Santé-Lhoutellier V (2014) Application to proteomics to understand and modify meat quality. Meat Sci 98:539–543
10. Bassols A, Turk R, Roncada P (2014) A proteomics perspective: from animal welfare to food safety. Curr Protein Pept Sci 15:156–168

11. Piovesana S, Capriotti AL, Caruso G, Cavaliere C, La Barbera G, Chiozzi RZ et al (2016) Labeling and label free shotgun proteomics approaches to characterize muscle tissue from farmed and wild gilthead sea bream (*Sparus aurata*). J Chromatogr A 1428:193–201
12. Yang Y, Zheng N, Zhao X, Zhang Y, Han R, Yang J et al (2016) Metabolomic biomarkers identify differences in milk produced by Holstein cows and other minor dairy animals. J Proteomics 136:174–182
13. Colburn W, Degruttola VG, Demets DL, Downing GJ, Hoth DF, Oates JA et al (2001) Biomarkers and surrogate endpoints: preferred definitions and conceptual framework. Biomarkers definitions working group. Clin Pharmacol Ther 69:89–95
14. Ceciliani F, Ceron JJ, Eckersall PD, Sauerwein H (2012) Acute phase proteins in ruminants. J Proteomics 75:4207–4231
15. Henry CJ (2010) Biomarkers in veterinary cancer screening: applications, limitations and expectations. Vet J 185:10–14
16. Cox J, Mann M (2007) Is proteomics the new genomics? Cell 130:395–398
17. Rogowska-Wrzesinska A, Le Bihan M-C, Thaysen-Andersen M, Roepstorff P (2013) 2D gels still have a niche in proteomics. J Proteomics 88:4–13
18. Albalat A, Husi H, Siwy J, E Nally J, Mclauglin M, D Eckersall P et al (2014) Capillary electrophoresis interfaced with a mass spectrometer (CE-MS): technical considerations and applicability for biomarker studies in animals. Curr Protein Pept Sci 15:23–35
19. Mullen W, Albalat A, Gonzalez J, Zerefos P, Siwy J, Franke J et al (2012) Performance of different separation methods interfaced in the same MS-reflection TOF detector: a comparison of performance between CE versus HPLC for biomarker analysis. Electrophoresis 33:567–574
20. Klein J, Papadopoulos T, Mischak H, Mullen W (2014) Comparison of CE-MS/MS and LC-MS/MS sequencing demonstrates significant complementarity in natural peptide identification in human urine. Electrophoresis 35:1060–1064
21. Nally JE, Mullen W, Callanan JJ, Mischak H, Albalat A (2015) Detection of urinary biomarkers in reservoir hosts of leptospirosis by capillary electrophoresis-mass spectrometry. Proteomics-Clin Appl 9:543–551
22. Albalat A, Husi H, Stalmach A, Schanstra JP, Mischak H (2014) Classical MALDI-MS versus CE-based ESI-MS proteomic profiling in urine for clinical applications. Bioanalysis 6:247–266
23. Mansor R, Mullen W, Albalat A, Zerefos P, Mischak H, Barrett DC et al (2013) A peptidomic approach to biomarker discovery for bovine mastitis. J Proteomics 85:89–98
24. Di Luca A, Elia G, Hamill R, Mullen AM (2013) 2D DIGE proteomic analysis of early post mortem muscle exudate highlights the importance of the stress response for improved water-holding capacity of fresh pork meat. Proteomics 13:1528–1544
25. Janjanam J, Singh S, Jena MK, Varshney N, Kola S, Kumar S et al (2014) Comparative 2D-DIGE proteomic analysis of bovine mammary epithelial cells during lactation reveals protein signatures for lactation persistency and milk yield. PLoS ONE 9:e102515
26. Almeida AM, Plowman JE, Harland DP, Thomas A, Kilminster T, Scanlon T et al (2014) Influence of feed restriction on the wool proteome: a combined iTRAQ and fiber structural study. J Proteomics 103:170–177
27. Huang J, Luo G, Zhang Z, Wang X, Ju Z, Qi C et al (2014) iTRAQ-proteomics and bioinformatics analyses of mammary tissue from cows with clinical mastitis due to natural infection with *Staphylococci aureus*. BMC Genomics 15:1
28. Long M, Zhao J, Li T, Tafalla C, Zhang Q, Wang X et al (2015) Transcriptomic and proteomic analyses of splenic immune mechanisms of rainbow trout (*Oncorhynchus mykiss*) infected by *Aeromonas salmonicida* subsp. salmonicida. J Proteomics 122:41–54
29. Thingholm TE, Palmisano G, Kjeldsen F, Larsen MR (2010) Undesirable charge-enhancement of isobaric tagged phosphopeptides leads to reduced identification efficiency. J Proteome Res 9:4045–4052
30. Burkhart JM, Vaudel M, Zahedi RP, Martens L, Sickmann A (2011) iTRAQ protein quantification: a quality-controlled workflow. Proteomics 11:1125–1134

31. Jiang X, Zeng T, Zhang S, Zhang Y (2013) Comparative proteomic and bioinformatic analysis of the effects of a high-grain diet on the hepatic metabolism in lactating dairy goats. PLoS ONE 8:e80698
32. Romao JM, He ML, Mcallister TA, Guan LL (2014) Effect of age on bovine subcutaneous fat proteome: molecular mechanisms of physiological variations during beef cattle growth. J Anim Sci 92:3316–3327
33. Phongpa-Ngan P, Grider A, Mulligan JH, Aggrey SE, Wicker L (2011) Proteomic analysis and differential expression in protein extracted from chicken with a varying growth rate and water-holding capacity. J Agric Food Chem 59:13181–13187
34. Doherty MK, Mclean L, Hayter JR, Pratt JM, Robertson DHL, El-Shafei A et al (2004) The proteome of chicken skeletal muscle: changes in soluble protein expression during growth in a layer strain. Proteomics 4:2082–2093
35. Teltathum T, Mekchay S (2009) Proteome changes in Thai indigenous chicken muscle during growth period. Int J Biol Sci 5:679
36. Lamy E, Da Costa G, Santos R, Capela ESF, Potes J, Pereira A et al (2009) Sheep and goat saliva proteome analysis: a useful tool for ingestive behavior research? Physiol Behav 98:393–401
37. Gutiérrez AM, Miller I, Hummel K, Nöbauer K, Martínez-Subiela S, Razzazi-Fazeli E et al (2011) Proteomic analysis of porcine saliva. Vet J 187:356–362
38. Ang CS, Binos S, Knight MI, Moate PJ, Cocks BG, Mcdonagh MB (2011) Global survey of the bovine salivary proteome: integrating multidimensional prefractionation, targeted, and glycocapture strategies. J Proteome Res 10:5059–5069
39. Lamy E, Da Costa G, Santos R, Capela E Silva F, Potes J, Pereira A et al (2011) Effect of condensed tannin ingestion in sheep and goat parotid saliva proteome. J Anim Physiol Anim Nutr 95:304–312
40. Mau M, Kaiser TM, Sudekum KH (2009) Evidence for the presence of carbonic anhydrase 29-kDa isoenzyme in salivary secretions of three ruminating species and the gelada baboon. Arch Oral Biol 54:354–360
41. Mau M, Kaiser TM, Sudekum KH (2010) Carbonic anhydrase II is secreted from bovine parotid glands. Histol Histopathol 25:321–329
42. Martin SAM, Vilhelmsson O, Médale F, Watt P, Kaushik S, Houlihan DF (2003) Proteomic sensitivity to dietary manipulations in rainbow trout. Biochimica et Biophysica Acta (BBA) Proteins Proteomics 1651:17–29
43. Vilhelmsson OT, Martin SA, Médale F, Kaushik SJ, Houlihan DF (2004) Dietary plant-protein substitution affects hepatic metabolism in rainbow trout (*Oncorhynchus mykiss*). Br J Nutr 92:71–80
44. Brunt J, Hansen R, Jamieson DJ, Austin B (2008) Proteomic analysis of rainbow trout (*Oncorhynchus mykiss*, Walbaum) serum after administration of probiotics in diets. Vet Immunol Immunopathol 121:199–205
45. Sveinsdóttir H, Steinarsson A, Gudmundsdóttir Á (2009) Differential protein expression in early Atlantic cod larvae (*Gadus morhua*) in response to treatment with probiotic bacteria. Comp Biochem Physiol D Genomics Proteomics 4:249–254
46. Agrawal A, Karim S, Kumar R, Sahoo A, John P (2014) Sheep and goat production: basic differences, impact on climate and molecular tools for rumen microbiome study. Int J Curr Microbiol App Sci 3:684–706
47. Yang Y, Wang J, Yuan T, Bu D, Yang J, Sun P (2013) Proteome profile of bovine ruminal epithelial tissue based on GeLC–MS/MS. Biotechnol Lett 35:1831–1838
48. Hollmann M, Miller I, Hummel K, Sabitzer S, Metzler-Zebeli BU, Razzazi-Fazeli E et al (2013) Downregulation of cellular protective factors of Rumen Epithelium in goats fed high energy diet. PLoS ONE 8:e81602
49. Luo J, Zheng A, Meng K, Chang W, Bai Y, Li K et al (2013) Proteome changes in the intestinal mucosa of broiler (*Gallus gallus*) activated by probiotic Enterococcus faecium. J Proteomics 91:226–241

50. Tang Y, Underwood A, Gielbert A, Woodward MJ, Petrovska L (2014) Metaproteomics analysis reveals the adaptation process for the Chicken Gut Microbiota. Appl Environ Microbiol 80:478–485
51. Deusch S, Seifert J (2015) Catching the tip of the iceberg—evaluation of sample preparation protocols for metaproteomic studies of the rumen microbiota. Proteomics 15:3590–3595
52. Bouley J, Chambon C, Picard B (2004) Mapping of bovine skeletal muscle proteins using two-dimensional gel electrophoresis and mass spectrometry. Proteomics 4:1811–1824
53. Picard B, Berri C, Lefaucheur L, Molette C, Sayd T, Terlouw C (2010) Skeletal muscle proteomics in livestock production. Briefings Funct Genomics: elq005
54. Jia X, Ekman M, Grove H, Færgestad EM, Aass L, Hildrum KI et al (2007) Proteome changes in bovine longissimus thoracis muscle during the early postmortem storage period. J Proteome Res 6:2720–2731
55. Ouali A, Gagaoua M, Boudida Y, Becila S, Boudjellal A, Herrera-Mendez CH et al (2013) Biomarkers of meat tenderness: present knowledge and perspectives in regards to our current understanding of the mechanisms involved. Meat Sci 95:854–870
56. Herrera-Mendez CH, Becila S, Boudjellal A, Ouali A (2006) Meat ageing: reconsideration of the current concept. Trends Food Sci Technol 17:394–405
57. Kim NK, Cho S, Lee SH, Park HR, Lee CS, Cho YM et al (2008) Proteins in longissimus muscle of Korean native cattle and their relationship to meat quality. Meat Sci 80:1068–1073
58. Laville E, Sayd T, Morzel M, Blinet S, Chambon C, Lepetit J et al (2009) Proteome changes during meat aging in tough and tender beef suggest the importance of apoptosis and protein solubility for beef aging and tenderization. J Agric Food Chem 57:10755–10764
59. Polati R, Menini M, Robotti E, Millioni R, Marengo E, Novelli E et al (2012) Proteomic changes involved in tenderization of bovine Longissimus dorsi muscle during prolonged ageing. Food Chem 135:2052–2069
60. Joseph P, Suman SP, Rentfrow G, Li S, Beach CM (2012) Proteomics of muscle-specific beef color stability. J Agric Food Chem 60:3196–3203
61. Sayd T, Morzel M, Chambon C, Franck M, Figwer P, Larzul C et al (2006) Proteome analysis of the sarcoplasmic fraction of pig semimembranosus muscle: implications on meat color development. J Agric Food Chem 54:2732–2737
62. Kim N-K, Lee S-H, Cho Y-M, Son E-S, Kim K-Y, Lee C-S et al (2009) Proteome analysis of the m. longissimus dorsi between fattening stages in Hanwoo steer. BMB Rep 42:433–438
63. De Liu X, Jayasena DD, Jung Y, Jung S, Kang BS, Heo KN et al (2012) Differential proteome analysis of breast and thigh muscles between korean native chickens and commercial broilers. Asian-Australas J Anim Sci 25:895–902
64. Zanetti E, Molette C, Chambon C, Pinguet J, Rémignon H, Cassandro M (2011) Using 2-DE for the differentiation of local chicken breeds. Proteomics 11:2613–2619
65. Sentandreu MA, Fraser PD, Halket J, Patel R, Bramley PM (2010) A proteomic-based approach for detection of chicken in meat mixes. J Proteome Res 9:3374–3383
66. Kinkead R, Elliott C, Cannizzo F, Biolatti B, Mooney M (2015) Proteomic identification of plasma proteins as markers of growth promoter abuse in cattle. Anal Bioanal Chem 407:4495–4507
67. Affolter M, Grass L, Vanrobaeys F, Casado B, Kussmann M (2010) Qualitative and quantitative profiling of the bovine milk fat globule membrane proteome. J Proteomics 73:1079–1088
68. Tacoma R, Fields J, Ebenstein DB et al (2015) Characterization of the bovine milk proteome in early-lactation Holstein and Jersey breeds of dairy cows. J Proteomics 130:200–210
69. Roncada P, Piras C, Soggiu A, Turk R, Urbani A, Bonizzi L (2012) Farm animal milk proteomics. J Proteomics 75:4259–4274
70. Sui S, Zhao J, Wang J et al (2014) Comparative proteomics of milk fat globule membrane proteins from transgenic cloned cattle. PLoS ONE 9:1–12
71. Abd El-Salam MH (2014) Application of proteomics to the areas of milk production, processing and quality control—a review. Int J Dairy Technol 67:153–166

72. Bendixen E, Danielsen M, Hollung K, Gianazza E, Miller I (2011) Farm animal proteomics —a review. J Proteomics 74:282–293
73. Caroli A, Rizzi R, Lühken G, Erhardt G (2010) Short communication: milk protein genetic variation and casein haplotype structure in the original Pinzgauer cattle. J Dairy Sci 93: 1260–1265
74. Singh K, Erdman RA, Swanson KM, Molenaar AJ, Maqbool NJ, Wheeler TT et al (2010) Epigenetic regulation of milk production in dairy cows. J Mammary Gland Biol Neoplasia 15:101–112
75. Lu J, Antunes Fernandes E, Páez Cano AE, Vinitwatanakhun J, Boeren S, Van Hooijdonk T et al (2013) Changes in milk proteome and metabolome associated with dry period length, energy balance, and lactation stage in postparturient dairy cows. J Proteome Res 12: 3288–3296
76. Raemy A, Meylan M, Casati S, Gaia V, Berchtold B, Boss R et al (2013) Phenotypic and genotypic identification of streptococci and related bacteria isolated from bovine intramammary infections. Acta Vet Scand 55:53
77. Sospedra I, Soler C, Mañes J, Soriano JM (2011) Analysis of staphylococcal enterotoxin A in milk by matrix-assisted laser desorption/ionization-time of flight mass spectrometry. Anal Bioanal Chem 400:1525–1531
78. Dušková M, Šedo O, Kšicová K, Zdráhal Z, Karpíšková R (2012) Identification of lactobacilli isolated from food by genotypic methods and MALDI-TOF MS. Int J Food Microbiol 159:107–114
79. Jadhav S, Sevior D, Bhave M, Palombo EA (2014) Detection of Listeria monocytogenes from selective enrichment broth using MALDI-TOF mass spectrometry. J Proteomics 97:100–106
80. Jadhav S, Gulati V, Fox EM, Karpe A, Beale DJ, Sevior D et al (2015) Rapid identification and source-tracking of Listeria monocytogenes using MALDI-TOF mass spectrometry. Int J Food Microbiol 202:1–9
81. Halasa T, Huijps K, Østerås O, Hogeveen H (2007) Economic effects of bovine mastitis and mastitis management: a review. Vet Q 29:18–31
82. Miller G, Bartlett P, Lance S, Anderson J, Heider LE (1993) Costs of clinical mastitis and mastitis prevention in dairy herds. J Am Vet Med Assoc 202:1230–1236
83. De Mol R, Ouweltjes W (2001) Detection model for mastitis in cows milked in an automatic milking system. Prev Vet Med 49:71–82
84. Kamphuis C, Sherlock R, Jago J, Mein G, Hogeveen H (2008) Automatic detection of clinical mastitis is improved by in-line monitoring of somatic cell count. J Dairy Sci 91:4560–4570
85. Chagunda M, Friggens N, Rasmussen MD, Larsen T (2006) A model for detection of individual cow mastitis based on an indicator measured in milk. J Dairy Sci 89:2980–2998
86. Ibeagha-Awemu EM, Ibeagha AE, Messier S, Zhao X (2010) Proteomics, genomics, and pathway analyses of *Escherichia coli* and *Staphylococcus aureus* infected milk whey reveal molecular pathways and networks involved in mastitis. J Proteome Res 9:4604–4619
87. Larsen L, Hinz K, Jørgensen A, Møller H, Wellnitz O, Bruckmaier R et al (2010) Proteomic and peptidomic study of proteolysis in quarter milk after infusion with lipoteichoic acid from *Staphylococcus aureus*. J Dairy Sci 93:5613–5626
88. Anton M, Nau F, Nys Y (2006) Bioactive egg components and their potential uses. World's Poult Sci J 62:429–438
89. Mine Y, Kovacs-Nolan J (2006) New insights in biologically active proteins and peptides derived from hen egg. World's Poult Sci J 62:87–96
90. Mine Y (2007) Egg proteins and peptides in human health-chemistry, bioactivity and production. Curr Pharm Des 13:875–884
91. Mann K, Mann M (2008) The chicken egg yolk plasma and granule proteomes. Proteomics 8:178–191

92. Farinazzo A, Restuccia U, Bachi A, Guerrier L, Fortis F, Boschetti E et al (2009) Chicken egg yolk cytoplasmic proteome, mined via combinatorial peptide ligand libraries. J Chromatogr A 1216:1241–1252
93. Guérin-Dubiard C, Pasco M, Mollé D, Désert C, Croguennec T, Nau F (2006) Proteomic analysis of hen egg white. J Agric Food Chem 54:3901–3910
94. Mann K, Mann M (2011) In-depth analysis of the chicken egg white proteome using an LTQ orbitrap velos. Proteome Sci 9:1–6
95. Omana DA, Liang Y, Kav NNV, Wu J (2011) Proteomic analysis of egg white proteins during storage. Proteomics 11:144–153
96. Qiu N, Ma M, Zhao L, Liu W, Li Y, Mine Y (2012) Comparative proteomic analysis of egg white proteins under various storage temperatures. J Agric Food Chem 60:7746–7753
97. Mikšík I, Sedláková P, Lacinová K, Pataridis S, Eckhardt A (2010) Determination of insoluble avian eggshell matrix proteins. Anal Bioanal Chem 397:205–214
98. Rose-Martel M, Du J, Hincke MT (2012) Proteomic analysis provides new insight into the chicken eggshell cuticle. J Proteomics 75:2697–2706
99. Marco-Ramell A, De Almeida A, Cristobal S et al (2016) Proteomics and the search for welfare and stress biomarkers in animal production in the one-health context. Mol BioSyst 12:2024–2035
100. Bonizzi L, Roncada P (2007) Welfare and immune response. Vet Res Commun 31:97–102
101. Mackenzie S (2013) Behaviour, individual variation and immunity. Fish Shellfish Immunol 6:1663
102. Glaser R, Kiecolt-Glaser JK (2005) Stress-induced immune dysfunction: implications for health. Nat Rev Immunol 5:243–251
103. Christensen L, Ertbjerg P, Løje H, Risbo J, Van Den Berg FW, Christensen M (2013) Relationship between meat toughness and properties of connective tissue from cows and young bulls heat treated at low temperatures for prolonged times. Meat Sci 93:787–795
104. Aguayo-Ulloa L, Pascual-Alonso M, Campo M, Olleta J, Villarroel M, Pizarro D et al (2014) Effects of an enriched housing environment on sensory aspects and fatty-acid composition of the longissimus muscle of light-weight finished lambs. Meat Sci 97:490–496
105. Mormède P, Andanson S, Aupérin B, Beerda B, Guémené D, Malmkvist J et al (2007) Exploration of the hypothalamic–pituitary–adrenal function as a tool to evaluate animal welfare. Physiol Behav 92:317–339
106. Marco-Ramell A, Pato R, Peña R, Saco Y, Manteca X, Ruiz De La Torre JL et al (2011) Identification of serum stress biomarkers in pigs housed at different stocking densities. Vet J 190:e66–e71
107. Cordeiro OD, Silva TS, Alves RN, Costas B, Wulff T, Richard N et al (2012) Changes in liver proteome expression of Senegalese sole (Solea senegalensis) in response to repeated handling stress. Mar Biotechnol 14:714–729
108. Cruzen S, Pearce S, Baumgard L, Gabler N, Huff-Lonergan E, Lonergan S (2015) Proteomic changes to the sarcoplasmic fraction of predominantly red or white muscle following acute heat stress. J Proteomics 128:141–153
109. Marco-Ramell A, Arroyo L, Saco Y, García-Heredia A, Camps J, Fina M et al (2012) Proteomic analysis reveals oxidative stress response as the main adaptative physiological mechanism in cows under different production systems. J Proteomics 75:4399–4411
110. Ibarz A, Martín-Pérez M, Blasco J, Bellido D, De Oliveira E, Fernández-Borràs J (2010) Gilthead sea bream liver proteome altered at low temperatures by oxidative stress. Proteomics 10:963–975
111. Alves RN, Cordeiro O, Silva TS, Richard N, De Vareilles M, Marino G et al (2010) Metabolic molecular indicators of chronic stress in gilthead seabream (*Sparus aurata*) using comparative proteomics. Aquaculture 299:57–66

Chapter 10
Applications of Proteomics in Aquaculture

Pedro M. Rodrigues, Denise Schrama, Alexandre Campos, Hugo Osório and Marisa Freitas

Abstract Aquaculture is one of the fastest growing world industries due to the increased demand of fishery products for human consumption and capture restrictions as a result of aquatic ecosystems exploitation. Aquaculture is therefore an extremely competitive business with major challenges to keep a high quality farmed fish through a sustainable production system. These challenges imposed quite important changes in this more traditional market, namely at the level of integrating scientific knowledge and research. Proteomics presents itself as a powerful tool not only for a better understanding of the marine organisms biology but also to provide solutions to deal with changes and the increasing demand in the system's

---

P.M. Rodrigues (✉) · D. Schrama
Departamento de Química e Farmácia, Universidade do Algarve, CCMar,
Edifício 7, Campus de Gambelas, 8005-139 Faro, Portugal
e-mail: pmrodrig@ualg.pt

A. Campos
Interdisciplinary Centre of Marine and Environmental Research (CIIMAR/CIMAR),
University of Porto, Rua dos Bragas 289, 4050-123 Porto, Portugal

H. Osório
Instituto de Investigação e Inovação em Saúde - i3S (Institute for Research and Innovation in Health), University of Porto, Rua Alfredo Allen, 208, 4200-135 Porto, Portugal

H. Osório
Institute of Molecular Pathology and Immunology of the University of Porto (IPATIMUP), Rua Júlio Amaral de Carvalho, 45, 4200-135 Porto, Portugal

H. Osório
Faculty of Medicine of the University of Porto, Alameda Prof. Hernâni Monteiro, 4200-319 Porto, Portugal

M. Freitas
Department of Environmental Health, Escola Superior de Tecnologia da Saúde do Porto, Polytechnic Institute of Porto, CISA/Research Center in Environment and Health, Rua de Valente Perfeito, 322, 4400-330 Gaia, Portugal

© Springer International Publishing Switzerland 2016
G.H. Salekdeh (ed.), *Agricultural Proteomics Volume 1*,
DOI 10.1007/978-3-319-43275-5_10

production line to ensure the required supply. In this book chapter we will give an overview of aquaculture nowadays, its challenges and describe relevant proteomics studies in several areas of this industry. A brief description of the proteomics technical approaches applied to aquaculture will also be addressed.

Keywords Proteomics · Fish · Aquaculture · Fish proteomics · Fish biology

10.1 Introduction

Aquaculture refers to all forms of active culturing of aquatic animals and plants, occurring in marine, brackish, or fresh waters. In recent decades the once almost exclusive practice of Easter Asia countries (namely China) has spread around the world. According to the Global Aquaculture Production statistics database from FAO, this sector continues to grow until the date of 2013, reaching a value of about 97.2 million tons, of which around 70.0 million tons of the total food fish and 27.0 million tons of aquatic plants [1]. The tremendous growth of this industry has been stimulated by the intrinsic limitations to the productivity of the wild, unmanaged aquatic ecosystems overexploited by humans as sources of fish, aquatic invertebrates and seaweeds, with harvest yields declining substantially over the last decades.

Opposite to traditional aquaculture invented by the Chinese many thousands of years ago, modern aquaculture is driven by competitiveness in the industry based on scientific and technical knowledge that focuses on improving fish health and nutrition, welfare assessment and stress reduction, diseases and the use of antibiotics and vaccines.

The main quality attributes of aquaculture products is the starting point to a perspective of the current consumer expectations, followed by a critical review of the main sensory, physical, chemical and microbial methods used until the present for quality evaluations of aquatic farmed products.

The increase in demand for fishery products for human consumption during the last decade made aquaculture an extremely competitive market and more recent approaches using emerging technologies like proteomics and other "Omics" (genomics, metabolomics) have been used in order to improve the knowledge on the impact of nutritional and welfare factors on aquaculture products, quality, health and safety, therefore contributing towards obtaining higher quality of farmed species through a sustainable production system (Fig. 10.1). Proteomics is seen nowadays as a complementary science to genomics and transcriptomics and the number of aquaculture research studies using proteomics has grown almost exponentially over the last decade [2]. The proteome is defined as the entire set of proteins expressed in an organism, cell, tissue or fluid at a given time under defined conditions and unlikely the genome, the proteome has the ability of a dynamic response to an external/environmental stimulus. Proteomics can also be used as a novel way of understanding biological mechanisms since protein posttranslational

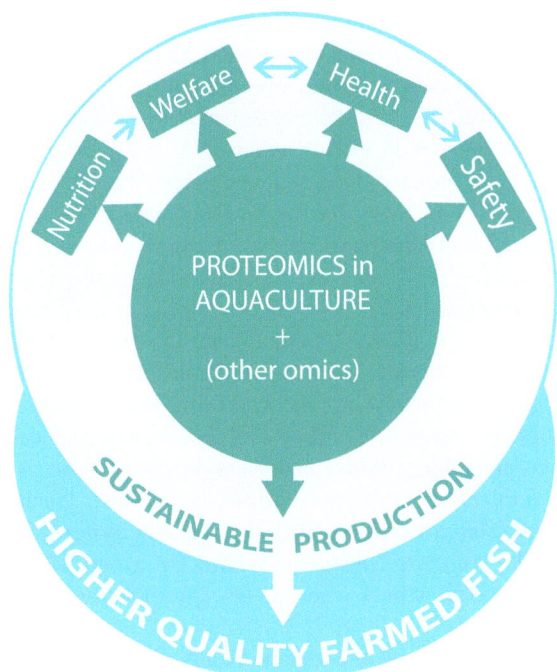

Fig. 10.1 Aquaculture sustainable production system

modifications that play a main role in protein function, can be characterized using this type of approach. Nevertheless there is a major drawback that is related to the lack of available information at the genome level in the vast majority of the cultured species [3]. We believe that this problem will be partially overcome in a near future, based on recent developments in genome sequence technologies that will trigger the availability of genome and EST data of aquaculture organisms [4].

In this review we will address the different applications of proteomics in aquaculture, with a focus on food fish species, with a brief overview on fish model organisms and the latest proteomic technologies and their contribution to the knowledge of species biology, welfare, and to the safety and quality of aquaculture products.

10.2 Fish Species in Aquaculture

The major utilization of cultivated food fishery species is for human consumption, being recognized as a valuable and complementary source of protein, or as fish and animal feed. In 2013 the contribution of aquaculture to the world total fish production reached 43.1 % [1]. A total of 567 species have been referenced by the FAO, including finfishes (354 species, with 5 hybrids), molluscs (102), crustaceans

(59), amphibians and reptiles (6), aquatic invertebrates (9) and marine and freshwater algae (37). Finfish represent about 63.5 % of world production of farmed fishery products followed by crustaceans (22.4 %), molluscs (11.5 %) and other species (2.5 %). Freshwater aquaculture is mostly focused in the production of finfish whereas in marine aquaculture the main production comes from the cultivation of mollusc species [5]. Despite the increased number of fishery species cultivated nowadays, the largest percentage of the global production is represented by a reduced number of species. A list of 25 most cultivated species is presented in Table 10.1. Fish species of Cyprinidae family (e.g. *Ctenopharyngodon idellus*, *Hypophthalmichthys molitrix*, *Cyprinus carpio*), tilapias and other cichlids (e.g. *Oreochromis niloticus*) predominate in world aquaculture. Other most representative fish species are the Atlantic salmon (*Salmo salar*), rainbow trout (*Oncorhynchus mykiss*) and milkfish (*Chanos chanos*) (Table 10.1). Other species worth of mention regarding the high economical value are the European seabass (*Dicentrarchus labrax*) and Atlantic cod (*Gadus morhua*). The clam *Ruditapes philippinarum* is the most cultivated mollusc species, followed by the Agemaki clam (*Sinonovacula constricta*), Pacific oyster (*Crassostrea gigas*) and blood cockle (*Anadara granosa*). Within the group of Shrimps and prawns the most cultivated are whiteleg shrimp (*Penaeus vannamei*) and giant tiger prawn (*Penaeus monodon*).

With respect to the aquatic plants the Japanese kelp (Laminaria japonica) and Wakame and Porphyra seaweeds are the most cultivated with the purpose of human consumption. Eucheuma seaweeds (*Kappaphycus alvarezii* and *Eucheuma* spp.) are also extensively cultivated but mainly for extraction of high economic value compounds such as agar and the polysaccharide carrageenan utilized for instance in food industry [6]. Microalgae are rich sources of essential polyunsaturated fatty acids (PUFAs), such as eicosapentaenoic acid (EPA) and docosahexaenoic acid (DHA), as well as vitamins and several antioxidants [7]. Species such as *Chlorella vulgaris*, *Haematococcus pluvialis*, *Dunaliella salina* and cyanobacteria *Spirulina maxima* are thus widely cultivated and commercialized as nutritional supplements for humans and as animal feed additives [8]. In addition microalgae could have an important role in the supply of novel natural bioactive compounds with a variety of technological applications and can be used for the production of biodiesel [9–11].

The most cultivated fishery species have been object of intensive research, involving genomics, proteomics and metabolomics. This work is enabling to understand in detail for some of these species their biology and physiology or to identify major environmental threats. Of notice regarding future research developments in aquaculture is the sequence of the genome of some of the most representative species such as the common carp, Nile tilapia, Atlantic salmon, Pacific oyster, European seabass, Atlantic cod, Japanese eel, among others. The genome information has a great impact in the understanding of the functions in biological systems. This information also has application in aquaculture. On the other hand the analysis of the transcriptome has become extremely useful in research particularly in species without sequenced genomes. Complementary global transcriptome analysis, using technologies like microarrays and deep RNA sequencing, has

Table 10.1 25 principal food fish species in aquaculture regarding their production values (t = tonnes) [1]

Species	UniProtKB entries	Common name	Family	Production (t)[a]	Species group	Environment[b]	Genome (Mb)[c]	Transcriptome[d]	Proteome[d]
Ctenopharyngodon idellus	859	Grass carp	Cyprinidae	5226,202	Carps, barbels, cyprinids	FW		1	0
Hypophthalmichthys molitrix	471	Silver carp	Cyprinidae	4591,852	Carps, barbels, cyprinids	FW		2	0
Cyprinus carpio	1874	Common carp	Cyprinidae	4080,045	Carps, barbels, cyprinids	FW	1380.1	8	11
Oreochromis niloticus	27,627	Nile tilapia	Cichlidae	3436,508	Tilapias and other cichlids	FW	927.696	6	3
Ruditapes philippinarum	282	Japanese carpet shell	Veneridae	3896,436	Clams, cockles, arkshells	BW, MW		7	4
Penaeus vannamei	579	Whiteleg shrimp	Penaeidae	3314,447	Shrimps, prawns	MW		18	4
Hypophthalmichthys nobilis	359	Bighead carp	Cyprinidae	3059,555	Carps, barbels, cyprinids	FW		2	0
Catla catla	175	Indian carp	Cyprinidae	2776,074	Carps, barbels, cyprinids	FW		0	1
Salmo salar	9944	Atlantic salmon	Salmonidae	2087,111	Salmons, trouts, smelts	Anadromous	2966.89	44	15
Chanos chanos	96	Milkfish	Chanidae	1043,936	Marine fish	BW, MW		0	0
Oncorhynchus mykiss	49,988	Rainbow trout	Salmonidae	814,068	Salmons, trouts, smelts	Anadromous		35	27
Penaeus monodon	485	Giant tiger prawn	Penaeidae	803,783	Shrimps, prawns	MW		2	12
Eriocheir sinensis	289	Chinese mitten crab	Varunidae	729,969	Crabs	Catadromous		15	5

(continued)

Table 10.1 (continued)

Species	UniProtKB entries	Common name	Family	Production (t)[a]	Species group	Environment[b]	Genome (Mb)[c]	Transcriptome[d]	Proteome[d]
Sinonovacula constricta	57	Agemaki clam	Pharidae	720,804	Clams, cockles, arkshells	BW, MW		1	0
Procambarus clarkii	284	Red swamp crawfish	Cambaridae	652,051	Freshwater crustaceans	FW		6	4
Crassostrea gigas	26,901	Pacific oyster	Ostreidae	555,994	Oyster	MW	557.736	14	15
Channa argus	72	Northern snakehead	Channidae	510,116	Other FW fish	FW		0	0
Anadara granosa	115	Blood cockle	Arcidae	448,864	Clams, cockles, arkshells	BW		0	1
Silurus asotus	69	Amur catfish	Siluridae	438,736	Other FW fish	FW		0	0
Ictalurus punctatus	7412	Channel catfish	Ictaluridae	419,215	Other FW fish	FW		9	1
Trionyx sinensis	20,836	Chinese soft-shell turtle	Trionychidae	347,587	Turtle	BW	2202.48	0	1
Monopterus albus	337	Rice eel	Synbranchidae	346,143	River eels	FW, BW		1	1
Micropterus salmoides	227	Largemouth black bass	Centrarchidae	339,900	Other FW fish	FW		1	1
Misgurnus anguillicaudatus	335	Pond loach	Cobitidae	322,207	Other FW fish	FW		2	0
Pangasius hypophthalmus	96	Striped catfish	Pangasiidae	306,077	Other FW fish	FW		0	0

[a]Production values from FAO relative to 2013
[b]Freshwater (FW), brackish water (BW), marine water (MW)
[c]Information retrieved from genome database at NCBI
[d]Number of publications searched in each species within the fields of transcriptomics and proteomics

reached a broader number of cultivated species (Table 10.1) hence enabling a conjunct of novel opportunities in those species regarding to functional genomics. Proteomics approaches have also been also explored to investigate protein expression in a variety of fishery species (Table 10.1). New opportunities for high throughput approaches (mass spectrometry-based proteomics) and deep proteome coverage in aquaculture species were created due to the complementary information available at the level of the genome and transcriptome [12–15]. The OMICs technologies have been employed in many different research topics from nutrition [16, 17], to the disclosure of immune mechanisms and response to infection [12, 18, 19], characterization of the reproductive systems [15, 20–23], understanding of the effects of climate change [24, 25], characterization of neuropeptides and hormone like elements in crustaceans [26, 27], growth and development [28], risk assessment and toxicity of environmental contaminants [29–33], and mechanisms of detoxification of contaminants [34].

10.3 Challenges in Aquaculture; the Proteomics Approach

Over the last five decades global fishery production has grown steadily with world aquaculture expanding at an average annual rate of 6.2 % between 2000 and 2012 and fish remaining among the most traded commodities worldwide. The demand for fishery products as a way of accessing a "better" animal protein both in developed countries as well as in emerging economies has turned this industry into one of the most promising economical commercial trades. This poses enormous constrains especially at the production level where the main challenge is to produce more of a higher quality farmed fish through a sustainable production. This is targeted in the Blue Growth framework—one of the last initiatives by the Food and Agriculture Organization (FAO) that promotes responsible and sustainable aquaculture with the strengthening of policy environment, institutional arrangements and collaborative processes to empower fish farming. Within this initiative, science can play an extremely important role with a better knowledge of all the production chain in aquaculture not only focused on the product (fishery) but also with new technological advances created or adapted to achieve this goal. Is at this level that proteomics can present itself as an innovative and cutting edge technology to face the big challenge.

The proteome consists of all the proteins produced from the genome and offers a holistic and comprehensive information regarding the proteins expressed in a given organ, tissue, cell or fluid at a given time and under a specific condition or stimulus. Unlike the genome, the proteome defines a dynamic state subject to a multitude of changes of diverse nature and varies with time. Proteomics provides information not only regarding changes in protein expression (comparative proteomics), but also the characterization of protein post-translational modifications (PTMs). Quantitative proteomics approaches include the more traditional gel-based [2D-electrophoresis or 2D-DIGE (Difference gel electrophoresis)] and the gel-free based methods that might be separated further into label-free approaches (Spectral counting and absolute quantitation) and labelling approaches [stable isotope labelling (SILAC), isotope coded affinity tags (ICAT) and isobaric tagging (iTRAQ)]. Until recently, the majority of proteomics-based studies in aquaculture were gel-based approaches, mainly due to the lack of genomic database information on fishery species and also the economic factor, since gel-free approaches require more expensive equipment.

10.4 Proteomics Technologies in Aquaculture Research

Aquaculture research has benefited from the continuous and fast development of proteomics technologies in the last two decades. The most common proteomics workflow (Fig. 10.2) comprehend the following phases: (1) Protein extraction, clean-up and enrichment; (2) protein separation; (3) protein identification; (4) quantification of protein expression levels and; (5) characterization of protein post-translational modifications (PTMs).

Since a panoply of complementary approaches are available for each stage, a careful planning of the experimental protocol has to be performed with several steps of optimization for the best possible characterization of a given proteome. Some experimental approaches perform complementary proteomics approaches, for example, integrating gel-based and gel-free procedures as described below.

10.4.1 Sample Preparation

The sample preparation is the most critical stage of all the proteomics procedure workflow since the strategies and options to extract and enrich the proteomes will affect all the downstream processes.

The first step is to select a proper solubilisation buffer for protein extraction. It is essential to ensure protein solubilisation for further analysis. The following aspects should be considered for an optimal protein extraction:

(i) A proper pH is important to keep proteins soluble and stable. Most common buffers are Tris-HCl (pH range 7.0–9.0) or HEPES–NaOH (7.2–8.2) at a 20–50 mM concentration.
(ii) Detergents for solubilisation of poorly soluble/membrane proteins. For example Triton X-100 (non-ionic), CHAPS (zwitterionic), or SDS (anionic) at a concentration 0.1–1 %.
(iii) Salts for maintaining the ionic strength of the medium and to increase the protein solubility. Typical recommended salts are NaCl or KCl at 50–150 mM.
(iv) Reducing agents to reduce oxidation damage. It is recommended DTT (1–10 mM) or 2-mercaptoethanol (0.05 %).

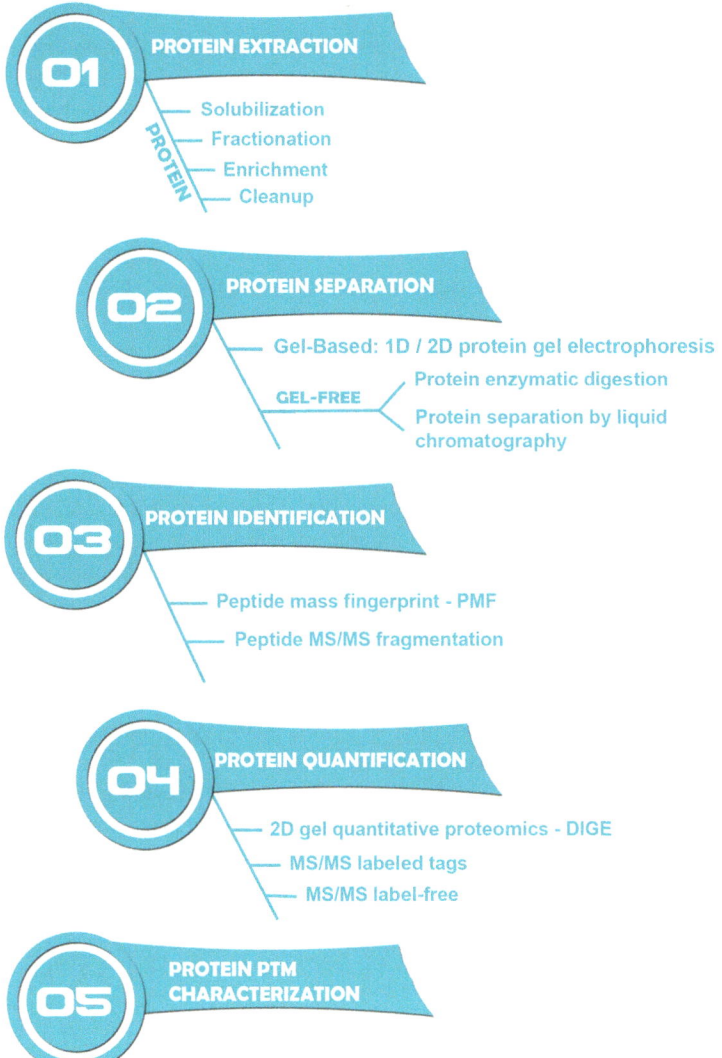

Fig. 10.2 Proteomics technologies in aquaculture research

(v) Metal chelators to chelate metal ions and reduce oxidation damage (ex. EGTA or EDTA at 1 mM).
(vi) Chaotropic reagents to disrupt the three-dimensional structure of proteins and to reduce protein aggregates. Examples include guanidine hydrochloride (0.5–6 M) or urea (0.5–8 M).
(vii) It is very important to add protease inhibitors since cell/tissue disruption will release proteases that will degrade proteins thus reducing total protein yield. Specific inhibitors should be used for the different proteolytic enzymes: aprotinin, PMSF or benzamidine for serine proteases; EDTA or EGTA for metallo proteases; pepstatin A for aspartic proteases; and leupeptin for cysteine and serine proteases.

Depending on cells or tissues samples, it may be necessary to enhance protein extraction using physical disruption methods of cell lysis: mechanical, using a polytron or beads, liquid homogenization with a French press or sonication.

A clean-up centrifugation step is necessary to remove non-solubilized compounds, intact cells and lipids from the protein solution.

Often, proteomes present a highly dynamic range of protein levels. In the same proteome, strongly expressed proteins may mask key regulatory proteins that are expressed in small quantities. Protein fractionation is thus recommended to normalize/decrease the protein dynamic range. Some examples of protein fractionation include: subcellular fractionation, including cytoplasmic, membrane and nuclear fractions; organelle enrichment including nuclei, mitochondria, lysosome or peroxisome; or proteome equalization/normalization using combinatorial peptide ligand libraries.

Liquid chromatography and mass spectrometry approaches have a low tolerance to salt and detergents. It may be necessary to further clean-up protein samples by dialysis, salt/detergent removal tips or a short SDS-PAGE gel run before further proteomics downstream analysis.

It is strongly recommended that all the described steps should be performed at 4 °C to prevent sample degradation. After the conclusion, protein extracts should be divided in single-use aliquots to avoid freeze-thaw cycles and stored preferentially at −80 °C.

10.4.2 Protein Separation

Protein gel electrophoresis became a routine technique in biochemistry laboratories decades ago for protein separation. It still remains today as a simple, fast and powerful technique used in several proteomics applications. The most common approach is based on protein separation by molecular mass—1D SDS-PAGE. However, the currently most effective protein gel-based separation methodology is 2D gel electrophoresis. This approach is based on protein separation by isoelectric focusing (IEF) followed by protein molecular mass separation. Protein bands from

1D gels or protein spots from 2D gels are selected for protein identification by mass spectrometry. After selection, the protein gel plugs are enzymatically "in gel" digested, usually with trypsin, and further analysed by mass spectrometry.

In the last years, high throughput proteomics approaches based on liquid-chromatography (LC) online coupled with a mass spectrometer have been progressively implemented. Usually, the protein extract is "in solution" enzymatically digested generally with trypsin. Further, peptides are separated by C18 reverse-phase LC chromatography. Multi-dimensional LC approaches may be performed combining reverse-phase with strong cation exchange (SCX) or hydrophobic interaction liquid chromatography (HILIC) for example.

10.4.3 Protein Identification

Mass spectrometry (MS) is currently the most efficient and informative approach for protein identification. The simplest protein identification method is based on Peptide Mass Fingerprint (PMF). In this approach, the experimental peptide masses obtained from the mass spectra of the enzymatically, usually tryptic and digested unknown proteins are compared with a database containing protein sequences. This analysis is performed by a software that digests "in silico" a set of proteins downloaded from a repository and further compares the experimental peptide peaks with the theoretical ones providing a statistical analysis. The success of this analysis fully depends on the information amount and quality of a specific protein sequence database. UniProt is currently one of the best publically available protein sequence databases. It is divided in two sections, Swiss-Prot and TrEMBL, in which the number of protein sequences has grown significantly in the last years. Swiss-Prot, manually reviewed and annotated, has grown from less than 100 thousand entries in 2000 to more than 500 thousand entries in 2015. More impressively, TrEMBL, the non reviewed and automatically annotated section, has grown from less than 0.5 million entries to more than 50 million available protein sequences for the same period. This increasingly available information is decisively contributing for the success of the PMF approach.

Peptide fragmentation and sequencing by MS/MS, often combined with PMF, is currently the most effective way to perform protein identification. The most common process is to perform a selective peptide collision against air or a gas. As a result of the collision, peptides will dissociate producing several related ions. The most common fragment ions are the N-terminal charged b-ions and the C-terminal charged y-ions. Using proper software tools that compare the MS/MS information with theoretical MS/MS data derived from "in silico" fragmentation of a protein database such as UniProt will allow performing high-throughput "shotgun" sequence analysis.

Cordero and collaborators established the *Dicentrarchus labrax* skin mucus proteome map using 2DE followed by LC-MS/MS [35].

10.4.4 Protein Quantification

2D gel electrophoresis is a simple and effective approach for evaluation of protein expression. It allows performing gel-to-gel comparison. Some experimental approaches use fluorescent dyes to label proteins, allowing run up to 3 samples in one gel using a technique called 2D-DIGE (two-dimensional difference in gel electrophoresis). Proper software that integrates statistical tools allowing the determination of protein expression levels performs this analysis. Dietrich and collaborators developed 1DE-LC-MS/MS and 2D-DIGE methodologies for quantitative *Cyprinus carpio* seminal proteome analysis [21, 36].

Isobaric labelling is performed in quantitative MS approaches. Peptides or proteins are chemically labelled in specific amino acids. At the MS/MS fragmentation, the labelled peptides will produce reporter tags allowing performing its quantification. The currently available isobaric tags include isobaric tags for relative and absolute quantitation, iTRAQ or tandem mass tags, TMT. Some mass spectrometers may also perform label-free proteomics approaches. An iTRAQ-based quantitative proteomics approach was used for instance to investigate sea cucumber *Apostichopus japonicus* infection by *Vibrio splendidus* [37]. Chemical dimethyl labelling and label-free quantitative proteomics approaches have been developed to analyse the *Sparus aurata* muscle tissue proteome [38]. Another gel-free proteomics approach was performed on stickleback's plasma. A label-free quantitative proteome profiling allowed identifying 45 population-specific plasma proteins [39]. *Edwardsiella tarda* proteome, a pathogen prejudicial for aquaculture was analysed by a gel-based approach. A comparison between a virulent isolate and an avirulent strain was performed. More than one hundred differentially expressed proteins have been identified [40].

To better understand the freshwater crayfish reproduction the *Astacus astacus* spermatophore proteome has been characterized. A gel-free based approach has been developed for protein identification including protein label-free quantification [41]. Long designed a combined transcriptomic and proteomic approach were applied to study *Oncorhynchus mykiss* (rainbow trout) infected by *Aeromonas salmonicida* [12]. The quantitative protein expression levels were accessed using the iTRAQ methodology.

Other study focused in the detection of cryo-preservation-induced alterations in protein composition of trout semen with complementary proteomics approaches including 1DE SDS-PAGE-pre-fractionation combined with LC-MS/MS and 2DE followed by MALDI-TOF/TOF identification [42].

One increasingly popular MS-based protein quantification approach relies on selective quantification of surrogate peptide(s) in a digested protein sample being referred as multiple or selective reaction monitoring (MRM or SRM). Groh and collaborators developed a targeted proteomics method based on SRM to analyse protein expression in zebrafish during gonad differentiation [43].

10.4.5 Protein PTMs Characterization

Protein PTMs can be relatively simple such acetylation, formylation or phosphorylation or highly complex such as glycosylation. The major challenges on PTMs proteomic approaches rely on the location determination of the amino acid site where the PTM is sited and how to unravel intricate PTM structures such as glycans being MS/MS fragmentation critical to characterize complex PTMs. Unlike protein identification analysis, it still remains a huge challenge to develop sophisticated software tools and proper PTM databases to perform an efficient and automated MS/MS ion fragment assignment required for proteomics PTM characterization. Jin and collaborators characterized mucin O-glycosylation on freshwater acclimated Atlantic salmon by LC-MS/MS mass spectrometry [44].

10.5 Welfare and Nutrition in Fish Farming

Fish welfare and stress are important issues to aquaculture mainly because of public perception, marketing, product acceptance, and production efficiency, quality and quantity [45–47]. Fish are vertebrates and though they share many traits in common with the more familiar intensively farmed animals such as pigs, chickens or cows. However, due to the separate evolutionary history and different adaptation needs, they have a number of special features that are relevant to the way welfare is approached. Fish welfare is in that sense a complex concept, which underlines the importance of a multidisciplinary and holistic approach in its study. Proteomics can therefore be an important part of the toolset required for such studies, ensuring that marine animals are reared in an environment that optimizes their capacity to cope with unavoidable challenges/stress [2]. Also it is important to understand that stress responses will not provide all the necessary information about fish welfare, since in aquaculture this last one is largely associated with tertiary effects of stress response that are generally indicative of prolonged, repeated or unavoidable stress [47–49] that is mostly related to maladaptive effects on growth, reproductive function, immune function and disease resistance [50].

There is a significant number of proteomics studies in aquaculture targeting areas of research related to welfare. These are mostly focused on health aspects in fish organs/tissues/fluids like the liver, brain, skeletal muscle, blood plasma and osmoregulatory and immune-related organs and tissues. Of these, the liver due to its central role in key metabolic processes and the blood plasma due to its non-evasive collection, are by far the most used ones.

Fish production is a factor of stress induction as it involves management practices and environmental sources with influence in fish welfare. Proteome changes in several tissues have been reported in fish submitted to high stocking densities [51, 52], handling [52, 53], pre-slaughter stress [54, 55], hypoxia [56, 57], anoxia [58–60], hyperoxygenation [61], osmotic [62–67] and temperature changes [25, 68–70]. Also proteomic approaches have been used as an attempt to establish welfare biomarkers in farmed fish [52, 71–73] with several proteins identified.

Another important field of aquaculture research covered by proteomics is fish nutrition. Due to increasing demand for fish oil and fishmeal in aquafeeds, more sustainable alternatives to the traditional diet formulations, such as plant-derived oils and proteins are emerging. Few papers interestingly report the fish proteome response to these new fish meals, as a new insight into the response of fish metabolism to dietary substitutions [17, 74–79]. Furthermore, the impact on the fish proteome of specific diet formulations, formulated to mitigate disease effects has been reported [80].

10.6 Proteomics Applied to Safety and Quality of Aquaculture Products

Aquaculture products have a great importance concerning food security and according to human health perspective there is an extensive evidence of several benefits associated to their regular consumption. However, along with the benefits, there are several potential risks, especially related to food-borne diseases and authenticity, which emphasizes the challenge to guarantee the safety and quality of aquaculture products.

Proteomic approaches have proven to be promising to evaluate the safety and quality of aquaculture products, being applied either as a monitoring tool or for discovery of potential biomarkers [2, 81–84]. Indeed, the identification of proteins and their function in a specific physiological condition may unequivocally reveal the mechanisms underlying the safety and quality changes, which can be crucial for guarantee the consumer satisfaction and to protect the aquaculture industry from financial damages. The hazard analysis carried out to ensure food safety presupposes the identification and characterization of chemical, biological or physical agents capable of causing quality loss. In this field, proteomics has been applied to the identification and biomarker exploration of (1) biotoxins and other environmental contaminants (industrial organic contaminants, environmental inorganic contaminants and veterinary drugs); (2) microbial contaminants (pathogenic and spoilage bacteria, virus and protozoan); and (3) allergens (Table 10.2).

Filter-feeding bivalve molluscs are often contaminated by biotoxins and proteomics techniques have been used to study the changes in protein expression and potential biomarkers either by the toxin producers or mechanisms of tissue response [85–87]. Furthermore, the identification and characterization of dinoflagellates proteome have also been explored to early discriminate toxic and nontoxic strains [88, 89]. Polychlorinated biphenyls (PCBs), polycyclic aromatic hydrocarbons (PAHs) and alkyl phenols are industrial organic compounds well studied, since they can cause severe effects on aquatic organisms. The identification of distinct protein expression signatures (PES) provided the knowledge of key proteins possibly

Table 10.2 Studies performed with application of proteomics tools on issues related to the safety of aquaculture products

Parameters analyzed	Focus of the study	Main discovery	Aquaculture genus/species (Common name)	Class (Group)	References
Biotoxins and other environmental contaminants	Differential protein expression in *Mytilus galloprovincialis* exposed to *Cylindrospermopsis raciborskii* cells	Alterations in actin and tubulin of mussel gills and digestive gland when exposed to cells containing or not cylindrospermopsin (CYN), suggesting the induction of physiological stress and tissue injury. The presence of CYN may lead to additional toxic effects	*Mytilus galloprovincialis* (Mediterranean mussel)	Bivalves (Molluscs)	[85]
	Proteomics analysis of digestive gland extracts of mussels employed to identify biomarkers of contamination due to okadaic acid-group toxins	Potential detection and identification of biomarkers of biotoxin contamination in shellfish, including both proteins expressed by the toxin producers and components that participate to the tissue response to the okadaic acid-group toxins	*Mytilus* (Mussels)	Bivalves (Molluscs)	[86]
	Application of proteomics for further identification of biomarkers in *Mytilus edulis* contaminated with azaspiracid toxins (AZA)	The identity of AZA biomarkers will support the investigations of the mechanisms underlying the toxin–protein interactions. AZAs bind weakly to a protein with a molecular weight of 45 kDa and a 22 kDa protein was also present only in contaminated mussel samples	*Mytilus edulis* (Blue mussel)	Bivalves (Molluscs)	[87]
	A proteomics approach was applied to discriminate toxic and nontoxic strains of *Alexandrium minutum*, a paralytic shellfish poisoning toxin-producing dinoflagellate	Pronounced differences were detected between toxic and nontoxic strains. The most notable differences between these strains were several abundant proteins with pIs ranging from 4.8 to 5.3 and apparent molecular masses between 17.5 and 21.5 kDa	–	–	[88]
	A proteomics approach was applied to identify and characterize a "biomarker of toxicity" from the proteome of *Alexandrium tamarense*, a paralytic shellfish toxin-producing dinoflagellate	Identification and characterization of *Alexandrium tamarense* proteome and determination of biomarkers of toxicity namely AT-T1, AT-T2 and AT-T3 that were present in all toxic strains of this algae	–	–	[89]

(continued)

Table 10.2 (continued)

Parameters analyzed	Focus of the study	Main discovery	Aquaculture genus/species (Common name)	Class (Group)	References
	The study of *Mytilus edulis* proteome exposed to copper (70 ppb), polychlorinated biphenyl (PCB) (Aroclor 1248 1 ppb) and lowered salinity	Each stressor produced distinct protein expression signatures in blue mussels. Monitoring of environments for specific proteins using proteome analysis may allow for detection and identification of biological effects	*Mytilus edulis* (Blue mussel)	Bivalves (Molluscs)	[90]
	A proteomics approach was applied to investigate protein changes in plasma of juvenile *Gadus morhua* induced by crude North Sea oil and North Sea oil spiked with alkyl phenols and polycyclic aromatic hydrocarbons (PAHs)	The protein changes observed in this study represent a first screening for potential biomarkers in cod plasma reflecting potential effects of crude oil and produced water exposure on fish	*Gadus morhua* (Atlantic cod)	Bony fish (Fish)	[71]
	Proteomics study in *Chamaelea gallina* as a preliminary screening of changes in protein expression caused by to four model contaminants, Aroclor 1254, copper (II), tributyltin (TBT), and arsenic (III), potentially useful as new biomarkers	Aroclor 1254 and Cu (II) upregulated putative isoforms of tropomyosin and light chain of myosin. Actin was downregulated by Aroclor and Cu (II) but upregulated by TBT and As (III). The exclusive identification of cytoskeletal proteins could reflect their role as major targets of pollutant-related oxidative stress	*Chamaelea gallina* (Clams)	Bivalves (Molluscs)	[91]
	Investigation of Cadmium (Cd) effects in the gill and digestive gland protein expression profiles (PEPs) of the sentinel species *Ruditapes decussatus* exposed to 40 ppb, during 21 days	Cd induces major changes in proteins involved in cytoskeletal structure maintenance, cell maintenance and metabolism, suggesting potential energetic change. They provide a valuable knowledge of Cd effects at biochemical and molecular levels in the gill and digestive gland of clams	*Ruditapes decussatus* (Clams)	Bivalves (Molluscs)	[92]
	Proteomics study to assess the effects of chemotherapeutics and production management strategy on *Penaeus monodon*	The study demonstrates that tested antibiotics on patterns of haemolymph protein expression, which were overwhelmed by the effects of the conditions in different production management systems, caused very subtle effects	*Penaeus monodon* (Giant tiger shrimp)	Crustaceans (Arthropoda)	[93]

(continued)

10 Applications of Proteomics in Aquaculture

Table 10.2 (continued)

Parameters analyzed	Focus of the study	Main discovery	Aquaculture genus/species (Common name)	Class (Group)	References
Microbial contamination	Proteome changes of *Bacillus cereus* due to high hydrostatic pressure (HHP) treatment, a technique of food preservation	The proteomics study highlights as a specific HHP treatment at 700 MP may reduce the virulence and protective response against oxidative stress	–	–	[99]
	Development of a whole-cell MALDI-TOF MS method for the distinction of *V. Parahaemolyticus* (an important causative agent of bacterial seafood-borne gastroenteritis) from other *Vibrio* spp.	The first use of whole-cell MALDI-TOF MS analysis for the rapid identification of *V. parahaemolyticus*. In addition, potential peaks that could be further developed into biomarkers were identified	–	–	[94]
	Identification of membrane-associated proteins from *Campylobacter jejuni* strains using complementary proteomics technologies	The proteome of *C. jejuni* was analysed for the first time. The study provides a comprehensive analysis of membrane-associated proteins from *C. Jejuni*	–	–	[95]
	Isolation and characterization of *Streptococcus parauberis* (a bacterium producing streptococcosis in farmed fish) from vacuum packaging refrigerated seafood products	Proteomics analysis by MALDI-TOF MS allowed the identification of five mass peaks specific to the species *S. parauberis* and the rapid identification with respect to other pathogenic and spoilage bacteria potentially present	–	–	[96]
	Proteomics analysis of head kidney tissue from high and low susceptibility families of channel catfish following challenge with *Edwardsiella ictaluri* [Enteric Septicemia of Catfish (ESC)]	The study demonstrates that 2-D gel electrophoresis coupled with PMF MALDI-TOF-MSMS can be applied successfully to channel catfish and that proteomics analysis of protein expression profiles, in response to bacterial challenge, has the potential to contribute significantly to the search for genetic markers for disease resistance, despite the limited genomic sequence data available	*Ictalurus punctatus* (Channel catfish)	Bony fish (Fish)	[100]
	Species differentiation of seafood spoilage and pathogenic gram-negative bacteria by MALDI-TOF mass fingerprinting	Strains isolated from seafood were identified, demonstrating the applicability of the technique for the characterization of unknown strains	–	–	[98]

(continued)

Table 10.2 (continued)

Parameters analyzed	Focus of the study	Main discovery	Aquaculture genus/species (Common name)	Class (Group)	References
	Rapid species identification of seafood spoilage and pathogenic gram-positive bacteria by MALDI-TOF mass fingerprinting	The method was successfully applied to identify the six strains isolated from seafood (*Bacillus* spp., *Listeria* spp., *Clostridium* spp., *Staphylococcus* spp. and *Carnobacterium* spp.) by comparison with the reference library	–	–	[97]
	To characterize host protein expression changes in *Litopenaeus vannamei* stomach cells after white spot syndrome virus (WSSV) infection	The study increases the understanding of the molecular pathogenesis of this virus-associated shrimp disease, which should be useful both for identifying potential biomarkers and for developing antiviral measures	*Litopenaeus vannamei* (Shrimp)	Crustaceans (Arthropoda)	[101]
	A proteomics approach was applied to understand the mechanism of molecular responses of *Fenneropenaeus chinensis* to WSSV infection	Protein expression patterns of the infected shrimp were drastically altered by WSSV infection. These data may provide some information about shrimp proteins that participate in the WSSV infection process	*Fenneropenaeus chinensis* (Chinese white shrimp)	Crustaceans (Arthropoda)	[102]
	A comparative proteomics analysis was employed to identify altered proteins in the yellow head virus (YHV) infected lymphoid organ (LO) of *Penaeus monodon* (24 h post-infection)	Identification of altered proteins in the YHV-infected shrimps may provide novel insights into the molecular responses of *P. monodon* to YHV infection	*Penaeus monodon* (Giant tiger shrimp)	Crustaceans (Arthropoda)	[103]
	A proteomics approach was applied to investigate altered proteins in hemocytes of *Penaeus vannamei* during Taura syndrome virus (TSV) infection (24 h post-infection)	Identification of several proteins in hemocytes that were modulated during TSV infection. While roles of some of these alterations could be related to host responses during a viral infection, functional significance of other altered proteins remains unclear. Further investigation of these data may lead to better understanding of the molecular responses of crustacean hemocytes to TSV infection	*Penaeus vannamei* (Shrimp)	Crustaceans (Arthropoda)	[104]

(continued)

Table 10.2 (continued)

Parameters analyzed	Focus of the study	Main discovery	Aquaculture genus/species (Common name)	Class (Group)	References
	A proteomics approach was applied to identify and quantitate differentially expressed proteins in the liver and kidneys of diseased [infected with Hematopoietic Necrosis Virus (IHNV) or *Renibacterium salmoninarum* (BKD)] and healthy *Salmo salar*	The results provide significant insights into differential protein expression, allowing better molecular understanding of events attending disease and pointing the way to more precise experiments to reveal the role of significant molecules	*Salmo salar* (Atlantic salmon)	Bony fish (Fish)	[105]
	A proteomics approach was applied to compare the haemolymph protein profiles between the two oyster species and between *Ostrea edulis* stocks with different susceptibility to bonamiosis, a protozoan infection	To understand the oyster–*B. ostreae* interaction and to find the bases of tolerance/resistance to bonamiosis, which would be useful for genetic selection programs with the objective of oyster increased tolerance	*Ostrea edulis* (European flat oyster) and *Crassostrea gigas* (Pacific cupped oyster)	Bivalves (Molluscs)	[106]
Allergens	Comparative proteomics analysis of allergens from GH-transgenic and non-transgenic *Oncorhynchus masou ishikawae*	The results indicate that amago salmon endogenous allergen expression does not seem to be altered by genetic modification	*Oncorhynchus masou ishikawae* (Amago salmon)	Bony fish (Fish)	[110]
	Rapid direct detection of the major fish allergen, parvalbumin, by selected MS/MS ion monitoring mass spectrometry	The present strategy allowed the direct detection of the presence of fish parvalbumin beta (β-PRVBs) in any food product in less than 2 h	Several reference fish species	–	[107]
	Proteomics and immunological analysis of a novel shrimp allergen, Pen m 2	A novel allergen from *Penaeus monodon*, designated Pen m 2, was identified by two-dimensional immunoblotting using sera from subjects with shrimp allergy, followed by MALDI-TOF MS analysis of the peptide digest. This novel allergen could be useful in allergy diagnosis and in the treatment of Crustacea-derived allergic disorders	*Penaeus monodon* (Giant tiger shrimp)	Crustaceans (Arthropoda)	[109]
	Analysis of the allergenic proteins in *Penaeus monodon* and characterization of the major allergen tropomyosin using mass spectrometry	A signature peptide was assigned for future quantification of black tiger prawn tropomyosin levels in different matrices (i.e. water, air, food) in the seafood industry	*Penaeus monodon* (Black tiger prawn)	Crustaceans (Arthropoda)	[108]

involved in toxicity mechanisms of PCBs in bivalve molluscs [90], and the potential discovery of sensitive biomarkers in cod plasma after chronic exposure to crude oil and produced water [71]. Likewise, proteomic approaches have been applied to assess the effects and to find potential biomarkers in sentinel organisms (clams) exposed to inorganic environmental stressors such as copper, arsenic and cadmium [91, 92]. Although briefly mentioned in this work, aquatic plants, which are largely consumed in Asian countries and nowadays in European countries as a part of sushi, can accumulate high levels of heavy metals, thus future studies should employ proteomics techniques to identify these chemical contaminants in such organisms. On the other hand, the intensification of aquaculture has been related to the increasing use of chemotherapeutics. Although the marked improvements in laboratory methods to detect residues of antibiotics, according to our knowledge only a study relying on proteomics technologies in antibiotic identification has been developed in shrimps [93].

The identification of spoilage and pathogenic bacteria isolated from fishery products is another issue of great importance regarding to aquaculture proteomics, where MALDI-TOF MS/MS has been successfully applied [94–98]. Studies on the bacterial proteome to test the effectiveness of food preservation techniques and the determination of potential markers for bacterial disease resistance in fish have also been developed [99, 100]. Proteomics has allowed further novel insights into the molecular pathogenesis of virus-associated shrimp and fish diseases [101–105] as well as the understanding of the susceptibility of two oyster species to a protozoan infection [106].

Allergic reactions are frequently reported as a consequence of seafood consumption. The major fish and crustacean allergens are the proteins parvalbumins beta and tropomyosin, respectively. Proteomic investigations towards identification and quantification of allergens in aquaculture products are still scarce [107, 108], however efforts have been made to identify novel allergens [109] and characterize the potential allergenicity of transgenic and non-transgenic fish [110]. Innovative is the approach proposed by Rodrigues and collaborators with the use of proteomics to track the modulation of parvalbumin expression in European seabass exposed to diets aimed to reduced its allergen content with the purpose of a low allergen farmed fish as the end product. In this field, studies regarding the stability of the allergenic proteins subjected to food processing and digestion are still missing.

Proteomics technologies have also proved to be very useful in the quality assessment of aquaculture products, and research studies have been developed on the following subjects: (1) traceability and authentication; (2) production process; (3) storage and processing; (4) sensorial and nutritional quality; and (5) functionality (Table 10.3).

So far, several studies have shown the effectiveness of proteomics approaches for discriminate and guarantee the traceability and authentication of aquaculture products, especially fish and crustaceans [111–118]. Furthermore, the use of protein biomarkers in cultured fish, particularly in edible *ante mortem* tissues, is an issue of high significance. Pre-slaughter conditions may influence the *post mortem* muscle

Table 10.3 Studies performed with application of proteomics tools on issues related to the quality of aquaculture products

Parameters analyzed	Focus of the study	Main discovery	Aquaculture species (Common name)	Class and group	References
Traceability and authentication	Characterization and partial sequencing of species-specific sarcoplasmic polypeptides from commercial hake species by MS following 2-DE	The work opens the way to the application of proteomics to the differential characterization of commercial hake species at the molecular level	*Merluccius merluccius* (European hake), *M. australis* (Southern hake), *M. hubbsi* (Argentinian hake), *M. gayi* (Chilean hake), and *M. capensis* (Cape hake)	Bony fish (Fish)	[111]
	Fish authentication by gel-free proteomics strategies based on MALDI-TOF MS	Highly specific mass spectrometric profiles from 25 different fish species were obtained. This method is suitable for verifying commercial product authenticity and to rapidly discriminate species subjected to fraudulent substitutions, such as those belonging to Gadidae and Pleuronectiformes	Several reference fish species	Bony fish (Fish)	[112]
	Discrimination of freshwater fish species by MALDI-TOF MS: a pilot study	The study allowed discriminating three freshwater fish species using both muscle and liver tissues. The technology enables to analyze tissues after a simple single-step extraction procedure without any further purification	*Alosa agone* (The shad); *Coregonus macrophthalmus* (Whitefish); *Rutilus rutilus* (The roach)	Bony fish (Fish)	[118]
	Application of proteome analysis to seafood authentication	Identification of species and muscle tissues, namely white muscle from Arctic charr, Hammerfest strain, and skeletal muscle of an Arctic charr from a lake called Sila	*Salvelinus alpinus* (Arctic charr)	Bony fish (Fish)	[113]
	Identification of commercial hake and grenadier species by proteomics analysis of the parvalbumin fraction	Analysis of parvalbumin fractions through proteomics methodologies allowed a clearly individual classification of ten commercial, closely related species of the family Merlucciidae	*Merluccius* and *Macruronus* (Hake)	Bony fish (Fish)	[114]
	High-resolution 2-DE as a tool to differentiate wild from farmed *Gadus morhua* and to assess the protein composition of klipfish	The protein patterns of wild and farmed cod seemed to indicate that farmed cod muscle had a different protein expression and/or different *post mortem* degradation pattern than wild cod	*Gadus morhua* (Cod)	Bony fish (Fish)	[115]

(continued)

Table 10.3 (continued)

Parameters analyzed	Focus of the study	Main discovery	Aquaculture species (Common name)	Class and group	References
	Isoelectric focusing (IEF) and 2-DE were used to distinguish four freshwater fish species, which are sold under the generic label of "perch"	Native IEF of the water-soluble proteins extracted from the white muscle of the four species of fish sold under the generic name of "perch" serves to distinguish the species	*Perca fluviatilis* (European perch), *Lates niloticus* (Nile perch), *Stizostedion lucioperca* (European pikeperch) and *Morone chrysops x saxatilis* (Sunshine bass)	Bony fish (Fish)	[116]
	A shotgun proteomics approach was applied to the detection of previously characterized species-specific peptides from different seafood species	The study allowed the differential classification of seven commercial, closely related, species of Decapoda shrimps proving to be an excellent tool for seafood product authentication	Several commercially relevant shrimp species	Crustaceans (Arthropoda)	[117]
Production process	Modifications of *Oncorhynchus mykiss* muscle proteins by pre-slaughter activity	Persistent under-representation of desmin, a key cytoskeletal protein, in fish submitted to intense muscular activity suggests that a pre-slaughter treatment can have an effect on *post mortem* muscle integrity	*Oncorhynchus mykiss* (Rainbow trout)	Bony fish (Fish)	[54]
	Changes in muscle and blood plasma proteomes of *Salmo salar* induced by crowding	The changes observed in muscle and blood plasma proteomes not only correspond with the physiological stress response in fish, but also highlights the mechanisms causing an accelerated muscle pH decline and *rigor mortis* contraction in crowded salmon	*Salmo salar* (Atlantic salmon)	Bony fish (Fish)	[119]
	Metabolic molecular indicators of chronic stress in *Sparus aurata* using comparative proteomics	Proteins involved in lipid transport and antioxidant role, chaperoning, Ca^{2+} signaling, lipid oxidation, ammonia metabolism, cytoskeleton, hemoglobin and carbohydrate metabolism were differentially expressed in fish under chronic stress	*Sparus aurata* (Gilthead seabream)	Bony fish (Fish)	[52]

(continued)

Table 10.3 (continued)

Parameters analyzed	Focus of the study	Main discovery	Aquaculture species (Common name)	Class and group	References
	Comparison of muscle protein expression between wild and farmed (mariculture) sea bream	There were no differences between muscle protein expression of wild and farmed sea bream	*Sparus aurata* (Gilthead seabream)	Bony fish (Fish)	[69]
	Comparison of water-soluble muscle proteins from edible parts of wild and farmed sea bass	In aquaculture fish, the enzymes involved in carbohydrate metabolism were over-expressed, while the expression of the major fish allergen, parvalbumin, was reduced by 22 %. Aquaculture could induce significant chemical and biochemical differences in fish muscle that may have an impact on food quality	*Dicentrarchus labrax* (Sea bass)	Bony fish (Fish)	[120]
	Two-dimensional difference gel electrophoresis (2D DIGE) was applied to investigate the impact of three different slaughtering techniques on the *post mortem* integrity of muscle tissue proteins in *Dicentrarchus labrax*	The study showed that slaughtering by spinal cord severance preserves protein integrity better than death by asphyxia, either in ice or in air	*Dicentrarchus labrax* (European sea bass)	Bony fish (Fish)	[121]
	Proteomics signature of muscle atrophy in rainbow trout. Characterization of proteomics profile in degenerating muscle of rainbow trout in relation to the female reproductive cycle	The data will help to identify genes associated with muscle degeneration and superior flesh quality in rainbow trout, facilitating identification of genetic markers for muscle growth and quality	*Oncorhynchus mykiss* (Rainbow trout)	Bony fish (Fish)	[122]
Storage and processing	Proteomics, mainly 2-DE, has been used to characterize *post mortem* changes during cold storage in *Dicentrarchus labrax* muscle	The study allowed the classification of fish samples according to *post mortem* time. Three spots of interest, which disappeared progressively, were identified on the 2-DE patterns	*Dicentrarchus labrax* (Sea bass)	Bony fish (Fish)	[123]

(continued)

Table 10.3 (continued)

Parameters analyzed	Focus of the study	Main discovery	Aquaculture species (Common name)	Class and group	References
	Effects of *post mortem* storage temperature on *Dicentrarchus labrax* muscle protein degradation: Analysis by 2-D DIGE and MS	The study offer new knowledge on changes occurring in sea bass muscle proteins during *post mortem* storage at different temperatures and provide indications on protein degradation trends that might be useful for monitoring freshness of fish and quality of storage conditions	*Dicentrarchus labrax* (Sea bass)	Bony fish (Fish)	[124]
	Changes in cod muscle proteins during frozen storage revealed by proteome analysis and multivariate data analysis	Application of proteomics, multivariate data analysis and MS/MS to analyse protein changes in cod muscle proteins during storage has revealed new knowledge on the issue and enables a better understanding of biochemical processes occurring	*Gadus morhua* (Cod)	Bony fish (Fish)	[133]
	Effect of HHP treatment on proteins, lipids and nucleotides in chilled farmed *Oncorhynchus kisutch* muscle	HHP treatment led to an increased free fatty acid (FFA) formation (day 0 values); on the contrary, an inhibitory effect on FFA formation could be observed at the end of the storage (15–20 days) in T-3 treated fish as a result of microbial activity inhibition	*Oncorhynchus kisutch* (Coho salmon)	Bony fish (Fish)	[128]
	Two-dimensional electrophoretic analyses of *Gadus morhua* whole muscle proteins, water-soluble fraction and surimi. Effect of the addition of $CaCl_2$ and $MgCl_2$ during the washing procedure	The study showed that the main myofibrillar proteins, including myosin, actin and tropomyosin, remained in the surimi. Several other proteins were selectively removed during the washing procedure. 2-DE has proved to be a valuable tool to quickly and easily assess the effect of different processing conditions on the protein content of the products	*Gadus morhua* (Cod)	Bony fish (Fish)	[129]
					(continued)

Table 10.3 (continued)

Parameters analyzed	Focus of the study	Main discovery	Aquaculture species (Common name)	Class and group	References
	Use of 2-DE to evaluate proteolysis in *Salmo salar* muscle affected by a lactic fermentation	Protein fragments appeared increasingly with time in both samples (salmon fillets inoculated or not with the starter culture *Lactobacillus* sake LAD), indicating that the main quantitative changes were due to endogenous enzymes. In contrast, fermentation had a significant effect in the pH range 6.20–8.35, suggesting a specificity of the bacterial proteases of *L. sake* toward alkaline to slightly acidic proteins	*Salmo salar* (Atlantic salmon)	Bony fish (Fish)	[130]
	Peptides in *Oncorhynchus mykiss* muscle subjected to ice storage and cooking	MS analysis revealed a limited but highly reproducible appearance of small peptides in trout muscle during the ice storage and after cooking	*Oncorhynchus mykiss* (Rainbow trout)	Bony fish (Fish)	[127]
	Post mortem muscle protein degradation during ice-storage of Arctic (*Pandalus borealis*) and tropical (*Penaeus japonicus* and *Penaeus monodon*) shrimps: A comparative electrophoretic and immunological study	The *post mortem* protein degradation during ice-storage seemed to be somehow different in the Arctic than in the tropical species: while in the Arctic species important protein degradation suggests active muscle proteases, proteolytic degradation in the tropical species seemed to be of much less relevance and probably inhibited by the greater difference between the normal living environmental (28 °C) and storage (0 °C) temperatures	*Pandalus borealis* (Arctic shrimp) *Penaeus japonicus* and *Penaeus monodon* (Tropical shrimps)	Crustaceans (Arthropoda)	[126]
Sensorial and nutritional quality	Effect of antioxidants, citrate, and cryoprotectants on protein denaturation and texture of frozen *Gadus morhua*	Ice crystal formation and lipid oxidation products are the major factors that cause protein denaturation in lean frozen fish, and antioxidants in addition to cryoprotectants can be used to minimize toughness	*Gadus morhua* (Cod)	Bony fish (Fish)	[131]
	Proteomics studies on protein oxidation in *Katsuwonus pelamis* muscle during storage	The study suggested that the oxidative deterioration of proteins in bonito muscle had proceeded during a 4-day storage period	*Katsuwonus pelamis* (Bonito)	Bony fish (Fish)	[132]

(continued)

Table 10.3 (continued)

Parameters analyzed	Focus of the study	Main discovery	Aquaculture species (Common name)	Class and group	References
	2-DE detection of protein oxidation in fresh and tainted rainbow trout muscle	The study gives an estimate of the level of protein carbonylation in rainbow trout and reveals that oxidation increases for a distinct number of proteins during tainting	*Oncorhynchus mykiss* (Rainbow trout)	Bony fish (Fish)	[133]
	Identification of carbonylated protein in *Oncorhynchus mykiss* fillets and development of protein oxidation during frozen storage	Low-abundant proteins could be relatively more carbonylated than high-abundant proteins, thereby indicating that some proteins are more susceptible to oxidation than others, due to their cellular localization, amino acid sequence or biochemical function	*Oncorhynchus mykiss* (Rainbow trout)	Bony fish (Fish)	[125]
Functionality	Proteomics analysis of processing by-products from canned and fresh tuna: Identification of potentially functional food proteins	Application of proteomics technologies has revealed new knowledge on the composition of important by-products from tuna species, enabling a better evaluation of their potential applications	*Thunnus alalunga* (Fresh tuna)	Bony fish (Fish)	[135]

integrity and factors such as stress, physical activity, crowding, growing and slaughtering conditions and life cycle were already studied by using proteomic approaches [52, 54, 69, 119–122]. On the other hand, the processes occurring during *post mortem* metabolism also produce several changes in seafood freshness and quality. Proteomics has been used to characterize *post mortem* changes during cold storage [123–126] and processing (either by cooking, food preservation treatment, addition of Calcium and Magnesium or fermentation) [127–130]. Additionally, oxidation modifications (e.g. carbonylation, thiol oxidation and aromatic hydroxylation and Maillard glycation) during storage and processing treatments significantly affect the sensorial and nutritional quality of fishery products, thus proteomics studies on the protein oxidation have also been performed [131–134]. Finally, recently proteomics techniques have also been applied to investigate functional food proteins of fish [135]. Although proteomics has only been sparingly applied on this field, it can be of great interest for potential biotechnological applications.

According to the above studies, the increasingly application of proteomics may provide valuable information regarding safety and quality of aquaculture products, enabling the evaluation of benefits and risks to consumer and industrial process improvements. In the future, the optimization and development of more cost-effective and sensitive proteomic technologies may allow its routine use.

10.7 Final Remarks

This book chapter gives a brief overview of the use of proteomics in aquaculture nowadays and illustrates some of the main constrains in the aquaculture industry and how science and state of the art technologies, in particular proteomics can be considered a turning point in terms of addressing issues like welfare, nutrition, safety and quality of farmed fish.

There are nevertheless challenges inherent to a fast growing industry like aquaculture that need to be addressed. Limitations in the technology both at the level of gene annotation for most fish species and its integration with other Omics technologies like metabolomics or transcriptomics, tend to postpone some of the knowledge and answers to problems like the establishment of protein biomarkers for fish welfare, disease or condition. This is a common problem to other areas of research using proteomics and we believe that the technological advances and the awareness for more and easier financing opportunities in this area will push this industry into the next level in a near future.

Another main challenge where proteomics can play an important role is a target of the Blue Growth framework by FAO that promotes the sustainability of all the aquaculture production process focused both on the technology and the end product; the farmed fishery products. As a response to this challenge and a clear adaptation effort, new emerging systems like Aquaponics where conventional aquaculture and hydroponics live in a symbiotic environment, has been developed.

Also, the world population growing demand for fish as part of its daily diet mainly due to the emerging countries raising economy and the global awareness for a healthier and "cleaner" protein, leaves no option to aquaculture but a major effort mainly based on solid scientific knowledge to overcome these issues.

Acknowledgments D. Schrama work is supported by grant Ref[a] 31-03-05-FEP-0060 from Promar - Projetos Pilotos e a Transformação de Embarcações de Pesca.

A. Campos work is supported by postdoctoral grant (SFRH/BPD/103683/2014) from FCT.

The Institute of Molecular Pathology and Immunology of the University of Porto integrates the Institute for Research and Innovation in Health, which is partially supported by the Portuguese Foundation for Science and Technology (FCT). This work is funded by the European Regional Development Fund (FEDER) through the Operational Program for Competitiveness Factors (COMPETE) and by national funds through the FCT, under the project PEst-C/SAU/LA0003/2013.

References

1. FAO (2015) Global aquaculture production statistics database updated to 2013. Summary information 2015
2. Rodrigues PM, Silva TS, Dias J, Jessen F (2012) PROTEOMICS in aquaculture: applications and trends. J Proteomics 75:4325–4345. doi:10.1016/j.jprot.2012.03.042
3. FAO (2008) World fisheries and aquaculture
4. Mukhopadhyay R (2009) DNA sequencers: the next generation. Anal Chem 81:1736–1740. doi:10.1021/ac802712u
5. FAO (2014) The state of world fisheries and aquaculture 2014
6. Hilliou L (2014) Hybrid carrageenans: isolation, chemical structure, and gel properties. Adv Food Nutr Res 72:17–43. doi:10.1016/B978-0-12-800269-8.00002-6
7. Gladyshev MI, Sushchik NN, Makhutova ON (2013) Production of EPA and DHA in aquatic ecosystems and their transfer to the land. Prostaglandins Other Lipid Mediat 107:117–126. doi:10.1016/j.prostaglandins.2013.03.002
8. Priyadarshani I, Rath B (2012) Commercial and industrial applications of micro algae—a review. J Algal Biomass Utln 3:89–100
9. Meena DK, Das P, Kumar S et al (2013) Beta-glucan: an ideal immunostimulant in aquaculture (a review). Fish Physiol Biochem 39:431–457. doi:10.1007/s10695-012-9710-5
10. Caipang CMA, Lazado CC, Berg I et al (2011) Influence of alginic acid and fucoidan on the immune responses of head kidney leukocytes in cod. Fish Physiol Biochem 37:603–612
11. Hutson KS, Mata L, Paul NA, de Nys R (2012) Seaweed extracts as a natural control against the monogenean ectoparasite, *Neobenedenia* sp., infecting farmed barramundi (*Lates calcarifer*). Int J Parasitol 42:1135–1141. doi:10.1016/j.ijpara.2012.09.007
12. Long M, Zhao J, Li T et al (2015) Transcriptomic and proteomic analyses of splenic immune mechanisms of rainbow trout (*Oncorhynchus mykiss*) infected by *Aeromonas salmonicida* subsp. *salmonicida*. J Proteomics 122:41–54. doi:10.1016/j.jprot.2015.03.031
13. Nynca J, Arnold GJ, Fröhlich T et al (2014) Proteomic identification of rainbow trout sperm proteins. Proteomics 14:1569–1573. doi:10.1002/pmic.201300521
14. Campos A, Apraiz I, da Fonseca RR, Cristobal S (2015) Shotgun analysis of the marine mussel *Mytilus edulis* haemolymph proteome and mapping the innate immunity elements. Proteomics 15:4021–4029
15. Talakhun W, Phaonakrop N, Roytrakul S et al (2014) Proteomic analysis of ovarian proteins and characterization of thymosin-β and RAC-GTPase activating protein 1 of the giant tiger

shrimp *Penaeus monodon*. Comp Biochem Physiol Part D Genomics Proteomics 11:9–19. doi:10.1016/j.cbd.2014.05.002
16. De Santis C, Crampton VO, Bicskei B, Tocher DR (2015) Replacement of dietary soy- with air classified faba bean protein concentrate alters the hepatic transcriptome in Atlantic salmon (*Salmo salar*) parr. Comp Biochem Physiol Part D Genomics Proteomics 16:48–58. doi:10.1016/j.cbd.2015.07.005
17. Xue X, Hixson SM, Hori TS et al (2015) Atlantic salmon (*Salmo salar*) liver transcriptome response to diets containing *Camelina sativa* products. Comp Biochem Physiol Part D Genomics Proteomics 14:1–15. doi:10.1016/j.cbd.2015.01.005
18. Braceland M, Bickerdike R, Tinsley J et al (2013) The serum proteome of Atlantic salmon, *Salmo salar*, during pancreas disease (PD) following infection with salmonid alphavirus subtype 3 (SAV3). J Proteomics 94:423–436
19. Provan F, Jensen LB, Uleberg KE et al (2013) Proteomic analysis of epidermal mucus from sea lice-infected Atlantic salmon, *Salmo salar* L. J Fish Dis 36:311–321. doi:10.1111/jfd.12064
20. Klinbunga S, Petkorn S, Kittisenachai S et al (2012) Identification of reproduction-related proteins and characterization of proteasome alpha 3 and proteasome beta 6 cDNAs in testes of the giant tiger shrimp *Penaeus monodon*. Mol Cell Endocrinol 355:143–152. doi:10.1016/j.mce.2012.02.005
21. Dietrich MA, Arnold GJ, Fröhlich T et al (2015) Proteomic analysis of extracellular medium of cryopreserved carp (*Cyprinus carpio* L.) semen. Comp Biochem Physiol Part D Genomics Proteomics 15:49–57. doi:10.1016/j.cbd.2015.05.003
22. Dietrich MA, Arnold GJ, Fröhlich T, Ciereszko A (2014) In-depth proteomic analysis of carp (*Cyprinus carpio* L.) spermatozoa. Comp Biochem Physiol Part D Genomics Proteomics 12:10–15. doi:10.1016/j.cbd.2014.09.003
23. Breton TS, Berlinsky DL (2014) Characterizing ovarian gene expression during oocyte growth in Atlantic cod (*Gadus morhua*). Comp Biochem Physiol Part D Genomics Proteomics 9:1–10
24. Timmins-Schiffman E, Coffey WD, Hua W et al (2014) Shotgun proteomics reveals physiological response to ocean acidification in *Crassostrea gigas*. BMC Genom 15:951. doi:10.1186/1471-2164-15-951
25. McLean L, Young IS, Doherty MK et al (2007) Global cooling: cold acclimation and the expression of soluble proteins in carp skeletal muscle. Proteomics 7:2667–2681. doi:10.1002/pmic.200601004
26. Stewart MJ, Favrel P, Rotgans BA et al (2014) Neuropeptides encoded by the genomes of the Akoya pearl oyster *Pinctata fucata* and Pacific oyster *Crassostrea gigas*: a bioinformatic and peptidomic survey. BMC Genom 15:840. doi:10.1186/1471-2164-15-840
27. Bigot L, Beets I, Dubos M-P et al (2014) Functional characterization of a short neuropeptide F-related receptor in a lophotrochozoan, the mollusk *Crassostrea gigas*. J Exp Biol 217:2974–2982
28. Marie B, Zanella-Cléon I, Guichard N et al (2011) Novel proteins from the calcifying shell matrix of the Pacific oyster *Crassostrea gigas*. Mar Biotechnol (NY) 13:1159–1168. doi:10.1007/s10126-011-9379-2
29. Schultz IR, Nagler JJ, Swanson P et al (2013) Toxicokinetic, toxicodynamic, and toxicoproteomic aspects of short-term exposure to trenbolone in female fish. Toxicol Sci 136:413–429. doi:10.1093/toxsci/kft220
30. Li Z-H, Li P, Sulc M et al (2012) Hepatic proteome sensitivity in rainbow trout after chronically exposed to a human pharmaceutical verapamil. Mol Cell Proteomics 11: M111.008409–M111.008409. doi:10.1074/mcp.M111.008409
31. Hampel M, Alonso E, Aparicio I et al (2015) Hepatic proteome analysis of Atlantic Salmon (*Salmo salar*) after exposure to environmental concentrations of human pharmaceuticals. Mol Cell Proteomics 14:371–381. doi:10.1074/mcp.M114.045120

32. Eyckmans M, Benoot D, Van Raemdonck GAA et al (2012) Comparative proteomics of copper exposure and toxicity in rainbow trout, common carp and gibel carp. Comp Biochem Physiol—Part D Genomics Proteomics 7:220–232. doi:10.1016/j.cbd.2012.03.001
33. Karlsen OA, Bjørneklett S, Berg K et al (2011) Integrative environmental genomics of Cod (*Gadus morhua*): the proteomics approach. J Toxicol Environ Health A 74:494–507. doi:10.1080/15287394.2011.550559
34. Martins J, Campos A, Osório H et al (2014) Proteomic profiling of cytosolic glutathione transferases from three bivalve species: *Corbicula fluminea, Mytilus galloprovincialis* and *Anodonta cygnea*. Int J Mol Sci 15:1887–1900. doi:10.3390/ijms15021887
35. Cordero H, Brinchmann MF, Cuesta A et al (2015) Skin mucus proteome map of European sea bass (*Dicentrarchus labrax*). Proteomics 15:4007–4020
36. Dietrich MA, Arnold GJ, Nynca J et al (2014) Characterization of carp seminal plasma proteome in relation to blood plasma. J Proteomics 98:218–232. doi:10.1016/j.jprot.2014.01.005
37. Zhang P, Li C, Zhang P et al (2014) iTRAQ-based proteomics reveals novel members involved in pathogen challenge in sea cucumber *Apostichopus japonicus*. PLoS ONE 9:e100492. doi:10.1371/journal.pone.0100492
38. Piovesana S, Capriotti AL, Caruso G et al (2015) Labeling and label free shotgun proteomics approaches to characterize muscle tissue from farmed and wild gilthead sea bream (*Sparus aurata*). J Chromatogr A. doi:10.1016/j.chroma.2015.07.049
39. Kültz D, Li J, Zhang X et al (2015) Population-specific plasma proteomes of marine and freshwater three-spined sticklebacks (*Gasterosteus aculeatus*). Proteomics. doi:10.1002/pmic.201500132
40. Buján N, Hernández-Haro C, Monteoliva L et al (2015) Comparative proteomic study of *Edwardsiella tarda* strains with different degrees of virulence. J Proteomics 127:310–320
41. Niksirat H, James P, Andersson L et al (2015) Label-free protein quantification in freshly ejaculated versus post-mating spermatophores of the noble crayfish *Astacus astacus*. J Proteomics 123:70–77. doi:10.1016/j.jprot.2015.04.004
42. Nynca J, Arnold GJ, Fröhlich T, Ciereszko A (2015) Cryopreservation-induced alterations in protein composition of rainbow trout semen. Proteomics 15:2643–2654. doi:10.1002/pmic.201400525
43. Groh KJ, Schönenberger R, Eggen RIL et al (2013) Analysis of protein expression in zebrafish during gonad differentiation by targeted proteomics. Gen Comp Endocrinol 193:210–220. doi:10.1016/j.ygcen.2013.07.020
44. Jin C, Padra JT, Sundell K et al (2015) Atlantic salmon carries a range of novel O-Glycan structures differentially localized on skin and intestinal mucins. J Proteome Res 14:3239–3251. doi:10.1021/acs.jproteome.5b00232
45. Broom DM (2010) Animal welfare: an aspect of care, sustainability, and food quality required by the public. J Vet Med Educ 37:83–88
46. Southgate P, Wall T (2001) Welfare of farmed fish at slaughter. In Pract 23:277–284. doi:10.1136/inpract.23.5.277
47. Fish HOW, To R, Stressors N (2002) Fisheries Society of the British isles briefing paper 2, table of contents. Granta 44
48. Barton BA (2002) Stress in fishes: a diversity of responses with particular reference to changes in circulating corticosteroids. Integr Comp Biol 42:517–525
49. Conte FS (2004) Stress and the welfare of cultured fish. Appl Anim Behav Sci 86:205–223
50. Ashley PJ (2007) Fish welfare: current issues in aquaculture. Appl Anim Behav Sci 104:199–235
51. Provan F, Bjornstad A, Pampanin DM et al (2006) Mass spectrometric profiling—a diagnostic tool in fish? Mar Environ Res 62:S105–S108. doi:10.1016/j.marenvres.2006.04.002 S0141-1136(06)00046-8 [pii]
52. Alves RN, Cordeiro O, Silva TS et al (2010) Metabolic molecular indicators of chronic stress in gilthead seabream (*Sparus aurata*) using comparative proteomics. Aquaculture 299:57–66

53. Cordeiro OD, Silva TS, Alves RN et al (2012) Changes in liver proteome expression of senegalese sole (*Solea senegalensis*) in response to repeated handling stress. Mar Biotechnol 14:714–729
54. Morzel M, Chambon C, Lefevre F et al (2006) Modifications of trout (*Oncorhynchus mykiss*) muscle proteins by preslaughter activity. J Agric Food Chem 54:2997–3001. doi:10.1021/jf0528759
55. Silva TS, Cordeiro OD, Matos ED et al (2012) Effects of preslaughter stress levels on the post-mortem sarcoplasmic proteomic profile of gilthead seabream muscle. J Agric Food Chem 60:9443–9453. doi:10.1021/jf301766e
56. Jiang H, Li F, Xie Y et al (2009) Comparative proteomic profiles of the hepatopancraes in *Fenneropenaeus chinensis* response to hypoxic stress. Proteomics 9:3353–3367. doi:10.1002/pmic.200800518
57. Sun L, Liu S, Bao L et al (2015) Claudin multigene family in channel catfish and their expression profiles in response to bacterial infection and hypoxia as revealed by meta-analysis of RNA-Seq datasets. Comp Biochem Physiol Part D Genomics Proteomics 13:60–69. doi:10.1016/j.cbd.2015.01.002
58. Wulff T, Hoffmann EK, Roepstorff P et al (2008) Comparison of two anoxia models in rainbow trout cells by a 2-DE and MS/MS-based proteome approach. Proteomics 8:2035–2044. doi:10.1002/pmic.200700944
59. Smith RW, Cash P, Ellefsen S, Nilsson GE (2009) Proteomic changes in the crucian carp brain during exposure to anoxia. Proteomics 9:2217–2229. doi:10.1002/pmic.200800662
60. Mendelsohn BA, Malone JP, Townsend RR, Gitlin JD (2009) Proteomic analysis of anoxia tolerance in the developing zebrafish embryo. Comp Biochem Physiol D-Genomics Proteomics 4:21–31
61. Salas-Leiton E, Canovas-Conesa B, Zerolo R et al (2009) Proteomics of juvenile senegal sole (*Solea senegalensis*) affected by gas bubble disease in hyperoxygenated ponds. Mar Biotechnol 11:473–487. doi:10.1007/s10126-008-9168-8
62. Lee J, Valkova N, White MP, Kultz D (2006) Proteomic identification of processes and pathways characteristic of osmoregulatory tissues in spiny dogfish shark (*Squalus acanthias*). Comp Biochem Physiol Part D Genomics Proteomics 1:328–343. doi:10.1016/j.cbd.2006.07.001 S1744-117X(06)00073-6 [pii]
63. Dowd WW, Harris BN, Cech Jr. JJ, Kultz D (2010) Proteomic and physiological responses of leopard sharks (*Triakis semifasciata*) to salinity change. J Exp Biol 213:210–224. doi:10.1242/jeb.031781 213/2/210 [pii]
64. Kultz D, Fiol D, Valkova N et al (2007) Functional genomics and proteomics of the cellular osmotic stress response in "non-model" organisms. J Exp Biol 210:1593–1601. doi:10.1242/jeb.000141 210/9/1593 [pii]
65. Ky CL, de Lorgeril J, Hirtz C et al (2007) The effect of environmental salinity on the proteome of the sea bass (*Dicentrarchus labrax* L.). Anim Genet 38:601–608. doi:10.1111/j.1365-2052.2007.01652.x AGE1652 [pii]
66. Lu XJ, Chen J, Huang ZA et al (2010) Proteomic analysis on the alteration of protein expression in gills of ayu (*Plecoglossus altivelis*) associated with salinity change. Comp Biochem Physiol Part D Genomics Proteomics 5:185–189. doi:10.1016/j.cbd.2010.03.002 S1744-117X(10)00024-9 [pii]
67. Papakostas S, Vasemägi A, Himberg M, Primmer CR (2014) Proteome variance differences within populations of European whitefish (*Coregonus lavaretus*) originating from contrasting salinity environments. J Proteomics 105:144–150. doi:10.1016/j.jprot.2013.12.019
68. Ibarz A, Martin-Perez M, Blasco J et al (2010) Gilthead sea bream liver proteome altered at low temperatures by oxidative stress. Proteomics 10:963–975. doi:10.1002/pmic.200900528
69. Addis MF, Cappuccinelli R, Tedde V et al (2010) Proteomic analysis of muscle tissue from gilthead sea bream (*Sparus aurata*, L.) farmed in offshore floating cages. Aquaculture 309:245–252

70. Bosworth CA IV, Chou CW, Cole RB, Rees BB (2005) Protein expression patterns in zebrafish skeletal muscle: initial characterization and the effects of hypoxic exposure. Proteomics 5:1362–1371
71. Bohne-Kjersem A, Skadsheim A, Goksoyr A, Grosvik BE (2009) Candidate biomarker discovery in plasma of juvenile cod (*Gadus morhua*) exposed to crude North Sea oil, alkyl phenols and polycyclic aromatic hydrocarbons (PAHs). Mar Environ Res 68:268–277
72. Russell S, Hayes MA, Simko E, Lumsden JS (2006) Plasma proteomic analysis of the acute phase response of rainbow trout (*Oncorhynchus mykiss*) to intraperitoneal inflammation and LPS injection. Dev Comp Immunol 30:393–406. doi:10.1016/j.dci.2005.06.002 S0145-305X (05)00117-5 [pii]
73. Brunt J, Hansen R, Jamieson DJ, Austin B (2008) Proteomic analysis of rainbow trout (*Oncorhynchus mykiss*, Walbaum) serum after administration of probiotics in diets. Vet Immunol Immunopathol 121:199–205
74. Ghisaura S, Anedda R, Pagnozzi D et al (2014) Impact of three commercial feed formulations on farmed gilthead sea bream (*Sparus aurata*, L.) metabolism as inferred from liver and blood serum proteomics. Proteome Sci 12:44. doi:10.1186/s12953-014-0044-3
75. Panserat S, Kaushik SJ (2010) Regulation of gene expression by nutritional factors in fish. Aquac Res 41:751–762
76. Martin SAM, Vilhelmsson O, Medale F et al (2003) Proteomic sensitivity to dietary manipulations in rainbow trout. Biochim Biophys Acta-Proteins Proteomics 1651:17–29
77. Vilhelmsson OT, Martin SAM, Medale F et al (2004) Dietary plant-protein substitution affects hepatic metabolism in rainbow trout (*Oncorhynchus mykiss*). Br J Nutr 92:71–80
78. Sissener NH, Martin SA, Cash P et al (2010) Proteomic profiling of liver from Atlantic salmon (*Salmo salar*) fed genetically modified soy compared to the near-isogenic non-GM line. Mar Biotechnol 12:273–281. doi:10.1007/s10126-009-9214-1
79. Keyvanshokooh S, Tahmasebi-Kohyani A (2012) Proteome modifications of fingerling rainbow trout (*Oncorhynchus mykiss*) muscle as an effect of dietary nucleotides. Aquaculture 324–325:79–84. doi:10.1016/j.aquaculture.2011.10.013
80. Silva TS, da Costa AMR, Conceição LEC et al (2014) Metabolic fingerprinting of gilthead seabream (*Sparus aurata*) liver to track interactions between dietary factors and seasonal temperature variations. Peer J 2:e527. doi:10.7717/peerj.527
81. Pineiro C, Barros-Velazquez J, Vazquez J et al (2003) Proteomics as a tool for the investigation of seafood and other marine products. J Proteome Res 2:127–135
82. Zhou X, Ding Y, Wang Y (2012) Proteomics: present and future in fish, shellfish and seafood. Rev Aquac 4:11–20. doi:10.1111/j.1753-5131.2012.01058.x
83. Tedesco S, Mullen W, Cristobal S et al (2014) High-throughput proteomics: a new tool for quality and safety in fishery products. Curr Protein Peptide Sci 15:118–133
84. Carrera M, Cañas B, Gallardo JM (2013) Proteomics for the assessment of quality and safety of fishery products. Food Res Int 54:972–979
85. Puerto M, Campos A, Prieto A et al (2011) Differential protein expression in two bivalve species; *Mytilus galloprovincialis* and *Corbicula fluminea*; exposed to *Cylindrospermopsis raciborskii* cells. Aquat Toxicol 101:109–116. doi:10.1016/j.aquatox.2010.09.009
86. Ronzitti G, Milandri A, Scortichini G et al (2008) Protein markers of algal toxin contamination in shellfish. Toxicon 52:705–713. doi:10.1016/j.toxicon.2008.08.007
87. Nzoughet KJ, Hamilton JTG, Floyd SD et al (2008) Azaspiracid: first evidence of protein binding in shellfish. Toxicon 51:1255–1263. doi:10.1016/j.toxicon.2008.02.016
88. Chan LL, Hodgkiss IJ, Lam PKS et al (2005) Use of two-dimensional gel electrophoresis to differentiate morphospecies of *Alexandrium minutum*, a paralytic shellfish poisoning toxin-producing dinoflagellate of harmful algal blooms. Proteomics 5:1580–1593
89. Chan LL, Sit W-H, Lam PK-S et al (2006) Identification and characterization of a "biomarker of toxicity" from the proteome of the paralytic shellfish toxin-producing dinoflagellate *Alexandrium tamarense* (Dinophyceae). Proteomics 6:654–666

90. Shepard JL, Olsson B, Tedengren M, Bradley BP (2000) Protein expression signatures indented in *Mytilus edulis* exposed to PCBs, copper and salinity stress. Mar Environ Res 50:337–340
91. Rodríguez-Ortega MJ, Grøsvik BE, Rodríguez-Ariza A et al (2003) Changes in protein expression profiles in bivalve molluscs (*Chamaelea gallina*) exposed to four model environmental pollutants. Proteomics 3:1535–1543. doi:10.1002/pmic.200300491
92. Chora S, Starita-Geribaldi M, Guigonis J-M et al (2009) Effect of cadmium in the clam *Ruditapes decussatus* assessed by proteomic analysis. Aquat Toxicol 94:300–308
93. Silvestre F, Huynh TT, Bernard A et al (2010) A differential proteomic approach to assess the effects of chemotherapeutics and production management strategy on giant tiger shrimp *Penaeus monodon*. Comp Biochem Physiol Part D Genomics Proteomics 5:227–233
94. Hazen TH, Martinez RJ, Chen Y et al (2009) Rapid identification of *Vibrio parahaemolyticus* by whole-cell matrix-assisted laser desorption ionization-time of flight mass spectrometry. Appl Environ Microbiol 75:6745–6756. doi:10.1128/AEM.01171-09
95. Cordwell SJ, Len ACL, Touma RG et al (2008) Identification of membrane-associated proteins from *Campylobacter jejuni* strains using complementary proteomics technologies. Proteomics 8:122–139
96. Fernández-No IC, Böhme K, Calo-Mata P et al (2012) Isolation and characterization of *Streptococcus parauberis* from vacuum-packaging refrigerated seafood products. Food Microbiol 30:91–97. doi:10.1016/j.fm.2011.10.012
97. Böhme K, Fernandez-No IC, Gallardo JM et al (2011) Safety assessment of fresh and processed seafood products by MALDI-TOF mass fingerprinting. Food Bioprocess Technol 4:907–918
98. Böhme K, Fernández-No IC, Barros-Velázquez J et al (2010) Species differentiation of seafood spoilage and pathogenic gram-negative bacteria by MALDI-TOF mass fingerprinting. J Proteome Res 9:3169–3183
99. Martínez-Gomariz M, Hernáez ML, Gutiérrez D et al (2009) Proteomic analysis by two-dimensional differential gel electrophoresis (2D DIGE) of a high-pressure effect in *Bacillus cereus*. J Agric Food Chem 57:3543–3549. doi:10.1021/jf803272a
100. Booth NJ, Bilodeau-Bourgeois AL (2009) Proteomic analysis of head kidney tissue from high and low susceptibility families of channel catfish following challenge with *Edwardsiella ictaluri*. Fish Shellfish Immunol 26:193–196
101. Wang H-C, Wang H-C, Leu J-H et al (2007) Protein expression profiling of the shrimp cellular response to white spot syndrome virus infection. Dev Comp Immunol 31:672–686. doi:10.1016/j.dci.2006.11.001
102. Chai Y-M, Yu S-S, Zhao X-F et al (2010) Comparative proteomic profiles of the hepatopancreas in *Fenneropenaeus chinensis* response to white spot syndrome virus. Fish Shellfish Immunol 29:480–486
103. Bourchookarn A, Havanapan PO, Thongboonkerd V, Krittanai C (2008) Proteomic analysis of altered proteins in lymphoid organ of yellow head virus infected *Penaeus monodon*. Biochim Biophys Acta 1784:504–511
104. Chongsatja PO, Bourchookarn A, Lo CF et al (2007) Proteomic analysis of differentially expressed proteins in *Penaeus vannamei* hemocytes upon Taura syndrome virus infection. Proteomics 7:3592–3601
105. Booy AT, Haddow JD, Ohlund LB et al (2005) Application of isotope coded affinity tag (ICAT) analysis for the identification of differentially expressed proteins following infection of atlantic salmon (*Salmo salar*) with infectious hematopoietic necrosis virus (IHNV) or *Renibacterium salmoninarum*. J Proteome Res 4:325–334
106. Cao A, Fuentes J, Comesaña P et al (2009) A proteomic approach envisaged to analyse the bases of oyster tolerance/resistance to bonamiosis. Aquaculture 295:149–156
107. Carrera M, Cañas B, Gallardo JM (2012) Rapid direct detection of the major fish allergen, parvalbumin, by selected MS/MS ion monitoring mass spectrometry. J Proteomics 75:3211–3220

108. Abdel Rahman AM, Rahman AMA, Kamath S et al (2010) Analysis of the allergenic proteins in black tiger prawn (*Penaeus monodon*) and characterization of the major allergen tropomyosin using mass spectrometry. Rapid Commun Mass Spectrom 24:2462–2470
109. Yu C-J, Lin Y-F, Chiang B-L, Chow L-P (2003) Proteomics and immunological analysis of a novel shrimp allergen, Pen m 2. J Immunol 170:445–453. doi:10.4049/jimmunol.170.1.445
110. Nakamura R, Satoh R, Nakajima Y et al (2009) Comparative study of GH-transgenic and non-transgenic amago salmon (*Oncorhynchus masou ishikawae*) allergenicity and proteomic analysis of amago salmon allergens. Regul Toxicol Pharmacol 55:300–308
111. Piñeiro C, Vázquez J, Marina AI et al (2001) Characterization and partial sequencing of species-specific sarcoplasmic polypeptides from commercial hake species by mass spectrometry following two-dimensional electrophoresis. Electrophoresis 22:1545–1552. doi:10.1002/1522-2683(200105)22:8<1545:AID-ELPS1545>3.0.CO;2-5
112. Mazzeo MF, De Giulio B, Guerriero G et al (2008) Fish authentication by MALDI-TOF mass spectrometry. J Agric Food Chem 56:11071–11076
113. Martinez I, Jakobsen Friis T (2004) Application of proteome analysis to seafood authentication. Proteomics 4:347–354. doi:10.1002/pmic.200300569
114. Carrera M, Cañas B, Piñeiro C et al (2006) Identification of commercial hake and grenadier species by proteomic analysis of the parvalbumin fraction. Proteomics 6:5278–5287
115. Martinez I, Slizyte R, Dauksas E (2007) High resolution two-dimensional electrophoresis as a tool to differentiate wild from farmed cod (*Gadus morhua*) and to assess the protein composition of klipfish. Food Chem 102:504–510. doi:10.1016/j.foodchem.2006.03.037
116. Berrini A, Tepedino V, Borromeo V, Secchi C (2006) Identification of freshwater fish commercially labelled "perch" by isoelectric focusing and two-dimensional electrophoresis. Food Chem 96:163–168
117. Ortea I, Canas B, Gallardo JM (2011) Selected tandem mass spectrometry ion monitoring for the fast identification of seafood species. J Chromatogr A 1218:4445–4451
118. Volta P, Riccardi N, Lauceri R, Tonolla M (2012) Discrimination of freshwater fish species by matrix-assisted laser desorption/ionization—time of flight mass spectrometry (MALDI-TOF MS): a pilot study. J Limnol 71:17. doi:10.4081/jlimnol.2012.e17
119. Veiseth-Kent E, Grove H, Faergestad EM, Fjaera SO (2010) Changes in muscle and blood plasma proteomes of Atlantic salmon (*Salmo salar*) induced by crowding. Aquaculture 309:272–279
120. Monti G, De Napoli L, Mainolfi P et al (2005) Monitoring food quality by microfluidic electrophoresis, gas chromatography, and mass spectrometry techniques: effects of aquaculture on the sea bass (*Dicentrarchus labrax*). Anal Chem 77:2587–2594. doi:10.1021/ac048337x
121. Addis MF, Pisanu S, Preziosa E et al (2012) 2D DIGE/MS to investigate the impact of slaughtering techniques on postmortem integrity of fish filet proteins. J Proteomics 75:3654–3664
122. Salem M, Kenney PB, Rexroad CE, Yao J (2010) Proteomic signature of muscle atrophy in rainbow trout. J Proteomics 73:778–789. doi:10.1016/j.jprot.2009.10.014
123. Verrez-Bagnis V, Ladrat C, Morzel M et al (2001) Protein changes in post mortem sea bass (*Dicentrarchus labrax*) muscle monitored by one- and two-dimensional gel electrophoresis. Electrophoresis 22:1539–1544
124. Terova G, Addis MF, Preziosa E et al (2011) Effects of postmortem storage temperature on sea bass (*Dicentrarchus labrax*) muscle protein degradation: analysis by 2-D DIGE and MS. Proteomics 11:2901–2910
125. Kjaersgard IVH, Norrelykke MR, Baron CP, Jessen F (2006) Identification of carbonylated protein in frozen rainbow trout (*Oncorhynchus mykiss*) fillets and development of protein oxidation during frozen storage. J Agric Food Chem 54:9437–9446
126. Martinez I, Jakobsen Friis T, Careche M (2001) Post mortem muscle protein degradation during ice-storage of Arctic (*Pandalus borealis*) and tropical (*Penaeus japonicus* and *Penaeus monodon*) shrimps: a comparative electrophoretic and immunological study. J Sci Food Agric 81:1199–1208. doi:10.1002/jsfa.931

127. Bauchart C, Chambon C, Mirand PP et al (2007) Peptides in rainbow trout (*Oncorhynchus mykiss*) muscle subjected to ice storage and cooking. Food Chem 100:1566–1572
128. Ortea I, Rodríguez A, Tabilo-Munizaga G et al (2010) Effect of hydrostatic high-pressure treatment on proteins, lipids and nucleotides in chilled farmed salmon (*Oncorhynchus kisutch*) muscle. Eur Food Res Technol 230:925–934. doi:10.1007/s00217-010-1239-1
129. Martinez I, Solberg C, Lauritzen K, Ofstad R (1992) Two-dimensional electrophoretic analyses of cod (*Gadus morhua*, L.) whole muscle proteins, water-soluble fraction and surimi. Effect of the addition of $CaCl_2$ and $MgCl_2$ during the washing procedure. Appl Theor Electrophor 2:201–206
130. Morzel M, Verrez-Bagnis V, Arendt EK, Fleurence J (2000) Use of two-dimensional electrophoresis to evaluate proteolysis in salmon (*Salmo salar*) muscle as affected by a lactic fermentation. J Agric Food Chem 48:239–244
131. Badii F, Howell NK (2002) Effect of antioxidants, citrate, and cryoprotectants on protein denaturation and texture of frozen cod (*Gadus morhua*). J Agric Food Chem 50:2053–2061
132. Kinoshita Y, Sato T, Naitou H et al (2007) Proteomic studies on protein oxidation in bonito (*Katsuwonus pelamis*) muscle. Food Sci Technol Res 13:133–138
133. Kjaersgard IVH, Jessen F (2004) Two-dimensional gel electrophoresis detection of protein oxidation in fresh and tainted rainbow trout muscle. J Agric Food Chem 52:7101–7107
134. Baron CP, Kjaersgård IVH, Jessen F, Jacobsen C (2007) Protein and lipid oxidation during frozen storage of rainbow trout (*Oncorhynchus mykiss*). J Agric Food Chem 55:8118–8125
135. Sanmartín E, Arboleya JC, Iloro I et al (2012) Proteomic analysis of processing by-products from canned and fresh tuna: identification of potentially functional food proteins. Food Chem 134:1211–1219. doi:10.1016/j.foodchem.2012.02.177

Chapter 11
Wool Proteomics

Jeffrey E. Plowman and Santanu Deb-Choudhury

Abstract The recent sequencing of the wool genome has considerably assisted the characterisation of the proteome of wool fibres. This has been achieved by the coupling of mass spectral sequence identification to the traditional two-dimensional gel electrophoresis, resulting in the identification of most of the protein spots on the two-dimensional map. More recently, gel-free proteomic analysis of whole wool protein digests has resulted in improved sequence coverage of some proteins and the identification of novel isoforms of some of these proteins.

Keywords Keratin · Keratin associated protein · Electrophoresis · Mass spectrometry

Abbreviations

KAP	Keratin Associated Protein
HSP	High Sulphur Protein
UHSP	Ultra-High Sulphur Protein
HGTP	High Glycine-Tyrosine Protein
2DE	Two-Dimensional Electrophoresis
MALDI	Matrix Assisted Adsorption/Desorption Ionisation
TOF	Time of Flight
MS	Mass Spectrometry
LC	Liquid Chromatography
ESI	Electro-Spin Ionisation

11.1 Introduction

Proteins are the major component of wool contributing up to 98 % of the total mass of some fibres [1]. Of these, the keratinous proteins constitute 85 % of the protein material of the fibre [2]. They are further divided into two classes: keratins, and keratin associated

---

J.E. Plowman (✉) · S. Deb-Choudhury
Food & Bio-Based Products, AgResearch Lincoln Research Centre, Lincoln, New Zealand
e-mail: jeff.plowman@agresearch.co.nz

proteins (KAPs), of which the former make up 58 % of the total [3]. Keratins, with their α-helical core, make a significant contribution to the structural and mechanical properties of the fibre [4]. The primary building block is a coiled-coil heterodimer formed from the association of one acidic Type I keratin with a neutral-basic Type II keratin. Two of these heterodimers then associate in an anti-parallel manner to form a tetramer; which then undergo further associations to ultimately produce the macrofibrillar structures that occupy the major part of the cortical cells in wool [5].

The mechanical properties of wool are also dependent on the KAPs [6], of which at least three different classes have been classified. These are the high sulphur proteins (HSPs) with cysteine contents of less than 30 mol%; the ultra-high sulphur proteins (UHSPs) with cysteine contents higher that 30 mol%; and the high glycine-tyrosine proteins (HGTPs) [6]. In reality this classification system is somewhat arbitrary, because some UHSPs have cysteine contents just below 30 mol%, while other KAPs have been identified that do not fit into any of these classes. The whole proteome of wool also includes a number of non-keratinous proteins, including the keratin anchor proteins desmoplakin and desmoglein, and proteins found in the intermacrofibrillar material between the keratin macrofibrils, in the cellular remnants of the cortex and cuticle, and in the medulla in some fibres.

The wool proteome has been analysed by a number of approaches, of which two-dimensional gel electrophoresis (2DE) has been at the fore. Initially non-equilibrium approaches were used, whereby iodoacetic acid alkylated wool protein extracts were separated in the first dimension at a fixed pH and identified by amino acid profiling [7]. This was superseded by the use of isoelectric focusing of wool proteins alkylated with iodoacetic acid [8] and later iodoacetamide [9]. More recently there has been a move towards identification of proteins first by mass spectrometry (MS) peptide mass fingerprinting and later by matching peptide sequences by MS/MS. More recently gel-free proteomic approaches have come to the fore [10] whereby tryptic digests of whole or partial wool extracts are analysed directly by mass spectrometry. In this chapter, we briefly summarize the use of these different proteomics approaches to assess the protein content of wool, and how these have led to the current understanding of the wool proteome.

11.2 Keratin Protein Diversity

Successful identification of proteins using proteomics is entirely dependent on the sequences available in international sequence databases. In that respect wool keratin proteins were, for a long time, poorly served. Initial studies indicated that there were four Type I and four Type II keratins [11, 12] but more recent studies have revealed a total of 10 Type I and 7 Type II keratins (Table 11.1) [13, 14]. The KAPs are more complex than the keratins and initial studies indicated that there were about eight or nine families [6] but now 17 families have been identified in the wool proteome. Within these families there are 89 known individual KAPs, and an additional 44 polymorphic variants of these (Table 11.2) [15–21].

Table 11.1 Known human hair and wool keratins and their location in the fibre

	Wool	Hair	Location
Type I	K31	K31	Cortex
	K32	K32	Cuticle
	K33a	K33a	Cortex
	K33b	K33b	Cortex
	K34	K34	Cortex
	K35	K35	Cuticle and Cortex
	K36	K36	Cortex
		K37	Cortex—vellus hair (humans only)
	K38	K38	Cortex
	K39	K39	Cortex (sheep); cuticle and cortex (humans)
	K40	K40	Cuticle (sheep); cuticle and cortex (humans)
Type II	K81	K81	Cortex
	K82	K82	Cuticle
	K83	K83	Cortex
	K84	K84	Cuticle (wool); tongue (humans)
	K85	K85	Cuticle and cortex
	K86	K86	Cortex
	K87		Cortex

Another issue regarding keratin proteomics is that of nomenclature, in particular the continued use of the early nomenclature in public sequence databases. The original names given to the keratins were based on their separation by electrophoresis into three components (namely 5, 7 and 8) of which 7 was found to consist of three sub-components and 8 of four sub-components [11, 12]. Since then their nomenclature has been revised; in 1997 to the KRT and K prefix for the gene and protein respectively, and more recently a universal system encompassing both the trichocyte and epithelial keratins was proposed [22].

Likewise, the nomenclature of the KAPs was originally based on their fractional separation. The four members of the B2 family were resolved by DEAE-cellulose chromatography after the isolation of two components, B1 and B2, by ammonium sulphate precipitation [11]. In contrast column electrophoresis of the HSP fraction of wool saw it separated into two poorly resolved bands, I and III, the latter ultimately being further fractionated on Sephadex G100 into BIIIA and BIIIB [11]. This nomenclature persisted until quite recently when the KAP nomenclature was first introduced [23] and subsequently refined [24].

Nevertheless, elements of the old nomenclatures still exist in international databases, meaning that care and vigilance is required when interpreting protein matches in database searches of mass spectral data.

Table 11.2 Known human ovine and caprine keratin associated proteins [24] (Yu, Personal communication)

Class	Family	Sheep	Human	Location
High sulfur	KAP1	4 (29)	4 (2)	Cortex
	KAP2	3	5 + 1	Cortex
	KAP3	3 + 1	3 + 1	Cortex
	KAP10	1	11 + 1 (10)	Cortex
	KAP11	1	1	Cortex
	KAP12	1	4 + 1	Cuticle
	KAP13	2	4 + 2	Cortex/cuticle
	KAP15	1	1	Cortex/cuticle
	KAP16		1	
	KAP23	–	1	Cortex/cuticle
Ultra-high sulfur	KAP4	27	11 + 1	Cortex
	KAP5	4 + 1 (6)	12 + 2	Cuticle
	KAP9	7	7 + 1	Cortex
	KAP17		1	Cuticle
High glycine-tyrosine	KAP6	4 (4)	3	Cortex
	KAP7	1 (2)	1	Cortex
	KAP8	2 (3)	1 + 2	Cortex
	KAP16	4		Cortex
	KAP19	4	7 + 4	Cortex/cuticle
	KAP20	–	2	Cortex
	KAP21	2	2 + 1	Cortex/cuticle
	KAP22	–	1	
Other	KAP24	1	1	Cuticle
	KAP25	–	1	
	KAP26	1	1	
	KAP27	1	1	
Total		74 + 2 (44)	88 + 17 (12)	

Notes
1. Numbers after "+" represent pseudogenes
2. Numbers in brackets are genetic variants of family members
3. Human KAP16.1 is a HSP, whereas the goat and sheep KAP16.1s are HGTPs

11.3 Gel Proteomic Analysis of Wool

11.3.1 Proteomic Analysis of the Keratins

Analysis of the wool proteome using the 2DE approach has its own technical challenges because keratin proteins overlap considerably on 2DE maps. Type II keratins resolve over a broad pH range and migrate to the same molecular weight region in the second dimension. Resolution of Type I keratins is even lower than the Type II keratins because they tend to migrate to a narrow region in a 2DE map. It is of considerable importance that individual keratins can be unequivocally located and identified in 2DE maps because their expression levels can differ between breeds of animals and also across the surface of the skin. Keratins in wool exhibit strong sequence homology with 92–93 % sequence identity between some of the Type Is and Type IIs. Characterisation of these proteins therefore requires methods of identifying unique peptides belonging to these proteins. The conventional approach, using spot excision followed by enzymatic digestion and mass spectrometry identification, while useful has proved inadequate because of the high homology between the keratins. It is also difficult to excise spots from a tight cluster of proteins, especially from the region of the gel where Type I proteins migrate (Fig. 11.1). This is further confounded by various sample handling steps that result in sample losses.

To circumvent these problems a complimentary technique to the conventional approach using matrix assisted laser desorption ionisation (MALDI) imaging has been used. This technique produces peptide masses that provide additional information on the localisation of these spots via an order of magnitude increase in analytical resolution (from a gel spot scale ~ 1 mm diameter to a laser beam scale ~ 80 μm diameter) [25].

Briefly, a 2DE gel region with the highest degree of overlapping Type I keratins was excised intact and in-gel digested with trypsin. The peptides thus produced were blotted through an UltraBind membrane (US450 affinity membrane, Pall Life Sciences, USA) incorporating immobilised trypsin on to a PVDF membrane using a semi-dry transfer procedure. This treated membrane was then sprayed with the MALDI matrix (α-cyano-4-hydroxy-*trans*-cinnamic acid) and then covered with a thin layer of conductive gold (~ 5 nm) to reduce the charging effect encountered during MALDI imaging and the MALDI image acquired (Fig. 11.2). The spatial distribution of the peptide masses corresponding to K31, K34, K35, K33a and K33b are shown in Fig. 11.1. Peptide masses both unique and common to Type I keratins identified through imaging are listed in Table 11.3. MALDI imaging of the gel region provided interesting data in that the distribution of the Type I keratins

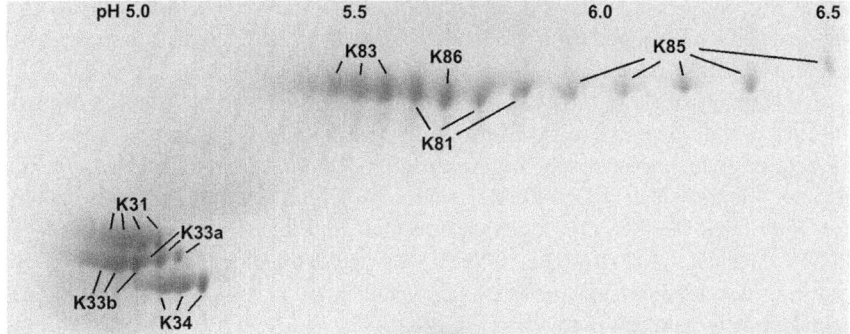

Fig. 11.1 A 2DE map of wool keratins separated over the pH range 4–7

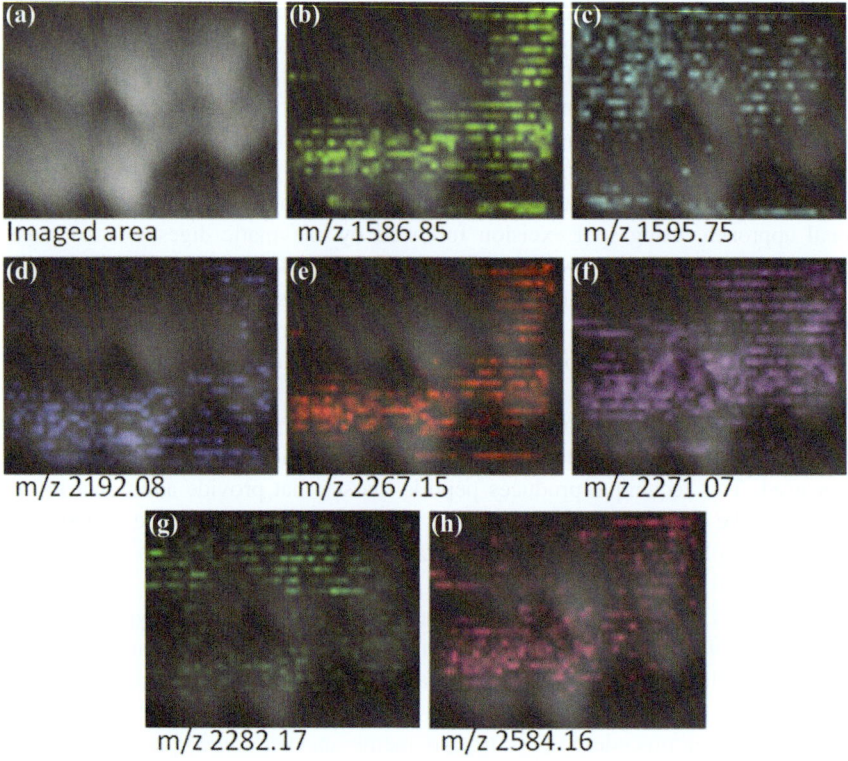

Fig. 11.2 The distribution of the ions across the PVDF membrane related to peptide sequences from K31, K34, K35, K33a and K33b. Images b-i show the distribution of peptides across this area: **a** imaged area of the 2-DE gel; **b** distribution of the peptide with m/z 1586.85, **c** distribution of the peptide with m/z 1595.75, **d** distribution of the peptide with m/z 2192.08, **e** distribution of the peptide with m/z 2267.15, **f** distribution of the peptide with m/z 2271.07 **g** distribution of the peptide with m/z 2282.17, **h** distribution of the peptide with m/z 2584.16

were determined to be more diffuse compared to their distribution implied by the manual excision of spots method. MALDI imaging clearly showed that the distribution of K34 overlapped with K33a, K33b and K35. K35 was also seen to be present in the K31 region.

An advantage of using the MALDI imaging approach is the observation of high molecular weight peptides (ranging from *m/z* 1500 to *m/z* 3400) and the release of new peptides which are not seen by the traditional approach. Avoidance of organic solvents to extract the tryptic peptides coupled with the electroblotting step could have contributed to this. One drawback to this procedure is the lack of useable MS/MS data, due to low signal intensity of precursor ions. However, it is an interesting development and serves as a complementary technique to the traditional manual excision and in-gel digest procedure.

Table 11.3 Type I IFP sequences obtained from MALDI imaging. Both common and unique sequences are shown

Protein	Experimental m/z (monoisotopic)	Calculated [M+H]+ (monoisotopic)	Peptide position	Peptide	Unique sequence
K31, K33b, K34	2282.17	2282.17	–	SQLGDRLNVEVDAAPTVDLNR	No
K33a, K33b	2267.14	2267.15	–	EELICLKQNHEQEVNTLR	No
K34	1586.85	1586.85	138–151	SENSRLVIQIDNAK	Yes
	2584.16	2584.16	383–405	LPSNPCATTNASSNFCRSSSQNR	Yes
K35	1595.75	1595.75	270–281	CQYETLVENNRR	Yes
	2192.08	2192.08	370–386	CDLERQNQEYQVLLDVR	Yes
	2271.07	2271.07	185–203	YETEVTMRQLVESDMNGLR	Yes

11.3.2 Proteomic Analysis of the KAPs

While keratins and some KAPs focus in the acidic region, other KAPs focus at alkaline pHs [26]. Keratins also dominate 2DE gels and to maintain good resolution of keratins while adequately visualising KAPs 2 mm thick second dimension gels overloaded with extract are necessary. Nevertheless, selective enhancement of the KAPs can be achieved through modifications to the standard extraction conditions [26, 27]. Keratin extractability is pH dependent and can be reduced to almost zero by lowering the pH from 9.3 to 5.5. The extractability of the KAPs is also affected by this but not as severely. Keratins can also be removed by lowering the urea concentration from 8 to 3 or 4 M, although this results in the loss of KAP bands below 10 kDa and the lowered extraction of another between 15 and 20 kDa. Lowering the dithiothreitol concentration from 50 to 2.5 mM also removes keratins from the extract.

Reducing the extractability of keratins by these procedures has enabled the identification of the KAP1 and KAP3 HSP families in the acidic regions by MALDI-TOF mass spectrometry. Up to four members of the KAP1 family have been found to separate between 20 and 30 kDa in the 2DE map (Fig. 11.2) but there is considerable variation in the expression of the members of this family both between and within breeds. Studies have shown that KAP1-3 and KAP1-4 are always present in wool fibres but the expression of the other proteins is variable [28]. The KAP3 family appears as a short train of spots between 10 and 15 kDa, with KAP3-2 at a slightly lower molecular weight than KAP3-3 and KAP3-4. The latter two proteins can also be separated using narrow range immobilised pH gradient strips. The other KAPs focus between pH 9 and 10, and the HSPs at higher molecular weight than the HGTPs (Fig. 11.3).

11.4 Gel-Free Wool Proteomics

11.4.1 Proteomic Analysis of the a-Layer

Both keratin and KAPs have been detected in the cuticle although it lacks the organised structure of the cortex. However, the cuticle has proved relatively intractable to the standard extraction conditions until recently, when it was found that it was possible to extract the exocuticle layer by replacing dithiothreitol with tris(2-carboxyethyl) phosphine in the extraction buffer. This opened the potential for analysis of the exocuticle layer by a gel electrophoretic approach [29]. The endocuticle can also be digested by enzymes such as Pronase E or trypsin, and it is possible to isolate the cuticle a-layer by tris(2-carboxyethyl) phosphine extraction followed by Pronase E digestion.

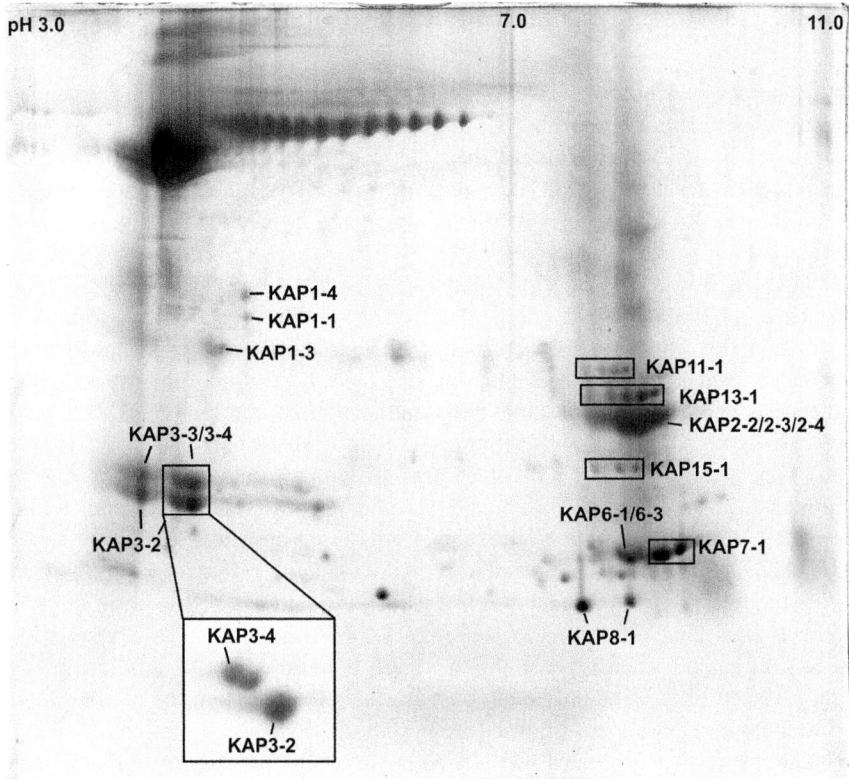

Fig. 11.3 The 2DE map of wool keratins and KAPs over the pH range 3–11, with an inset showing the further resolution of the KAP3 family using pH 5–6 immobilised pH gradient

Table 11.4 Amino acid residue sequences derived from ESI-MS/MS analysis of peptides isolated from the 2-nito-5-thiocyanobenzoic acid digest of the a-layer

Observed m/z	Primary structure (MS/MS)	Sequence homology with sheep ptoteins (BLAST)
442.02	CVPV.C	KAP5-5; CVPV.C (46–50, 58–90, 141–145, 188–192)
494.01	CACSS.C	KAP5-5; CSCSS.C (151–156)
562.03	CEPSC.C	KAP1-1; CEPSC.C (152–157)

The a-layer is the most intractable part of the wool fibre, proving relatively immune to enzymic or acidic reduction, however 2-nito-5-thiocyanobenzoic acid has proved to be capable of partially digesting it [29]. From this approach it has been possible to establish that the a-layer contains the UHSP KAP5-5 and proteins similar to the HSP KAP1 family (Table 11.4).

11.4.2 Gel-Free Proteomics of Whole Wool

The properties of wool make it a difficult substrate for proteomic investigations. Complications arise from a limited number of basic residues in KAPs and the presence of prolines after some of these basic residues result in low proteolytic digestion yield for downstream mass spectrometry analysis. Identification of wool keratin proteins from 2DE gels can be complex and a viable alternative is to use gel-free proteomic approaches.

Wool protein extracts, typically obtained using chaotropes such as urea in the presence of a reducing agent, can be analysed using mass spectrometry following pre-fractionation chromatography that may include various ion-exchangers. Four approaches using gel-free proteomic analysis have been used for increasing the sequence coverage and improving the identification of wool proteins. The first approach involved 1D-LC MS/MS runs of the whole wool protein enzymatic digest. The second approach involved fractionation of enzymatic wool protein digests using strong cation exchange chromatography. The third and fourth approaches involved separation of the intact solubilised proteins using either strong anion exchange or strong anion exchange-hydroxyapatite chromatography prior to enzymatic digestion of the fractions. The whole wool protein digest, as in the first approach, was analysed using LC-ESI and LC-MALDI runs. The digests were resolved by nanoflow HPLC with an analytical column (C18, 30 cm, 75 μm ID) using reversed phase gradient conditions. Digests obtained from the second, third and fourth approaches were analysed using only LC-ESI.

These approaches generated partially complementary, partially overlapping data sets allowing high significance identifications along with optimal sequence coverage. However, when working with multiple data sets, combining MS and MS/MS data across experiments for protein identification can result in extremely large peak list files. This results in searches taking an extremely long time to complete and the results become difficult to interpret. To circumvent this, a new approach was employed wherein data sets were searched individually using appropriate parameters and significant results subsequently combined into protein and peptide lists [10].

The high abundance of proteins such as keratins and the presence of lower abundance proteins such as KAPs in keratinous materials like wool contribute to a high dynamic range of protein mixtures. The greatest advantage of the multitude of MS/MS spectra obtained from the combined gel-free analyses is the increase in sequence coverage and specificity for lower abundance proteins. KAPs exhibit extensive sequence similarity and data from a single analysis is not enough to distinguish between isoforms. Combining data from multiple analyses has the added advantage of identifying discriminating peptides allowing unequivocal identification of certain isoforms of these proteins. Using these approaches, 76 proteins were identified using 1D-LC-MS/MS and LC-MALDI, 21 additional

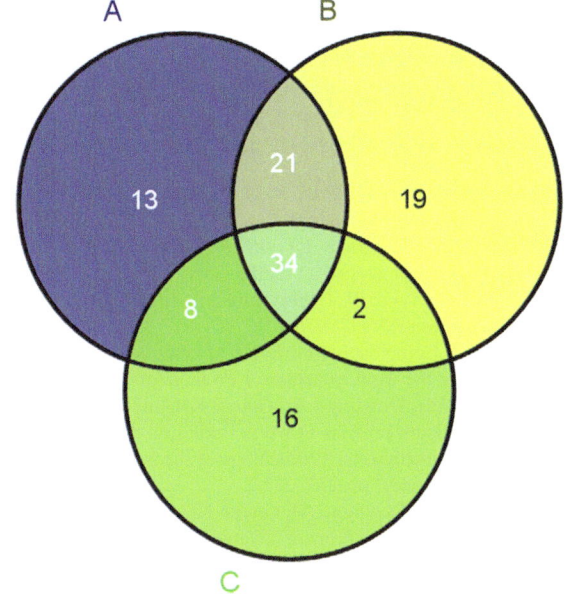

Fig. 11.4 Number of proteins identified with the three gel-free approaches. *A* 1D LC-MS/MS and LC-MALDI; *B* 2D LC-MS/MS; *C* Protein separation + LC-MS/MS

proteins by 2D-LC-MS/MS and another 16 unique proteins by protein separation followed by LC-MS/MS (Fig. 11.4).

11.5 Conclusion

Analysis of the proteome of wool is complicated by the high degree of homology among its constituent keratins and KAPs and it is therefore highly dependent on sequence information in international databases. Nevertheless good progress has been made using a combination of gel and gel-free proteomic approaches. To date most of the major cortical keratins and KAPs visible on 2DE gels have been identified and some progress has been made on identifying proteins in the cuticle, in particular the a-layer. In addition, gel-free approaches have extended this by increasing the sequence coverage of low abundance proteins, allowing the unequivocal identification of certain isoforms of these proteins.

References

1. Shorland FB, Gray JM (1970) The preparation of nutritious protein from wool. Br J Nutr 24:717
2. Gillespie JM, Goldsmith LA (1983) The structural proteins of hair: Isolation, characterization, and regulation of biosynthesis. In: Biochemistry and physiology of the skin. Oxford University Press, New York, pp 475–510

3. Maclaren JA, Milligan B (1981) Wool science—the chemical reactivity of the wool fibre. Science Press, Marrickville
4. Parry DAD, Steinert PM (1995) Intermediate filament structure. Springer, Heidelberg
5. Marshall RC, Orwin DFG, Gillespie JM (1991) Structure and biochemistry of mammalian hard keratin. Electron Microscope Rev 4:47–83
6. Powell BC (1996) The keratin proteins and genes of wool and hair. Wool Technol Sheep Breed 44:100–118
7. Marshall RC (1981) Analysis of the proteins from single wool fibers by two-dimensional polyacrylamide gel electrophoresis. Text Res J 51:106–108
8. Marshall RC, Blagrove RJ (1979) Successful isoelectric focusing of wool low-sulphur proteins. J Chromatogr A 172:351–356
9. Herbert BR, Woods JL (1994) Immobilised pH gradient isoelectric focusing of wool proteins. Electrophoresis 15:972–976
10. Clerens S, Cornellison CD, Deb-Choudhury S, Thomas A, Plowman JE, Dyer JM (2010) Developing the wool proteome. J Proteomics 73:1722–1731
11. Crewther WG, Dowling LM, Gough KH, Inglis AS, Mckern NM, Sparrow LG et al (1975) The low-sulphur proteins of wool: studies on their classification, characterization, primary and secondary structure. In: Proceedings of the 5th international wool textile research conference. Aachen, Germany, pp 233–242
12. Crewther WG, Dowling LM, Gough KH, Marshall RC, Sparrow LG, Parry DAD et al (1980) The microfibrillar proteins of α-keratin. In: Fibrous proteins: scientific, industrial and medical aspects. Academic Press Inc. (London) Ltd., London, pp 151–159
13. Yu Z, Gordon SW, Nixon AJ, Bawden CS, Rogers MA, Wildermoth JE et al (2009) Expression patterns of keratin intermediate filament and keratin associated protein genes in wool follicles. Differentiation 77:307–316
14. Yu Z, Wildermoth JE, Wallace OAM, Gordon SW, Maqbool NJ, Maclean P et al (2011) Annotations of sheep keratin intermediate filament genes and their patterns of expression. Exp Dermatol 20:582–588
15. Rogers GR, Hickford JG, Bickerstaffe R (1994) Polymorphism in two genes for B2 high sulfur proteins of wool. Anim Genet 25:407–415
16. Itenge-Mweza TO, Forrest RJH, Mckenzie G, Abbott J, Amoafo O, Hickford JG (2007) Polymorphism of the KAP1.1, KAP1.3 and KRT.1.2 genes in merino sheep. Mol Cell Probes 21:338–342
17. Gong H, Zhou H, Plowman JE, Dyer JM, Hickford JGH (2010) Analysis of variation in the ovine ultra-high sulphur keratin-associated protein KAP5-4 gene using PCR-SSCP technique. Electrophoresis 31:3545–3547
18. Gong H, Zhou H, Plowman JE, Dyer JM, Hickford JG (2011) Search for variation in the ovine KAP7-1 and KAP8-1 genes using PCR-SSCP. DNA Cell Biol 31:367–370
19. Gong H, Zhou H, Hickford JG (2010) Polymorphism of the ovine keratin-associated-protein 1-4 (KRTAP1-4) gene. Mol Biol Rep 37:3377–3380
20. Gong H, Zhou H, Dyer J, Hickford J (2011) Identification of the ovine KAP11-1 gene (KRTAP11-1) and genetic variation in its coding sequence. Mol Biol Rep 38:5429–5433
21. Gong H, Zhou H, Hickford JG (2011) Diversity of the glycine/tyrosine-rich keratin-associated protein 6 gene (KAP6) family in sheep. Mol Biol Rep 38:31–35
22. Schweizer J, Bowden PE, Coulombe PA, Langbein L, Lane EB, Magin TM et al (2006) New consensus nomenclature for mammalian keratins. J Cell Biol 174:169–174
23. Rogers MA, Langbein L, Winter H, Ehmann C, Praetzel S, Korn B et al (2001) Characterization of a cluster of human high/ultrahigh sulfur keratin-associated protein genes embedded in the type I keratin gene domain on chromosome 17q12-21. J Biol Chem 276:19440–19451
24. Gong H, Zhou H, Mckenzie GW, Yu Z, Clerens S, Dyer JM et al (2012) An updated nomenclature for keratin-associated proteins (KAPs). Int J Biol Sci 8:258–264

25. Deb-Choudhury S, Plowman JE, Thomas A, Krsinic GL, Dyer JM, Clerens S (2010) Electrophoretic mapping of highly homologous keratins: a novel marker peptide approach. Electrophoresis 31:2894–2902
26. Plowman JE, Deb-Choudhury S, Thomas A, Clerens S, Cornellison CD, Grosvenor AJ et al (2010) Characterisation of low abundance wool proteins through novel differential extraction techniques. Electrophoresis 31:1937–1946
27. Woods JL, Orwin DFG (1987) Wool proteins of New Zealand romney sheep. Aust J Biol Sci 40:1–14
28. Flanagan LM, Plowman JE, Bryson WG (2002) The high sulphur proteins of wool: towards an understanding of sheep breed diversity. Proteomics 2:1240–1246
29. Bringans SD, Plowman JE, Dyer JM, Clerens S, Vernon JA, Bryson WG (2007) Characterization of the exocuticle a-layer proteins of wool. Exp Dermatol 16:951–960

Chapter 12
Proteomic Research on Honeybee

Yue Hao and Jianke Li

Abstract Honeybee, as the most valuable insect pollinator, plays great roles for both ecological balance and agriculture. Although people started to investigate honeybee biology at the beginning of 20th century, it's not until the year 2006 that honeybees were largely explored at the molecular level, when the genome sequencing of *Apis mellifera* (*A. mellifera*) was finished. Since then and with the advances in relative protein technologies, proteomics has becoming one of the most efficient tools in addressing all aspects of honeybee biology. This chapter looks at recent developments in proteomic studies of honeybee using mass spectrometry (MS) and proteomic techniques on almost all the aspects of honeybee, including the growth stages, important tissues and organs, the caste differentiation, function transitions of worker bees, the reproduction of queen bees, and honeybee products such as royal jelly.

Keywords Honeybee · Growth stage · Development of organs · Reproduction · Hemolymph · Royal jelly

12.1 The Contribution of Honeybee to Agriculture

The honeybee is the most valuable insect in the world for their role as pollinator. The crops that benefit from honeybee pollination include cucumbers, blueberries, watermelons, apples, squash, strawberries, melons, peaches, alfalfa, cotton, peanuts, and soybeans, etc. According to a report by Cornell University, the value of honeybee pollination to United States agriculture is 11.68 billion US dollars in the year 2009, which accounts for near 80 % of the total value contributed by insect pollination [1]. In China, the average economic value contributed by honeybee pollination in the years 2006–2008, was estimated to be >300 billion Chinese Yuan, accounting for

---

Y. Hao · J. Li (✉)
Institute of Apicultural Research, Chinese Academy of Agricultural Sciences,
Beijing 100093, China
e-mail: apislijk@126.com

12.3 % of the gross output value of agriculture in China [2]. Scientists from France and Germany estimated that the worldwide economic value of the pollination by insect pollinators, mainly bees, was over 217 billion US dollars in the year 2005, which represented 9.5 % of the value of the world agricultural production used for human food that year [3]. Honeybee pollination has more than what can be estimated here. Without the production of fruit, the consequence of pollination, there will be no seeds to be spread out to sustain the plant reproduction. And the crops dependent on honeybee pollination are of vital important in providing food source for wild animals as well as economical animals. So the importance of honeybee goes far beyond what it plays in agriculture, it also plays great roles for both ecological balance and human life. If the bee disappeared off the surface of the globe then man would only have four years of life left, Elbert Einstein said. No more bees, no more pollination, no more plants, no more animals, no more man (Fig. 12.1).

Apart from the pollination service provided to ecosystem, human also benefits from a wide range of bee products. Honeybees convert floral nectar into honey and store it as their primary food source in their waxy honey combs. Although the specific composition, color and flavor of the honey depend on the flowers foraged by the bees, the main components of honey are fructose, glucose and maltose, which occupy almost 77 % of the dry weight. Besides sugars and other carbohydrates, honey also contains trace amounts of protein, vitamins and minerals. Human prefer it as sweetener over sugar not only because the distinctive flavor, but also for the health benefit. Honey possesses anti-bacterial activities [4]. Some evidence shows that honey is effective as a treatment of coughs, and other evidence show that honey can be used for wounds and burns. Million tons of honey is produced annually worldwide now. According to the statistic data from Food and Agriculture Organization of the United Nations Statistics Division (FAOSTAT), honeybee production of China accounted for about 28 % of the world total production in the year 2013. Regarding to other bee products such as royal jelly, bee pollen, bee wax and bee venom, they all have their high economic value and the importance for promoting human health.

Fig. 12.1 Honeybee forages pollen on the flowers and the flowers are pollinated (photographed by Dr. Jianke Li)

12.2 The Research History of Honeybee and General Proteomic Tools

Although honeybees play a vital role in agriculture as pollinators, it's not until the beginning of 20th century that people started to investigate honeybee biology. The first research paper on honeybee biology is about sex determination and was published on Science journal in the year 1904. Afterwards a wide cascade of biological issues related to the behaviors, reproduction and etiology were studied. For example, the artificial fertilization of queen bees, bees and perforated flowers, etiology of European foul-brood of bees, the division of labor of bees, and embryogenesis of bees. For a hundred years, because of the technical limitations, most studies of honeybee did not go beyond the description of the whole colonies or at the single animal level, and bees were rarely studied at the molecular, biochemical or cell biological level in the way that other systems have such as Drosophila. Despite regarded as essential model organism for understanding social behavior for a long period of time, honeybees are largely unexplored at the molecular level in developmental biology, neurobiology, genetics, immunology, and aging. This situation has enormously changed since the finish up of the genome sequencing of *A. mellifera* (western honeybee) in the year 2006, when the honeybee research entered the new era of functional genomics. As one of the most important functional genomic tools, proteomic has becoming the most emerging tool in addressing all the aspects of honeybee biology, including physiology, behavior and pathology.

With the advances in technologies, such as protein separation techniques, protein identification techniques, protein chemistry science and bioinformatics, the decipher of honeybee biology via proteomics has made great progress during the last ten years. At the beginning stage of 21 century, two demisional electrophoresis (2DE) combined with mass spectrometry (MS) was most often used approach to identify honeybee proteins. The laborious nature of 2DE hampers throughput and has only recently been minimized with advances in automation. Only the most abundant proteins were identified, as the dynamic range of most protein labels/stains is small compared to the dynamic range of protein levels in a given proteome; proteins with extreme pI (<4 or >9) and masses <15 or >200 kDa are hard to be resolved, and complex samples frequently require multiple gels to resolve the entire pI and Mr range; membrane proteins are still under-represented due to their poor solubility in the sample buffer and resolution in the gel. As a result, the 2DE-based the proteomics can only represent small fraction of protein in the honeybee protoeme. To improve the depth of proteome coverage, the MS based proteomics has becoming the most promising platform with the fast development of accuracy, sensitivity and high resolution. MS-based proteomics has been widely used in investigating large spectrum of the molecular underpinning regarding to the honeybee biology and bee products. These investigations help us gain significant new understanding of the physiology and behavior of honeybee both at qualitative and quantitative levels. Moreover, the proteomic research of protein modifications of honeybee biology and bee products have also been carried on at various levels, where significant findings yielded to address the questions about the regulatory mechanism and molecular interaction networks in honeybee life.

As it is known that proteomics can tell us all the proteins that exist in the specific cells, tissues and organs, at particular times, and this conceptual breakthrough has shifted our protein research from traditional "one at a time" to the "overall performance" and "systematic associated changes". For honeybee specifically, this novel concept help us "sidestep the difficulties" since that the gene editing tools were not well developed as the other well-studied organisms, but still surprised and have been continuing surprising us with various discoveries and implications on honeybee physiology, behavior, pathology, and many others.

Here we review the advances of proteomics in the research of all the aspects of honeybee, including the growth and organs, the caste differentiation, function transitions of worker bees and the reproduction of queen bees. We also summarize the progress of proteomic research on honeybee products.

12.3 Proteomic Research of Honeybee Physiology and Behavior

12.3.1 Growth Stages

A healthy honeybee hive usually has more than 20,000 bees in the summer, with its maximum population during the time of storing surplus honey. Almost all of these are female workers, along with a queen and hundreds of male drones. The queen lays as many as 1500 eggs each day during the active season, and start laying fewer eggs when temperature drops. During winters the queen stops laying eggs and the population reduces to around 10,000, which saves food from feeding young bees so as to make the colony lives longer. Both the queen and worker bees are developed from fertilized eggs. Drones are resulted from unfertilized eggs, so they only carry half of the chromosomes.

12.3.2 Egg Stage

Honeybees go through 4 important phases in their life in the cells of wax comb: egg, larva, pupa and adult. Honeybee embryo is rod-like shape with sizes varied markedly. For *A. mellifera*, the length of an egg is about 1.7 mm, and the width range is around 0.35 mm [5]. The egg stage is about 72 h, during which the weight of egg decreases about 30 % [5]. Although not fed, honeybees start their organ formation in egg stage. A tracheal network became visible even before hatching [6]. Honeybee embryo seems quiet while undergoing very active metabolism.

A study using 2-DE based proteomics revealed that a total number of 38 abundant proteins are present across the embryonic development of worker bees of *A. mellifera* [7]. Of these, 21 proteins are identified to be or be related with energy metabolism, heat shock, cytoskeleton, antioxidant protection and growth factors. Many of these proteins are found to be significantly up-regulated during the age of embryos. For example, F1 ATP synthase subunits, which is accounting for the ATP generation, tubulin, which constitutes the major component of cellular skeleton, thioredoxin peroxide, protecting against the oxidative stress, and lethal 37Cc, required for growth phase transitions. The fast accumulation of these proteins are called by cell divisions, tissue metamorphosis and self-saving, which might be of higher priority to the embryogenesis of honeybee. A similar research on drone embryogenesis also verified that proteins involved in carbohydrate metabolism, energy production and development are most abundant in 48 h old and 72 h old eggs, suggesting that organogenesis and cell differentiation of drone mainly occur at the middle to late stage of embryogenesis [8].

A recent MS-based in-depth proteomics with high resolution and high sensitivity was performed on the embryogenesis of eggs that intended to be worker bees and identified 1460 proteins of all the three ages [9]. The "core proteome" is composed of 585 proteins, which are mainly involved in translation, folding/degradation, carbohydrate metabolism, transporters and development [9]. The proteome of different ages are dynamically changed to coordinate the developmental events: the young (<24 h) eggs produce more proteins for nutrition storage and nucleic acid metabolism; the middle aged (24–48 h) eggs featured in enhanced levels of proteins for organogenesis, these proteins are involved in pathways of aminoacyl-tRNA biosynthesis, β-alanine metabolism, and protein exportation; for late-stage (48–72 h) eggs, biological pathways of fatty acid metabolism and RNA transport are highly activated, which matches with the physiological transition from egg to larva [8, 9].

The different developmental patterns of the embryos from worker bees and drones were also examined by MS-based proteomics (Fig. 12.2). It was found that the drones start their morphogenesis earlier than workers, and the levels of morphogenesis related proteins are higher in drone embryos across the embryogenesis. Drones also produce more cytoskeletal proteins to match their larger body size. It has also been found that drones and workers employ distinctive anti-oxidation mechanisms [10].

Chan et al. investigated the proteome profiling of honeybee cells in culture compared with freshly collected embryonic cells using MS-based proteomics, and revealed that 2/3 of all the detected proteins are up-regulated during culturing. The proteins involved in protein folding quality control and anti-oxidation are highly up-regulated, in keeping with the cultured cells' adaption to the unnatural environment. Moreover, the relatively high levels of lactate dehydrogenase and aldolase, related with anaerobic response and glycolysis, respectively, fits with the rich carbon source supplied in the medium [11].

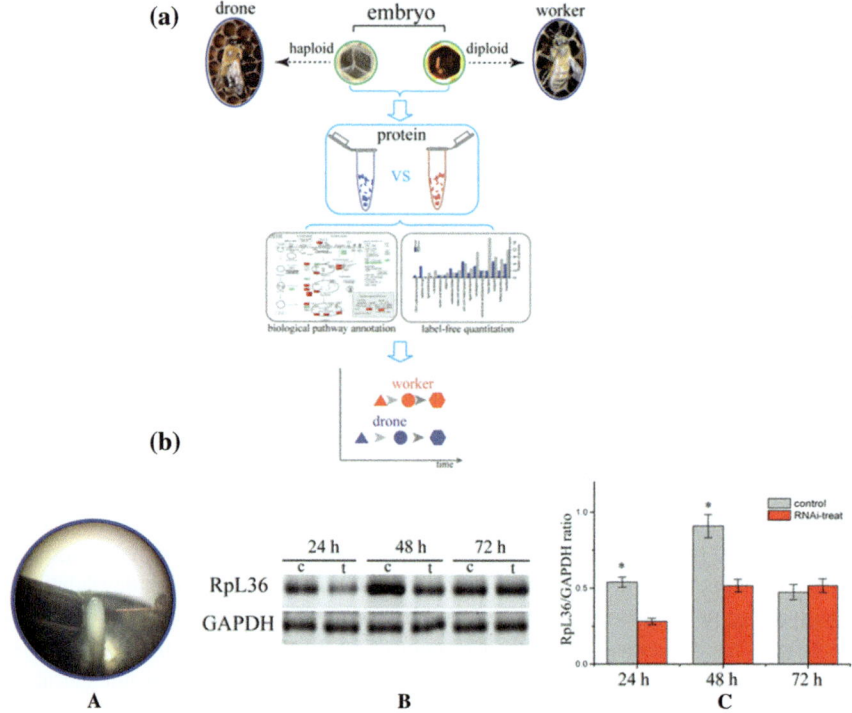

Fig. 12.2 a Shotgun proteome to investigate the mechanistic difference of embryogenesis between the drone and worker bees. **b** RNA interference induced knockdown of 60S ribosomal protein L36 (RpL36). (*A*) The dsRNA of gene RpL36 was injected at the concentration of 2.5 μg/μL in H_2O into freshly laid honeybee worker eggs (5 nL injected into each embryo), while an equal volume of sterile water was injected as a control. The injected embryos were incubated at 34 °C and 80 % humidity, and harvested at 24, 48 and 72 h. (*B*) Western-Blotting showing the RpL36 protein level in the worker embryo at three stages. (*C*) Quantitative real time PCR showing the transcript level RpL36 in the worker embryo at three stages. Letters "c" and "t" represent control and RNAi-treat. The error bar is standard deviation. The *asterisks* show significant differences ($p < 0.05$) [10]

12.3.3 Larva Stage

During larval stage, bees grow exponentially and female bees differentiate into queens and workers in response to their diet [12]. The most striking thing of larval stage is that the weight of the honeybee increases about 1500-fold over just 6 days [13]. The heavy demand of nutrient and energy is correlated with the proteomic observation of the constitutive accumulation of enolase, aldehyde dehydrogenase and phosphoglyceromutase and fatty acid synthase. While the storage protein-larval serum protein 2 only expressed on sixth instar, indicating that the larva stores amino acids for the following metamorphosis [13, 14]. Moreover, Proteomic profiling of hemolymph reveals that the levels of the immunity factors, prophenoloxidase and apismin, positively correlate with the aging of larvae for protection. However, other immunity factors are not up-regulated with age. These results suggested the maturity of the immune system during larval development and might give a explanation on why bees are susceptible to the age-associated bacterial infections such as American Foulbrood or fungal diseases [13].

Specific genes are activated for caste determination during larval development [15–18]. From the significant differences in protein expression of queen-intended larvae and worker-intended larvae at 72 and 120 h, it was found that the fate of the two castes is determined before 72 h [18]. By examining the nuclear proteome and mitochondrial proteome of queen and worker larvae at the 3rd, 4th and 5th instars, significant qualitative and quantitative protein expression differences between the two castes at the three developmental stages are found. In general, queen larvae are more active to produce physiometabolic-enriched proteins (metabolism of carbohydrate and energy, amino acid, and fatty acid) and nuclear function related proteins for their enhanced growth. The variations of proteins and enzymes between the two castes also provide us with insights of polymorphism of honeybee caste decision at the sub-cellular levels [19, 20]. These sub-cellular investigation are consistent with what had been seen from the whole larvae [18].

12.3.4 Pupa Stage

Bees are not fed during pupa stage, which lasts ~13 days. During pupation, bees undergo gradually body structure formation in the sealed wax cells, including the head, thorax, color, abdomen, and wings. Once the pupa develops into adult, it is ready to emerge. A research on pupa head development identified the differential expression patterns of 58 proteins on their 2-DE gels: 36 proteins involved in the head organogenesis are upregulated during early stages. However, 22 proteins involved in regulating the pupal head neuron and gland development are upregulated at later developmental stages [21]. These findings offered a possible explanation at molecular levels of the previous found correlation between the head weight of pupa and the production of royal jelly [22].

12.3.5 Adult Stage

Newly emerged worker bees first service as nurse bees to do cleaning, nursing, constructing the waxcomb, guide the hive entrance as well as ventilate the hive. Thereafter the worker bees leave the colony to visit flowers–forage for nectar, pollen, propolis and water. Comparative proteomics reveals that his behavioral change is linked to structural and biochemical alterations across the body of worker bees. 15 proteins diverged significantly between nurses and foragers are found in a whole-body proteomes and actin was used for standardization. Enzymes involved in ATP synthesis and consumption, sugar metabolism, such as fructose-1,6-bisphos-phate aldolase, glyceraldehyde-3-phosphate-dehydrogenase (GAPDH), enolase, glucosidase-like protein, are significantly elevated in foragers to support the higher metabolic rate during flight; in keeping with the nursing role, the jelly synthesis and transfer-related proteins and NADPH pathway are more important for nurse bees [23].

Nurse head (without glands) has been identified rich in proteins for olfactory system and major royal jelly proteins. Experienced foragers up-regulate proteins for energy production, iron binding, metabolic signaling and neurotransmitter [24].

According to a most large-scale mapping of the proteome from forage brain and nurse brain, 17 % of the total identified 2742 proteins are differentially expressed. Proteins related to cell adhesion, axonogenesis and gliogenesis, protein generation and modification, secretion and vesicular transport–related proteins have higher levels of expression in nurse, which is necessary for the cerebral maturation of the young workers. However, the experienced foragers are more active to produce synapses, enzymes for metabolism and energy generation, which is possibly triggered by the requirements of learning and memorization tasks [25]. Another research on the abdomen proteome of nest bees with different ages also reported a maturation-dependent adjustment of lipid metabolism, which is associated with the job transition of worker bees. Their RNAi data verify that the important changes in carbohydrate and lipid levels during transtition are regulated by *irs* gene, encoding a key regulator in insulin signaling network [26].

An in-depth comparison of the brain neuropeptidome from western bees (Aml) and high royal jelly producing bees (RJbs) that selected from western bees over 4 time points (newly emerged bees, 7 day-old bees, nurse bees and forager bees) identified 158 nonredundant neuropeptides (Fig. 12.3). A small number (14) of neuropeptide altered their expression during the labor division in both of the bee species. However, neuropeptides related to water homeostasis, brood pheromone recognition, foraging capacity, and pollen collection are significantly enriched in RJb. These findings indicate that both of the bee species employ a similar neruopeptidome during their age-dependent task transitions. Moreover, neuropeptides are related with the regulation of RJ secretion behavior [27].

12.4 Development of Organs

12.4.1 Antenna

As social insects living in a colony, honeybee has one of the most complex odor cues (pheromones) communication systems found in nature [28, 29]. As many as 15 glands are known to produce odor cues [30]. These chemical messengers emitted by queen, drone and worker bee induce immediate behavior responses in other bees. The queen bee produces unique odors to announce she's living, so as to avoid the

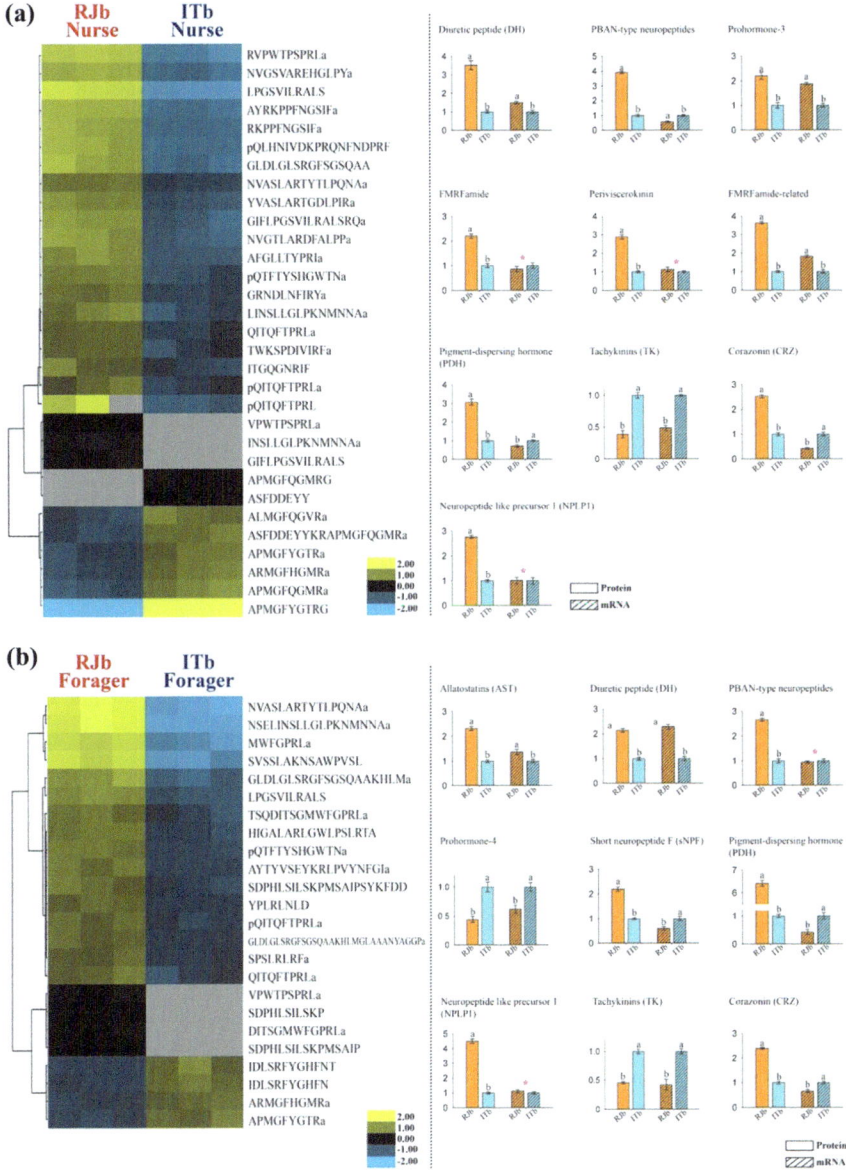

Fig. 12.3 Quantitative comparison of brain neuropeptides expression between the Italian bee (ITb) and the high royal jelly producing bee (RJb) in **a** nurse bees and **b** forager bees. Figures on the *left* are unsupervised hierarchical clustering of the differentially expressed (fold change >2 and $p < 0.05$) neuropeptides, the columns represent the two honeybee stains, and the rows represent the individual neuropeptides. The up- or down-regulated proteins are indicated by *yellow* and *blue* color code, respectively. The color intensity changes with the protein expressional level as noted on the key bar. The histograms on the *right* are the quantitative comparison of the expression trend of the precursor proteins and their mRNA between the ITb and the RJb. (*a*) is significantly higher than (*b*), and error bar is standard deviation [27]

Fig. 12.4 Antennal images of honeybee drone, worker and queen. Pane **a** is to compare whole antennae image of dorsum and ventral side of honeybee drone, worker and queen. Panel **b** is an enlarged segment where *A* an *D*, *B* and *E*, *C* and *F* independently represent the dorsum and ventral side of adult drone, worker and queen's flugellar segments. The letters on the image indicate the different kinds of sensilla; where AC is ampullaceal or coeloconic sensilla; B is basiconic sensilla; C is coelocapitular sensilla; ms is margin sensilla; P is poreplate sensilla; S is seta sensilla and T is trichoid sensilla [10]

introduction of a new queen by the hive. The vergin queen also emits odors to encourage the males to mate with her during mating flight, and to keep female workers from mating with males [31]. In addition, odor cues are also used to regulate the specific tasks of each individuals, such as nursing, alarming, defense and transmitting food information [32]. In insects such as honeybee, the reception of the odors mainly takes place in the antenna, via odorant binding proteins (OBPs) and chemosensory proteins (CSPs), to trigger series of neural responses and finally achieve social activities.

Early investigations on individual differences in the expression of OBPs and CSPs found that *A. mellifera* larva express 3 OBPs (OBP13, OBP14, OBP15) and a single CSP (CSP3), while forager antenna express 4 OBPs (OBP1, OBP2, OBP4 and OBP5) and 2 CSPs (CSP1 and CSP3). The ligand specificities to these organ and age specific binding proteins may support their roles in chemoreception, which are closely associated with the related social behavior [33].

Comparative investigations focusing on the whole proteome of antenna from worker bees and drones reveal that OBP21, OBP2 and OBP16 are exclusively upregulated in the worker, and OBP14 and OBP5 are significantly enriched in drone, suggesting the specific roles of OBPs for different castes. Although different in protein composition and abundance, both castes enrich biological pathways as energy production, carbohydrate and fatty acid metabolism, antioxidation, and molecular transportation. In particular, forager antenna expresses more enzymes for the adaption of their food foraging task, including searching for food sources, returning to nest, brood recognition and transmitting information. In contrast, these proteins are relatively low level of abundance in drone antenna, due to their less involvement in foraging activities as worker bees do. In keeping with the demand of responding to sex pheromones, drones antenna-expressing proteins play important roles in the metabolism of pheromone compounds, for example, aldh, a group of enzymes that catalyze the oxidation of aldehydes to carboxylic acids, is suggested to be involved in the transformation of male pheromone, which consequently promotes the recognition by queen during mating flight. Such examples include but not limit to jhe, carboxylesterases and FABPs [34]. Interestingly, an antenna-specific protein 3c in hemolymph has increased level in drones and workers than in queens [35]. A more comprehensive proteome comparison among drone, worker and queen confirm again that the differential expressions of antenna proteins are well associated with the different requirements of caste-dependent olfactory activities (Fig. 12.4).

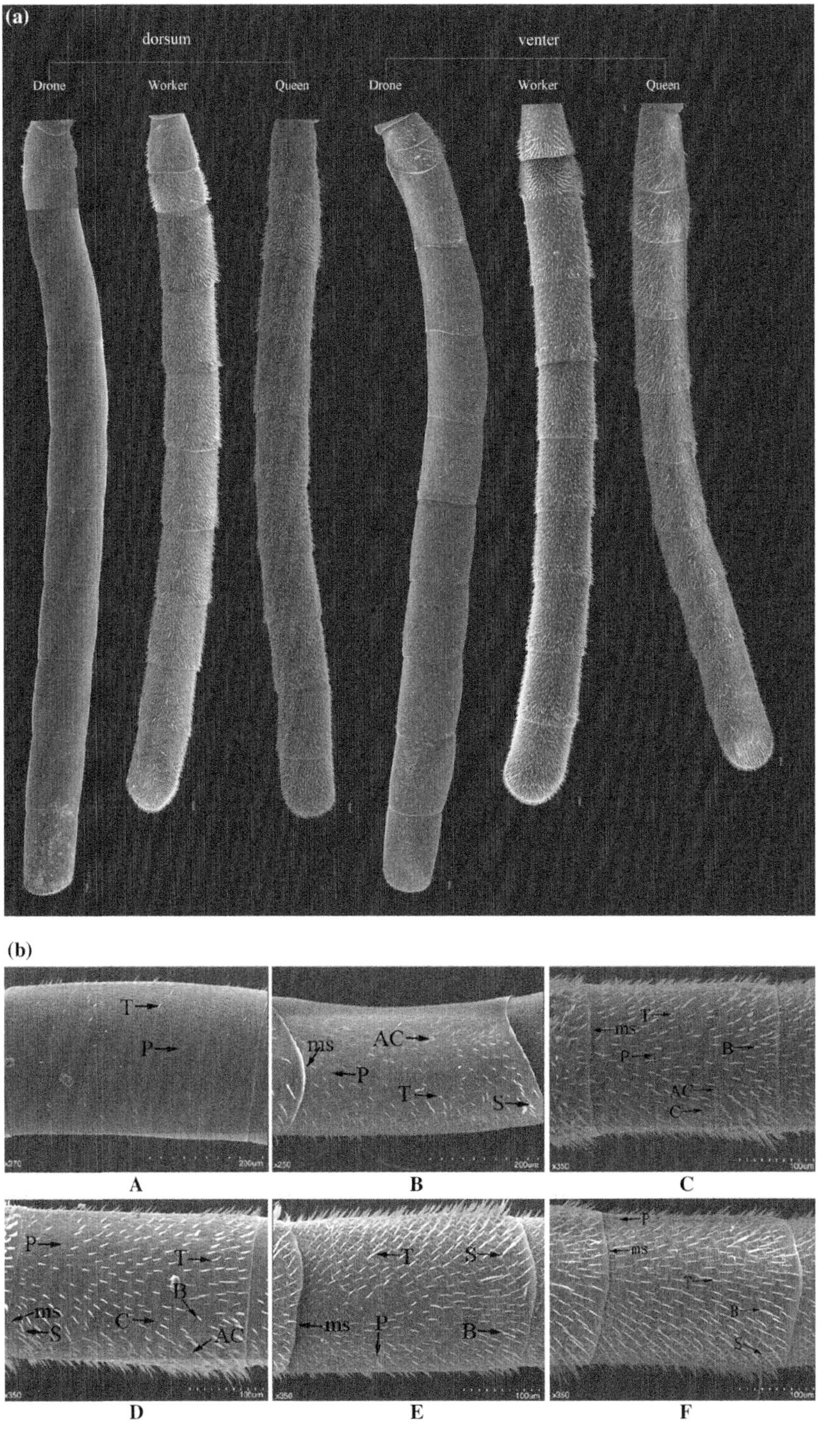

12 Proteomic Research on Honeybee 235

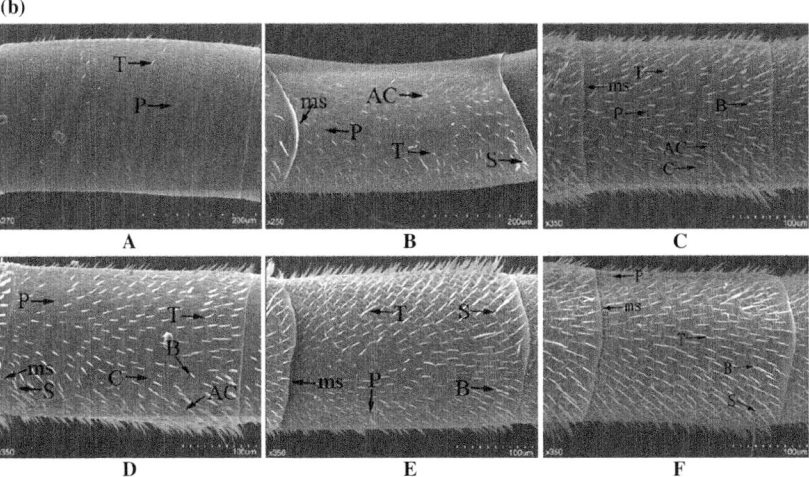

A following study comparing antenna proteome between *Apis mellifera ligustica* (Aml) and *Apis cerana cerana* (Acc) further confirm that the Aml drones are more active in carbohydrate metabolism, energy production, molecular transportation and antioxidation to deal with their respective evolutionary process. In addition to these biological pathways, Aml worker bees are more active in fatty acid metabolism [36], manifesting the fact that these two species have developed their own olfactory mechanism during the period of evolution.

12.4.2 Mandibular Gland

Mandibular glands are of special importance to queen bees. Many compounds in queen retinue pheromone (QRP), such as queen mandibular pheromone (QMP) and coniferyl alcohol, have been found in the queen mandibular glands. These chemicals keep the retinue attraction of the other bees around their queen, and also affect many other social behaviors and activities in both short and long term in the bee hive, including the maintenance of the hive, mating, inhibiting the ovary development of workers, and swarming, etc. [37]. Moreover, mandibular glands of worker bees release a highly volatile alarm pheromone when the hive is in danger from potential enemies and robber bees [38].

To illuminate the molecular mechanistic functionality driven the mandibular glands, proteome analysis revealed that the molecular compositions of OBPs and CSPs in the mandibular glands are caste and age-dependent. OBP13 is of high level of abundance in both virgin queens an newly emerged workers; OBP21 and CSP3 are up-regulated in experienced workers, drones and 2-years old queens; in addition, drones also express OBP18, whose level relatively low in females [39]. This castes-selective manner of mandibular gland proteome signature has also been confirmed by the investigations of aldehyde and fatty acid metabolism in worker and queen bees [40].

12.4.3 Hypopharyngeal Gland

The hypopharyngeal Gland (HG) is an important organ to synthesize and secrete royal Jelly (RJ), which is not only the major food for larvae and queen but also plays a crucial role in caste determination. In parallel with the age-dependent assignment of worker bees, HG is most active in nurse worker bees (adult bee age 5–15 days) and the secretion of the HG is rich in proteins, due to their nursing task. Older worker bees are usually lack of bee-milk proteins in their HGs, but this can be evoked when nurse worker bees are absent [41]. The HG of foragers contains sucrose hydrolysis enzymes [42]. Thus HG seems to have discrete differentiation states according to the roles of workers. The function of HG in different bee castes

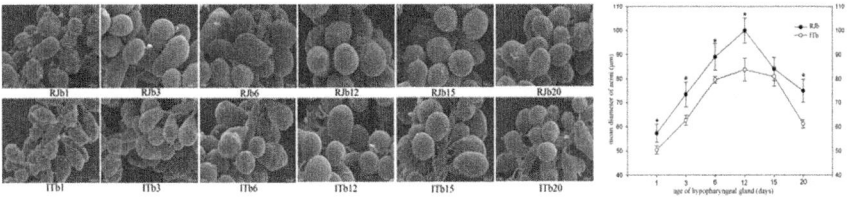

Fig. 12.5 Morphological comparison of hypopharyngeal gland between royal jelly producing bees and Italian bees [45]

and different bee species has been investigated in detail at molecular levels by means of proteomic analysis. The proteome comparison between Italian bees and the strain selected for high RJ production from Italian bees (whose production of RJ is 10 times more than that of control lines) revealed how proteome changes to help the RJ bee to achieve the high yield of RJ production.

To sustain the central functionality of producing RJ to feed the young larvae and queens, age-dependent proteome setting changes are observed. Generally, the proteins expressed in HG at young age are mainly implicated in promoting the gland growth. Proteins expressed by nurse HG are mainly major royal jelly proteins (MRJPs) that are important nutrients for RJ. To fit with the job switching as forager with age growing, the levels of MRJPs in HG decreased in the forager worker. However, the up-regulated proteins related to the metabolism of nectar in the HG of this stage are to help the bees translate the nectar into honey [43, 44].

Comparing with the control stock Italian honeybee (*A. m.ligustica*), the HG morphology of RJ bee shows a significant increase in acini size (Fig. 12.5). To maintain the RJ quality with the quantity increase, the abundance level of (MRJP) in RJ produced by RJ bees significantly increased. This is important for feeding the young larva as an efficient nutrition source. To consolidate the enhanced protein biothythsis, wide array of biological category and pathway are induced. For example, protein biosynthesis and protein folding activities are strengthened, protein related to carbohydrate metabolism and energy production are escalated to providing biological fuel for the protein synthesis. Protein implicated in development, metabolism of nucleotide, amino acid and fatty acid, transporter, cytoskeleton, and antioxidation are enhanced to ensure the gland development [45].

More recently, a phosphorproteome study of worker HGs across ages revealed that most proteins are regulated by phosphorylation independent of their expression levels. Proteins in key biological processes and pathways are dynamically phosphorylated with age development, including the centrosome cycle, mitotic spindle elongation, macromolecular complex disassembly, and ribosome, indicating that phosphorylation tunes protein activity in order to optimize cellular behavior of the HG over time. Moreover, both complementary protein and phosphoprotein expressions are required to support the unique physiology of secretory activity in the HG [46] (Fig. 12.6).

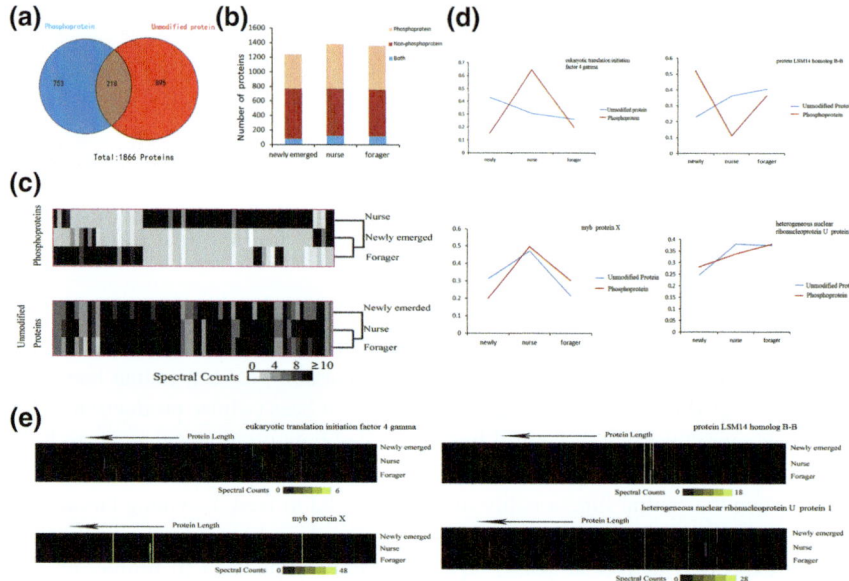

Fig. 12.6 Cross-age comparison of phosphoprotein and non-phosphoprotein expression in hypopharyngeal gland of honeybee workers. Overlap among phosphoproteins and unmodified proteins, overall (**a**); and in each age (**b**); Clustering of proteins based on spectral counts of phosphoproteins and non-phosphoproteins. Columns indicate 218 over-lapped proteins, and rows represent different ages of hypopharyngeal gland. Upper and lower panels represents expression profiles of 218 over-lapped phosphoproteins and non-phosphoproteins (**c**); Phosphorylated and non-phosphorylated abundance profiles for selected proteins (**d**); Heat maps depicting spectral counts observed across different ages of hypopharyngeal gland for sites along the length of each protein, reflecting variable phosphorylation within proteins (**e**) [46]

12.4.4 Salivary Gland

The honeybee salivary system is composed of two secretory glands: cephalic SG (HSG) and thoracic SG (TSG). Both HSG and TSG have their own protein expression profile, suggesting of their different roles in the salivary system of honeybee [47, 48]. The HSG activates the labor switching form in-hive work to field work via the modulation of juvenile hormone and ethyl oleate titer. The TSG emphasize the expression of proteins related to carbohydrate and energy metabolism, protein folding, protein metabolism, cellular homeostasis and cytoskeleton, to facilitate honey processing via the synthesis and secretion of saliva into nectar [47].

12.4.5 Brain

Social insects such as honeybee are featured in their advanced nervous system, which is necessary for the survival as well as the development of not only the individuals, but also the entire colony. Proteomic studies on honeybee nervous system, in particular the brain, promote our in-depth understanding of the molecular and neural bases that underlie the diverse social behaviors of honeybee.

Mushroom Bodies (MBs) are higher regions in the brain of insect and play crucial roles in learning and memory. Honeybee MBs have a complex layered architecture [49], with the sub-regions of MBs are specialized to receive different input signals, such as olfactory and visual signals [50].

A study using 2-DE based proteomics has found that 5 and 3 proteins are selectively expressed in worker bee's MB and Optical lobe (OL), respectively. Of these proteins juvenile hormone diol kinase (JHDK) and glyceraldehyde-3-phosphate dehydrogenase (GAPDH), are identified to be selectively expressed in MB and OL. JHDK catalyze the formation of JH diol phosphate, suggesting its activity to participate Ca^{2+} signaling in the MB; GAPDH, involved in sugar metabolism, is up-regulated in OL but not MB. In situ hybridization analysis further confirms that the expression patterns and localizations of these two proteins are cell-type and structure specific [51]. A following study by the same group has found that endoplasmic reticulum (ER)—related genes differentially expressed in honeybee MB. Reticulocalbin, an ER Ca^{2+} transporter, and ryanodine receptor, contributes to ER Ca^{2+} channel, are preferentially expressed in the large-type kenyon cells of MB. Whereas in MB and OL, no dramatic difference is found for the genes not related to Ca^{2+} signaling pathways. These findings highlight the role of Ca^{2+} signal pathway in honeybee brain in aspects of learning and memory [51].

Besides the expression in HGs, the *MRJP* transcripts [52, 53] as well as the protein products [54] have also been found differentially expressed in multiple sub-regions of the worker bee's brain. Comparative studies between nurse brain and forager brain show the presence of different MRJPs is age/task-related [24, 25, 55, 56]. MRJP1 and MRJP3, represented by peptide p57 and p70, respectively, overexpressed in nurse brain but not forager brain [56]. Differential expression patterns of MRJPs in honeybee brain and HGs have also been verified [56]. Given the dominant role of MRJPs in RJ, modulating the caste hierarchy, it is possible that MRJPs work as endogenous participants of diverse brain activities [56].

Brain function declines with age in forager bees. With increasing days spent as a forager—so called foraging duration, workers show mechanical senescence including the symptoms of reduced flight frequency [57], wing-wear from intense flight [58], oxidative damage to the brain [59], level-reductions of the brain proteins that are central to neuronal functions [60]. Calyx region of the MB is found to be intact in structure and robust in proteome profile in foraging duration, in spite of the significant level-decline of kinases, synaptic and neuronal growth-related proteins in the central brain (brain without OL). Thus the distinct brain areas are differently affected by the senescence of honeybee, and the calyx region is suggested not to be responsible for the foraging-dependent performance decline [60]. Foragers' brain function decline has also been characterized by their learning deficits, which can be overcome by reverting their job to nursing. Significant differences in the brain proteome of forager bees with or without learning ability reveals that the recovery is positively associated with the levels of stress response and cellular maintenance proteins in the central brain [55].

12.5 Reproduction

Honeybee queen mates with an average number of 19 drones in midair, and gathers millions of sperm she will need for her lifetime [61, 62]. The sperm has to survive for prolonged periods of time: It takes about 40 h for the sperm to be stored, during which over 90 % sperm will be lost; and it could take up to 8 years that the sperm is kept between mating and egg fertilization [62]. Thus the producing of high quality ejaculates and the beneficial environment supplied by secretions and storage organs are crucial to maximize the reproductive success. Studies in fruit fly indicate that the seminal fluid from male, with accessory gland secretions as the major component, possibly affect both sperm viability and survival [63]. Moreover, the secretions of spermathecae, the sperm storage organ of female, have been hypothesized to facilitate the survival of the stored sperm [64]. In honeybee, the detailed biochemical composition and physiological understanding of the sperm, seminal fluid, spermathecae and secretions from female have been studied in detail recently [65–70].

The proteome of honeybee sperm shares similarity with both fruit fly and human. Many important biological pathways that contributes to the reproduction of both human and fruit fly, have also been identified in the sperm proteome of honeybee, including energy and amino acid metabolism, cytoskeleton maintaining, protein folding and anti-oxidation. Besides the common set, 131 unique sperm proteins of *A. mellifera* have been verified as well, mainly involved in two biological processes: nucleic acid expression and protection; enzyme regulation, for example, GTPase regulators and kinase regulators [69].

Comparing the proteomes of the secretions from the male accessory gland and female spermathecal gland indicates that the former contributed to the sperm survival, while the latter has positive effect on enhancing the sperm viability [68]. Applying 2-DE analysis of the sperm form fresh male ejaculates and the sperm stored in female spermathecae, it shows that the enzymes for carbohydrate, glycolytic or respiratory metabolism counted for the main difference between the two groups, in the aspect of protein abundance as well as enzymatic activity [71].

The seminal fluid of honeybee is full of enzymes, regulators and structural proteins for energy production, antioxidation, maintaining the stability and viability of sperm [66, 67, 68, 71]. Some are also identified to have biological activity that affect female physiology [67]. To current stage, we have identified the specific proteome composition in the honeybee seminal fluid. First the fluid is quite different from the proteome of its haemolymph, though both are body fluid and full of metabolic enzymes [67]. Second, it has a complete different composition from sperm, especially that many glycosylations are found in seminal fluid, while no glycoproteins are detected in sperm [67]. Third, the seminal fluid has many more proteins that are absent from the accessory gland [67]. Interestingly, the comparison of the seminal fluid protein compositions between honeybee, *Drosophila* and human reveal that the bee is more close to human: among the 57 identified honeybee seminal fluid proteins, 23 have homologs in human set, and only 11 have homologs in the *Drosophila* set [67]. The dissimilarity in the male genetic background between honeybee and fruit fly had also been seen by other studies [69, 72].

Intra-species studies on seminal fluid proteome of *A. mellifera* suggest that evolutionary changes, including both protein abundance and modifications, happened for 16 % of all the identified proteins from three linages that had been bred for >20 generations. These proteins have different biological functions that are widely involved in male productive success, energy metabolism and cellular structural proteins, and immune defense [66].

In the honeybee hive, the queen bee is the mother to produce the most, if not all, of the offspring. The queen bee emits sex pheromones to encourage the males to mate with her during the nuptial flight, but also emits pheromones to reduce worker bees' ovary activation [31]. The queen bee has a more developed ovary than worker bees: usually consists of >150 ovarioles that allows the queen to produce thousands of eggs per day [73]. While the ovaries of worker bees have been suppressed early up to their larval stage, by means of apoptosis, resulting in inactivated ovaries with only a couple of ovarioles [73]. When the queen is present and healthy, this can be interpreted as an adaptive response. However, when the queen is absent, about 30 % of the worker bees' ovaries can be activated to lay unfertilized eggs, which will develop into males according to the haplo-diploid sex determination [74].

To identify the proximate factors involved in the regulation of worker bee's ovary activation/inactivation in honeybee, brain and ovary proteome from fertile and sterile worker bees were analyzed using 2-DE. A total number of 223 proteins have been identified in ovary proteome. Proteins implicated in the metabolism of various substrates are up-regulated in rudimentary ovaries. The coincidence of the oogenic process and the protein degradation in rudimentary ovaries suggested that the inhibition of ovary activity is a constant interplay between the two. Moreover, the identification of a batch of heat shock proteins, proteins with tetratricopeptide repeats and a steroid hormones binding protein in activated ovaries further pointed to the possibility that the activation of ovary is mediated through steroid and small neuropeptide signaling pathway. The most striking up-regulated proteins in sterile workers (in two of the three colonies that examined) are found to match the polyproteins of Kakugo virus (KV) and Deformed wing virus (DWV), in ovary and brain, respectively. On the other hand, the immune system component such as thioredoxin peroxidase is more abundant in the activated ovaries [75]. Similar viral load pattern has also been seen in the hemolymph in another comparative proteomic study [76]. These results indicated the possible correlation of virus loading, reproduction and the immunity of honeybee.

12.6 Hemolymph

Hemolymph in arthropods is the counterpart of blood as in the higher organisms. The major role of hemolymph is to distribute nutrients and immune components throughout the body. Thus, hemolymph is well deserved to be used as a powerful indicator of individual's physiological condition. The characterization of honeybee hemolymph proteome is of great interests to scientists ever since the accomplishment of the genome sequencing. The progress of understanding the honeybee hemolymph and immune system is also due to the efforts of the annotation of honeybee genome.

Key proteins in hemolymph include enzymes for the metabolism of carbohydrate and protein/peptide, transporters for nutrients and pheromones, structural proteins, immune-response proteins, MRJPs and others [35, 77]. Moreover, the identification of virus proteins in honeybee hemolymph creates the possibilities of its being used for biomarker research [35, 77], which will be reviewed in the second volume of this book.

The aim of the very first proteomic study of honeybee hemolymph was to identify the basic protein component using 1-DE in combination with MS. Among the identified 324 unique proteins, the overall overlap of adult queen, drone and worker is less than half [35]. Extremely high levels of vitellogenin, apolipophorin and hexamerins are presented only in the female bees. Adult workers have more immunity-associated proteins than larvae. For example, prophenoloxidase, prophenoloxidase-activating factor and β-1,3-glucan recognition protein are 30–50 folds more abundant in adult workers compared with larvae. These findings suggested a possible cause for some honeybee diseases that usually affect larvae, such as chalkbrood and American foulbrood.

The protein components changed with the development of honeybee. The hemolymph of young larvae emphasizes the importance of proteins for priming their rapid development and initial innate immune protection. During the honeybee's life transition from larva to pupa, the proteins involved in the metabolism have corresponding changes: from the metabolism of carbohydrates and MRJPs, to the metabolism of amino acids, nucleotides and fatty acids. The hemolymph of the young pupae expressed more number of proteins related to energy store for the preparation of their non-diet-driven pupation [78]. These protein profile changes of hemolymph match with the honeybee's normal life transition as nutrients and immune agents. In addition, the hemolymph proteome coverage was significantly increased using 2-DE and MS-based complimentary proteomics approaches [77].

The Western honeybees and Eastern honeybees have developed their own strategies of using hemolymph for nutrients deliver and immune defense. Compared with Eastern honeybees, the Western honeybees have stronger expressions of proteins related to energy production, protein folding, cytoskeleton, and development. These findings provide valuable resources for the future functional research on hemolymph with different phenotypes [43].

One of the most important physiological changes in worker bees is that their ovaries can be activated upon the loss of the queen and brood. To identify the relevant proteins for ovary activity/inactivity, a comparison between fertile and sterile worker bee's hemolymph proteome was carried using 2-DE based proteomics. Surprisingly, the sterile worker bees were found to be significantly enrich in polypeptides from DMV and KV, suggesting more viral infections on the sterile workers [76]. In comparison with the sterile worker bees, the hemolymph of the reproductive worker bees have stronger immune system. Reproductive worker bees are more rich in thioredoxin peroxidase, which is known to respond to oxidative stress [79] and functions as an apoptosis inhibitor [80]. Serine protease 8 (SP8) and a serine protease homolog (cSPH42), with the latter known to respond to microbial infection, are more abundant in reproductive worker bees. In addition, odorant binding protein 14 (OBP14) and a disc growth factor are also up-regulated in reproductive workers [76]. Some of these highly expressed proteins in reproductive workers are also present in queen spermathecae [70]. Aldolase involved in energy metabolism, which is essential for forager behaviors, is dramatically down-regulated in fertile workers. These findings match with the role switching of the worker bee from infertility to fertility [76].

12.7 Proteomic Research of Royal Jelly

Royal jelly (RJ), a secretion from hypopharyngeal gland (HG) of worker bees, is used as the food for all the larvae and the entire life of the queen in a colony (Fig. 12.7). Fresh RJ is a white-yellow colloid with a variety of biological compounds, including proteins, carbohydrates, lipids, minerals, vitamins, and free amino acids. It is believed that RJ is essential to castes determination and larvae development of honeybee [81]. RJ is also used as health supplements for its amazing health benefits, such as prolonging lifespan, anti-aging, reducing fatigue, enhancing immunity, control cholesterol, improving brain health. Studies suggested that RJ may also have pharmacological benefits. RJ has been found to prevent systemic lupus erythematosus [82], control the progression of sarcopenia [83], and restore the expression of vascular endothelial growth factor gene in rodents. It is suggested that RJ may be better than western medicine to treat presymptomatic disease [84]. However, the active substances in RJ that lead to these chemical properties are yet unknown. Since up to 50 % of the dry mass of RJ is protein, the proteome analysis of RJ is promising for addressing these questions.

MRJPs belong to a protein family consisting of nine members. However, in 2005, only MRJP1, 2 and 3 were identified in *A. mellifera* using 2-DE, MS and de novo sequencing [85]. MRJP4 was found for the first time in a comparison study of the RJ from Africanized honeybees and European honeybees (*A. mellifera*) using 2-DE and N-terminal amino acid sequence of protein spot. Besides, remarkable differences in heterogeneity of the MRJPs (especially MRJP3) and glucose oxidase were identified between the two species [86]. A proteome profiling of the secretion from HG of nurse bees identified 34 proteins, of which 27 proteins belong to the MRJPs family. The other proteins are ferritin-like protein, for iron storage, apisimin, for the regulation of MRJP1 oligomerization, and the enzymes involved in the metabolism of carbohydrates and energy, including alpha glucosidase, glucose oxidase and alpha amylase, alcohol dehydrogenase, aldehyde dehydrogenase. The identification of these proteins confirmed that RJ is synthesized and secreted in the HGs of nurse bees [87]. Since the sequencing of the honeybee genome

Fig. 12.7 A frame contains plastic queen cells to collect royal jelly. *Left panel* is a royal jelly frame drawing from the bee colony, and *right panel* is queen cell cups that remove the wax cap and larvae with the royal jelly left in queen cell cups (Photographed by Dr. Jianke Li)

accomplished in 2006, different methods have been applied for the identifications of MRJP1-7 and MRJP9 in RJ of *A. mellifera*. An phosphorylated venom protein was also found in RJ, which was previously known to be present only in bee venom [88]. Two proteins related to anti-oxidation, 1-peroxiredoxin and 1-gllutathione S-transferase S1 were identified in RJ samples from native Italian honeybees (Aml), Carnica honeybees (*A. mellifera carnica*, Amc) and high RJ producing honeybees (RJBs) that breeding selected from the native Italian honeybees using 2-DE based proteomics. MRJPs were found to be the major components in each of the species. Many MRJP heterogeneities were also identified, with MRJP3 the most remarkable. It was also found that the RJ protein compositions are identical within the three species, while the detected protein numbers are significantly lower in Amc [89]. Of all the MRJPs identified in Aml and Acc by 2-DE based and MS-based proteomics, MRJP1 is the most abundant. Peroxiredoxin 2540, glutathione S-transferase S1, and MRJP5 were detected only in the RJ of Aml. In contrast, MRJP7 was found only in the RJ of Acc. Glucose oxidase was identified for the first time in the RJ of Acc in this work. The RJ production and the levels of MRJPs are significantly higher in Aml than in Acc, suggesting that Aml needs more nutrients for its large body size [90].

19 novel proteins were found in the RJ of Aml using 2-DE-based and MS-based proteomics approaches. These proteins are involved in oxidation-reduction, protein binding and lipid transport [91]. Another 13 novel RJ proteins were identified using MS-based proteomics, with the majority of them are involved in metabolic processes (38.5 %), and health promotion activities (46.2 %) [92].

HG, postcerebral gland (PcG), and thoracic gland (TG) are believed to synthesize and secret RJ proteins. Fujita et al. investigated the proteomes of these glands as well as RJ using LC-MS/MS method. 9 novel RJ proteins, 22 putative secretory proteins and a protein functioning in insulin signaling pathway were identified [93]. MRJPs are derived from these glands, but are also present in other tissues, such as brain. 3 polypeptides—p50, p70 and p128, correlated with MRJP1 and MRJP3, have been found with differential expressions in the sub-regions of the brain from different castes. These findings contribute to the elucidation of the context-dependent roles of MRJP family [56].

According to the proteomic comparisons of RJ from two flower blooming seasons, where the RJ was collected at 3 time points (24, 48 and 72 h), clear differences between the two seasons were seen. However, there is no significant difference of the protein profiles at the 3 time points [94].

The proteome of RJ stored at different temperatures and for 80 days or one year were investigated (−20 °C, 4 °C and Room temperature) [95, 96]. Higher temperatures led to heavy degradation of protein, therefore it's better to store RJ under freezing conditions. MRJP4, MRJP5 and glucose oxidase, peroxiredoxin, and glutathione S-transferase, particularly MRJP5, are the most sensitive, indicating that these proteins could be used as the markers for RJ freshness [95, 96].

Proteins are synthesized by ribosomes through translation, may then undergo post-translational modification (PTM) to form the mature protein product. PTM plays a major role in the diversification of MRJPs in RJ. Methylation and deamidation are found in most of the MRJPs, indicating their essential roles in RJ function. Phosphorylation has characterized in MRJP1 (5 sites), MRJP2 (13 sites), apolipophorin-III-like protein (2 sites) and venom protein 2 (3 sites) [97]. Applying two complementary phosphopeptide enrichment materials and high-sensitivity MS [27], 16 phosphoproteins carrying 67 phosphorylation sites and 9 proteins carrying 71 phosphorylation sites are identified in the RJ from Aml and Acc, respectively. The overlap of these two species is composed of 8 phosphoproteins. However, these proteins are more abundant in Acc RJ than in Aml RJ, which is the opposite to the abundance level of nonphosphorylayted protein in the RJ samples [90]. Further experimental results indicated that the phosphorylation of RJ proteins is driven by an extracellular serine/threonine protein kinase (FAM20C-like protein) in the HG. Interestingly, Jelleine-II, an antimicrobial peptide phosphorylated at different sites in RJ of Aml and Acc, has different performance in antimicrobial activity. These differences suggest that these two species use different strategies for phosphorylation to tune their RJ proteins as efficient nutrients and immune agents [98] (Fig. 12.8).

Glycosylation is another major PTM that very common for RJ proteins, and is important for the functionality of RJ proteins. MRJP2a, carrying two fully occupied n-glycosylation sites, is found to inhibit the growth of *Paenibacillus larvae*, which is the primary pathogen of American foulbrood honeybee disease, with the similar effective inhibitory concentration of tetracycline hydrochloride. It is also found to fight against other Gram-positive bacteria such as *B. subtilis* and *E. coli*. In contrast, MRJP2 is inactive against these bacteria [99].

Recent efforts have been made in the in-depth and large-scale mappings of glycosylation sites of RJ proteins. In one study, 25 N-glycosylated proteins in Aml RJ are identified, most of which are related to the metabolic activities and the health improvement activities [92]. A more recent study found another 23 novel glycosylated proteins in Aml RJ and 43 proteins in Acc RJ. The absence of glycosylated apidaecin, hymenoptaecin and peritrophic matrix, which are related to anti-microbial, and the low inhibitory effect of N-glycosylated MRJP2 in Aml on

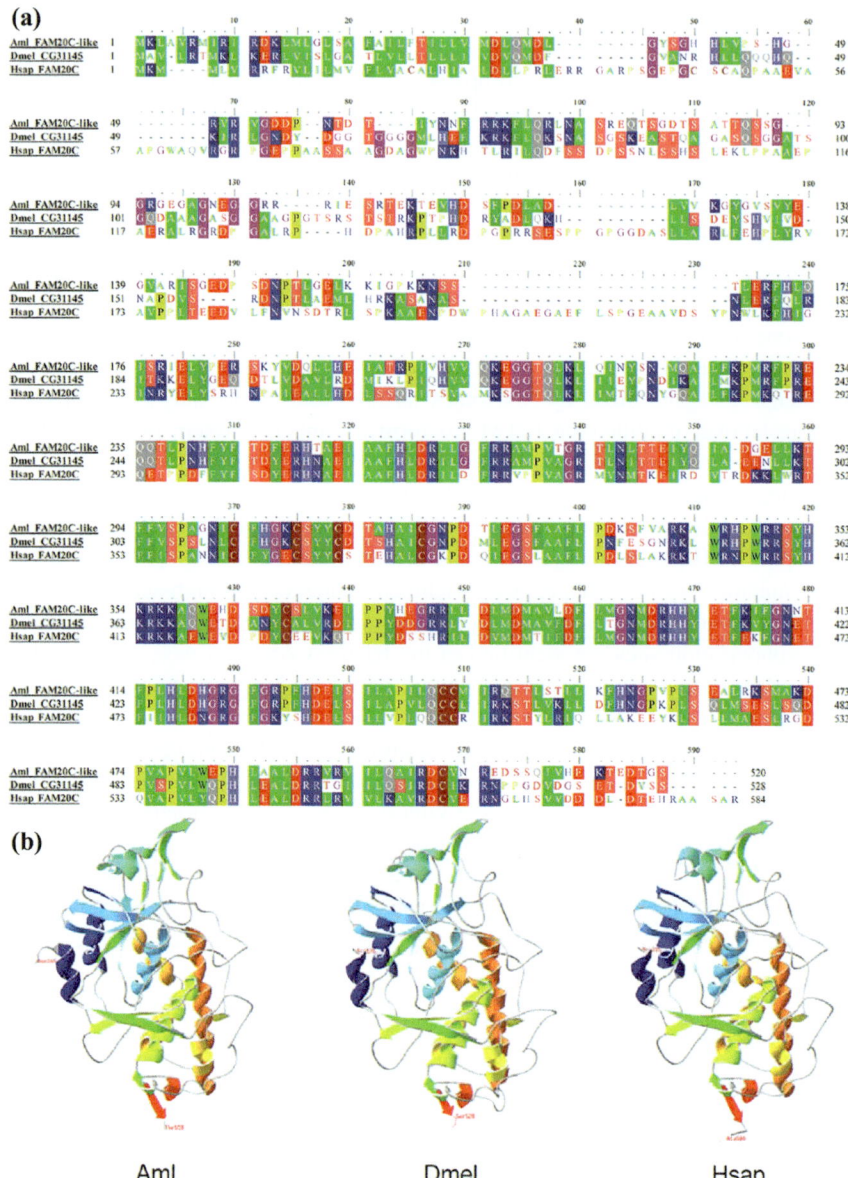

Fig. 12.8 Protein sequence alignment and three-dimensional structure comparison of FAM20C proteins. FAM20C protein of *Apis mellifera ligustica* (Aml FAM20C-like), *Drosophila melanogaster* (Dmel CG31145, isoform C), and Homo sapiens (Hsap FAM20C) were aligned by molecular evolutionary genetics analysis (MEGA) software (version 6.0). Three-dimensional structures were predicted by the Phyre2 online server and displayed using SPDBV software (version 4.1). **a** Protein sequence alignment. **b** Three-dimensional structure comparison [98]

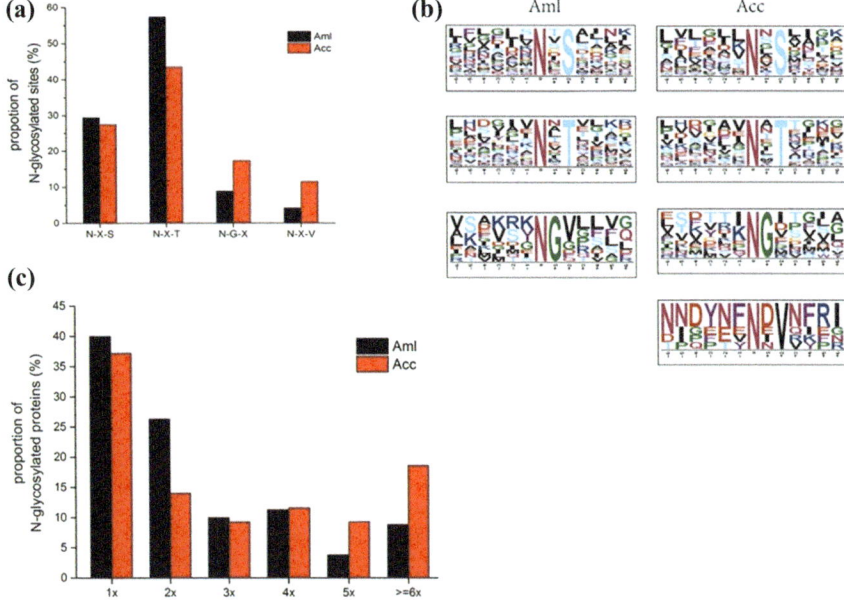

Fig. 12.9 Common characteristics of the identified N-glycoproteomes of RJ proteins produced by Aml and Acc. **a** The proportion of recognized sequence motifs from the N-glycopeptides, for N-X-S/T (X ≠ P/proline); for N-G-X and N-X-V, X can be any amino acid; **b** the extracted motifs using Motif-X algorithm. Only glycosites with a localization probability greater than 95 % are considered; **c** the distribution of N-glycosylated RJ proteins harboring single, double, and multiple glycosites [100]

P. larvae might be the reason why Aml is more susceptible to American foulbrood than Acc. It is also observed that N-glycosylated MRJP1 has an anti-hypertension activity. Again, the stronger activity has found in Acc than in Aml [100] (Fig. 12.9).

Many MRJPs and other RJ proteins have heterogeneities, which are most likely to be resulted from PTMs [86, 89]. The different PTM strategies employed by honeybee species reflect their biological characteristics of long-period evolution as Han et al. suggested in 2014. The characterization of PTM of RJ proteins will contribute to the in-depth understanding of the properties of RJ, and benefit the further development and utilization of RJ in various aspects.

References

1. Calderone NW (2012) Insect pollinated crops, insect pollinators and US agriculture: trend analysis of aggregate data for the period 1992–2009. PLoS ONE 7:e37235
2. Liu P, Wu J, Li H, Lin S (2011) Economic Values of Bee Pollination to China's Agriculture. Scientia Agricultura Sinica 44(24):5117–5123

3. Helmholtz Association of German Research Centres (2008) Economic value of insect pollination worldwide estimated At U.S. $217 Billion Science Daily. http://www.sciencedaily.com/releases/2008/09/080915122725.htm. Retrieved 14 July 2016
4. Molan PC (1992) The antibacterial activity of honey: 1. The nature of the antibacterial activity. Bee World 73:5–28
5. DuPraw EJ (1967) The honeybee embryo. In: Wilt F, Wessells NK (eds) Methods in developmental biology. Crowell Press, New York, pp 183–217
6. Collins AM (2004) Variation in time of egg hatch by the honey bee, apis mellifera (Hymenoptera: Apidae) no access. Ann Entomol Soc Am 97:140–146
7. Li J, Zhang L, Feng M, Zhang Z, Pan Y (2009) Identification of the proteome composition occurring during the course of embryonic development of bees (Apis mellifera). Insect Mol Biol 18:1–9
8. Li J, Fang Y, Zhang L, Begna D (2011) Honeybee (Apis mellifera ligustica) drone embryo proteomes. J Insect Physiol 57:372–384
9. Fang Y, Feng M, Han B, Lu X, Ramadan H, Li J (2014) In-depth proteomics characterization of embryogenesis of the honey bee worker (Apis mellifera ligustica). Mol Cell Proteomics 13:2306–2320
10. Fang Y, Feng M, Han B, Qi Y, Hu H, Fan P et al (2015) Proteome analysis unravels mechanism underling the embryogenesis of the honeybee drone and its divergence with the worker (Apis mellifera lingustica). J Proteome Res 14:4059–4071
11. Chan MMY, Choi SYC, Chan QWT, Li P, Guarna MM, Foster LJ (2010) Proteome profile and lentiviral transduction of cultured honey bee (Apis mellifera L.) cells. Insect Mol Biol 19:653–658
12. Allsopp MH, Calis JNM, Boot WJ (2003) Differential feeding of worker larvae affects caste characters in the cape honeybee, Apis mellifera capensis. Behav Ecol Sociobiol 54:555–561
13. Chan QWT, Foster LJ (2008) Changes in protein expression during honey bee larval development. Genome Biol 9:R156–R156
14. Li J, Li H, Zhang Z, Pan Y (2008) Identification of the proteome complement of high royal jelly producing bees (Apis mellifera) during worker larval development. Apidologie 38:545–557
15. Evans JD, Wheeler DE (1999) Differential gene expression between developing queens and workers in the honey bee, Apis mellifera. Proc Natl Acad Sci U S A 96:5575–5580
16. Hepperle C, Hartfelder K (2001) Differentially expressed regulatory genes in honey bee caste development. Naturwissenschaften 88:113–116
17. Wheeler DE, Buck N, Evans JD (2006) Expression of insulin pathway genes during the period of caste determination in the honey bee, Apis mellifera. Insect Mol Biol 15:597–602
18. Lefebvre B, Timmers T, Mbengue M, Moreau S, Hervé C, Tóth K et al (2010) A remorin protein interacts with symbiotic receptors and regulates bacterial infection. Proc Natl Acad Sci U S A 107:2343–2348
19. Begna D, Fang Y, Feng M, Li J (2011) Mitochondrial proteins differential expression during honeybee (Apis mellifera L.) queen and worker larvae caste determination. J Proteome Res 10:4263–4280
20. Begna D, Han B, Feng M, Fang Y, Li J (2012) Differential expressions of nuclear proteomes between honeybee (Apis mellifera L.) queen and worker larvae: a deep insight into caste pathway decisions. J Proteome Res 11:1317–1329
21. Zheng A, Li J, Begna D, Fang Y, Feng M, Song F (2011) Proteomic analysis of honeybee (Apis mellifera L.) pupae head development. PLoS ONE 6:e20428
22. Su SK, Chen SL (2003) Research on morphological genetic marker of honeybee (Apis mellifera ligustica) in royal jelly production performance. Yi Chuan 25:677–680
23. Wolschin F, Amdam GV (2007) Comparative proteomics reveal characteristics of life-history transitions in a social insect. Proteome Sci 5:10
24. Garcia L, Garcia CHS, Calabria LK, Da Cruz GCN, Puentes AS, Bao SN et al (2009) Proteomic analysis of honey bee brain upon ontogenetic and behavioral development. J Proteome Res 8:1464–1473

25. Hernández LG, Lu B, Da Cruz GCN, Calábria LK, Martins NF, Togawa R et al (2012) Worker honeybee brain proteome. J Proteome Res 11:1485–1493
26. Chan QW, Mutti NS, Foster LJ, Kocher SD, Amdam GV, Wolschin F (2011) The worker honeybee fat body proteome is extensively remodeled preceding a major life-history transition. PLoS ONE 6:e24794
27. Han B, Fang Y, Feng M, Hu H, Qi Y, Huo X et al (2015) Quantitative neuropeptidome analysis reveals neuropeptides are correlated with social behavior regulation of the honeybee workers. J Proteome Res 14:4382–4393
28. Slessor KN, Winston ML, Le Conte Y (2005) Pheromone communication in the honeybee (*Apis mellifera* L.). J Chem Ecol 31:2731–2745
29. Sato K, Pellegrino M, Nakagawa T, Nakagawa T, Vosshall LB, Touhara K (2008) Insect olfactory receptors are heteromeric ligand-gated ion channels. Nature 452:1002–1006
30. Free JB (1987) Pheromones of social bees. Chapman and Hall, London
31. Slessor KN, Kaminski L-A, King GGS, Borden JH, Winston ML (1988) Semiochemical basis of the retinue response to queen honey bees. Nature 332:354–356
32. Le Conte Y, Arnold G, Trouiller J, Masson C, Chappe B (1990) Identification of a brood pheromone in honeybees. Naturwissenschaften 77:334–336
33. Dani FR, Iovinella I, Felicioli A, Niccolini A, Calvello MA, Carucci MG et al (2010) Mapping the expression of soluble olfactory proteins in the honeybee. J Proteome Res 9:1822–1833
34. Feng M, Song F, Aleku DW, Han B, Fang Y, Li J (2011) Antennal proteome comparison of sexually mature drone and forager honeybees. J Proteome Res 10:3246–3260
35. Chan QW, Howes CG, Foster LJ (2006) Quantitative comparison of caste differences in honeybee hemolymph. Mol Cell Proteomics 5:2252–2262
36. Woltedji D, Song F, Zhang L, Gala A, Han B, Feng M et al (2012) Western honeybee drones and workers (*Apis mellifera ligustica*) have different olfactory mechanisms than eastern honeybees (*Apis cerana cerana*). J Proteome Res 11:4526–4540
37. Butler C, Callow R, Johnston NC (1962) The isolation and synthesis of queen substance, 9-oxodec-trans-2-enoic acid, a honeybee pheromone. Proc Royal Soc London B Biol Sci 155:417–432
38. Collins AM, Rinderer TE, Daly HV, Harbo JR, Pesante D (1989) Alarm pheromone production by two honeybee (*Apis mellifera*) types. J Chem Ecol 15:1747–1756
39. Iovinella I, Dani FR, Niccolini A, Sagona S, Michelucci E, Gazzano A et al (2011) Differential expression of odorant-binding proteins in the mandibular glands of the honey bee according to caste and age. J Proteome Res 10:3439–3449
40. Hasegawa M, Asanuma S, Fujiyuki T, Kiya T, Sasaki T, Endo D et al (2009) Differential gene expression in the mandibular glands of queen and worker honeybees, *Apis mellifera* L.: implications for caste-selective aldehyde and fatty acid metabolism. Insect Biochem Mol Biol 39:661–667
41. Ohashi K, Sasaki M, Sasagawa H, Nakamura J, Natori S, Kubo T (2000) Functional flexibility of the honey bee hypopharyngeal gland in a dequeened colony. Zool Sci 17:1089–1094
42. Kubo T, Sasaki M, Nakamura J, Sasagawa H, Ohashi K, Takeuchi H et al (1996) Change in the expression of hypopharyngeal-gland proteins of the worker honeybees (*Apis mellifera* L.) with age and/or role. J Biochem 119:291–295
43. Feng M, Ramadan H, Han B, Fang Y, Li J (2014) Hemolymph proteome changes during worker brood development match the biological divergences between western honey bees (*Apis mellifera*) and eastern honey bees (*Apis cerana*). BMC Genom 15:563
44. Ji T, Liu Z, Shen J, Shen F, Liang Q, Wu L et al (2014) Proteomics analysis reveals protein expression differences for hypopharyngeal gland activity in the honeybee. *Apis mellifera* carnica Pollmann. BMC genomics 15:665
45. Jianke L, Mao F, Begna D, Yu F, Aijuan Z (2010) Proteome comparison of hypopharyngeal gland development between Italian and royal jelly producing worker honeybees (*Apis mellifera* L.). J Proteome Res 9:6578–6594

46. Qi Y, Fan P, Hao Y, Han B, Fang Y, Feng M et al (2015) Phosphoproteomic analysis of protein phosphorylation networks in the hypopharyngeal gland of honeybee workers (*Apis mellifera ligustica*). J Proteome Res 14:4647–4661
47. Feng M, Fang Y, Han B, Zhang L, Lu X, Li J (2013) Novel aspects of understanding molecular working mechanisms of salivary glands of worker honeybees (*Apis mellifera*) investigated by proteomics and phosphoproteomics. J Proteomics 87:1–15
48. Fujita T, Kozuka-Hata H, Uno Y, Nishikori K, Morioka M, Oyama M, Kubo T (2010) Functional analysis of the honeybee (*Apis mellifera L.*) salivary system using proteomics. Biochem Biophys Res Commun 397(4):740–744
49. Sinakevitch I, Niwa M, Strausfeld NJ (2005) Octopamine-like immunoreactivity in the honey bee and cockroach: comparable organization in the brain and subesophageal ganglion. J Comp Neurol 488:233–254
50. Mobbs P (1982) The brain of the honeybee *Apis mellifera*. I. The connections and spatial organization of the mushroom bodies. Philos Trans Royal Soc London B Biol Sci 298: 309–354
51. Uno Y, Fujiyuki T, Morioka M, Kubo T (2013) Mushroom body-preferential expression of proteins/genes involved in endoplasmic reticulum Ca^{2+}-transport in the worker honeybee (*Apis mellifera L.*) brain. Insect Mol Biol 22:52–61
52. Kucharski R, Maleszka R (2002) Evaluation of differential gene expression during behavioral development in the honeybee using microarrays and northern blots. Genome Biol 3:1–0007
53. Kucharski R, Maleszka R, Hayward D, Ball E (1998) A royal jelly protein is expressed in a subset of Kenyon cells in the mushroom bodies of the honey bee brain. Naturwissenschaften 85:343–346
54. Garcia L, Saraiva Garcia CH, Calábria LK, Costa Nunes Da Cruz G, SáNchez Puentes A, Báo SN et al (2009) Proteomic analysis of honey bee brain upon ontogenetic and behavioral development. J Proteome Res 8:1464–1473
55. Baker N, Wolschin F, Amdam GV (2012) Age-related learning deficits can be reversible in honeybees *Apis mellifera*. Exp Gerontol 47:764–772
56. Peixoto LG, Calabria LK, Garcia L, Capparelli FE, Goulart LR, De Sousa MV et al (2009) Identification of major royal jelly proteins in the brain of the honeybee *Apis mellifera*. J Insect Physiol 55:671–677
57. Tofilski A (2000) Senescence and learning in honeybee (*Apis mellifera*) workers. Acta Neurobiol Exp 60:35–40
58. Finch C (1990) Longevity, senescence, and the genome. University of Chicago Press, Chicago
59. Seehuus S-C, Norberg K, Gimsa U, Krekling T, Amdam GV (2006) Reproductive protein protects functionally sterile honey bee workers from oxidative stress. Proc Natl Acad Sci U S A 103:962–967
60. Wolschin F, Münch D, Amdam G (2009) Structural and proteomic analyses reveal regional brain differences during honeybee aging. J Exp Biol 212:4027–4032
61. Koeniger N, Koeniger G (2000) Reproductive isolation among species of the genus Apis. Apidologie 31:313–339
62. Baer B (2005) Sexual selection in Apis bees. Apidologie 36:187–200
63. Chapman T, Davies SJ (2004) Functions and analysis of the seminal fluid proteins of male *Drosophila melanogaster* fruit flies. Peptides 25:1477–1490
64. Prokupek A, Hoffmann F, Eyun SI, Moriyama E, Zhou M, Harshman L (2008) An evolutionary expressed sequence tag analysis of *Drosophila spermatheca* genes. Evolution 62:2936–2947
65. Cardoen D, Wenseleers T, Ernst UR, Danneels EL, Laget D, Dc DEG et al (2011) Genome-wide analysis of alternative reproductive phenotypes in honeybee workers. Mol Ecol 20:4070–4084
66. Baer B, Zareie R, Paynter E, Poland V, Millar AH (2012) Seminal fluid proteins differ in abundance between genetic lineages of honeybees. J Proteomics 75:5646–5653

67. Baer B, Heazlewood JL, Taylor NL, Eubel H, Millar AH (2009) The seminal fluid proteome of the honeybee *Apis mellifera*. Proteomics 9:2085–2097
68. Den Boer SP, Boomsma JJ, Baer B (2009) Honey bee males and queens use glandular secretions to enhance sperm viability before and after storage. J Insect Physiol 55:538–543
69. Zareie R, Eubel H, Millar AH, Baer B (2013) Long-term survival of high quality sperm: insights into the sperm proteome of the honeybee *Apis mellifera*. J Proteome Res 12: 5180–5188
70. Baer B, Eubel H, Taylor NL, O'toole N, Millar AH (2009) Insights into female sperm storage from the spermathecal fluid proteome of the honeybee *Apis mellifera*. Genome Biol 10:R67
71. Poland V, Eubel H, King M, Solheim C, Harvey Millar A, Baer B (2011) Stored sperm differs from ejaculated sperm by proteome alterations associated with energy metabolism in the honeybee *Apis mellifera*. Mol Ecol 20:2643–2654
72. Collins AM, Caperna TJ, Williams V, Garrett WM, Evans JD (2006) Proteomic analyses of male contributions to honey bee sperm storage and mating. Insect Mol Biol 15:541–549
73. Snodgrass RE (1984) Anatomy of the honey bee. Comstock Publishing Associates, Ithaca, NY
74. Miller Iii DG, Ratnieks FLW (2001) The timing of worker reproduction and breakdown of policing behaviour in queenless honey bee (*Apis mellifera* L.) societies. Insectes Soc 48: 178–184
75. Cardoen D, Ernst UR, Boerjan B, Bogaerts A, Formesyn E, De Graaf DC et al (2012) Worker honeybee sterility: a proteomic analysis of suppressed ovary activation. J Proteome Res 11:2838–2850
76. Cardoen D, Ernst UR, Van Vaerenbergh M, Boerjan B, De Graaf DC, Wenseleers T et al (2011) Differential proteomics in dequeened honeybee colonies reveals lower viral load in hemolymph of fertile worker bees. PLoS ONE 6:e20043
77. Bogaerts A, Baggerman G, Vierstraete E, Schoofs L, Verleyen P (2009) The hemolymph proteome of the honeybee: gel-based or gel-free? Proteomics 9:3201–3208
78. Woltedji D, Fang Y, Han B, Feng M, Li R, Lu X et al (2013) Proteome analysis of hemolymph changes during the larval to pupal development stages of honeybee workers (*Apis mellifera ligustica*). J Proteome Res 12:5189–5198
79. Wang Y, Oberley LW, Murhammer DW (2001) Evidence of oxidative stress following the viral infection of two lepidopteran insect cell lines. Free Radic Biol Med 31:1448–1455
80. Zhang P, Liu B, Kang SW, Seo MS, Rhee SG, Obeid LM (1997) Thioredoxin peroxidase is a novel inhibitor of apoptosis with a mechanism distinct from that of Bcl-2. J Biol Chem 272:30615–30618
81. Kamakura M (2011) Royalactin induces queen differentiation in honeybees. Nature 473: 478–483
82. Mannoor M, Shimabukuro I, Tsukamotoa M, Watanabe H, Yamaguchi K, Sato Y (2009) Honeybee royal jelly inhibits autoimmunity in SLE-prone NZB × NZW F1 mice. Lupus 18:44–52
83. Niu K, Guo H, Guo Y, Ebihara S, Asada M, Ohrui T et al (2013) Royal jelly prevents the progression of sarcopenia in aged mice in vivo and in vitro. J Gerontol Series A: Biol Sci Med Sci 68:1482–1492
84. Miyata T (2007) Pharmacological basis of traditional medicines and health supplements as curatives. J Pharmacol Sci 103:127–131
85. Scarselli R, Donadio E, Giuffrida MG, Fortunato D, Conti A, Balestreri E et al (2005) Towards royal jelly proteome. Proteomics 5:769–776
86. Sano O, Kunikata T, Kohno K, Iwaki K, Ikeda M, Kurimoto M (2004) Characterization of royal jelly proteins in both africanized and european honeybees (*Apis mellifera*) by two-dimensional gel electrophoresis. J Agric Food Chem 52:15–20
87. Santos KS, Delazari Dos Santos L, Anita Mendes M, Monson De Souza B, Malaspina O, Palma MS (2005) Profiling the proteome complement of the secretion from hypopharyngeal gland of Africanized nurse-honeybees (*Apis mellifera* L.). Insect Biochem Mol Biol 35:85–91

88. Schönleben S, Sickmann A, Mueller M, Reinders J (2007) Proteome analysis of *Apis mellifera* royal jelly. Anal Bioanal Chem 389:1087–1093
89. Li J, Wang T, Zhang Z, Pan Y (2007) Proteomic analysis of royal jelly from three strains of western honeybees (*Apis mellifera*). J Agric Food Chem 55:8411–8422
90. Fang Y, Feng M, Li J (2010) Royal jelly proteome comparison between *A. mellifera ligustica* and *A. cerana cerana*. J Proteome Res 9:2207–2215
91. Han B, Li C, Zhang L, Fang Y, Feng M, Li J (2011) Novel royal jelly proteins identified by gel-based and gel-free proteomics. J Agric Food Chem 59:10346–10355
92. Zhang L, Han B, Li R, Lu X, Nie A, Guo L et al (2014) Comprehensive identification of novel proteins and N-glycosylation sites in royal jelly. BMC Genom 15:135
93. Fujita T, Kozuka-Hata H, Ao-Kondo H, Kunieda T, Oyama M, Kubo T (2013) Proteomic analysis of the royal jelly and characterization of the functions of its derivation glands in the honeybee. J Proteome Res 12:404–411
94. Furusawa T, Rakwal R, Nam HW, Shibato J, Agrawal GK, Kim YS et al (2008) Comprehensive royal jelly (RJ) proteomics using one- and two-dimensional proteomics platforms reveals novel RJ proteins and potential phospho/glycoproteins. J Proteome Res 7:3194–3229
95. Li J, Wang T, Peng W (2007) Cmparative analysis of the effects of different storage conditions on major royal jelly proteins. J Apic Res 46:73–81
96. Li JK, Feng M, Zhang L, Zhang ZH, Pan YH (2008) Proteomics analysis of major royal jelly protein changes under different storage conditions. J Proteome Res 7:3339–3353
97. Zhang L, Fang Y, Li R, Feng M, Han B, Zhou T et al (2012) Towards posttranslational modification proteome of royal jelly. J Proteomics 75:5327–5341
98. Han B, Fang Y, Feng M, Lu X, Huo X, Meng L et al (2014) In-depth phosphoproteomic analysis of royal jelly derived from western and eastern honeybee species. J Proteome Res 13:5928–5943
99. Bilikova K, Mirgorodskaya E, Bukovska G, Gobom J, Lehrach H, Simuth J (2009) Towards functional proteomics of minority component of honeybee royal jelly: the effect of post-translational modifications on the antimicrobial activity of apalbumin2. Proteomics 9:2131–2138
100. Feng M, Fang Y, Han B, Xu X, Fan P, Hao Y et al. (2015) In-depth N-Glycosylation reveals species-specific modifications and functions of the royal jelly protein from western (*Apis mellifera*) and eastern honeybees (*Apis cerana*). J Proteome Res 14(12):5327–5340

Erratum to: Proteomics in Energy Crops

Shiva Bakhtiari, Meisam Tabatabaei and Yusuf Chisti

Erratum to:
Chapter 6 in: G.H. Salekdeh (ed.), *Agricultural Proteomics Volume 1*, DOI 10.1007/978-3-319-43275-5_6

The original version of the book was inadvertently published with old version despite having received the revised manuscript. The erratum chapter and the book have been updated with the revised text.

The updated original online version for this chapter can be found at 10.1007/978-3-319-43275-5_6

S. Bakhtiari
Biology Department, Concordia University, 7141 Sherbrooke W., Montreal H4B 1R6, Canada

M. Tabatabaei (✉)
Microbial Biotechnology Department, Agricultural Biotechnology Research Institute of Iran (ABRII), Agricultural Research, Education, and Extension Organization (AREEO), Karaj, Iran
e-mail: meisam_tabatabaei@abrii.ac.ir; meisam_tab@yahoo.com

M. Tabatabaei
Biofuel Research Team (BRTeam), Karaj, Iran

Y. Chisti
School of Engineering, Massey University,
Private Bag 11 222, Palmerston North, New Zealand

© Springer International Publishing Switzerland 2016
G.H. Salekdeh (ed.), *Agricultural Proteomics Volume 1*,
DOI 10.1007/978-3-319-43275-5_13

Index

A
Acetylation, 12, 14, 15, 187
Adulteration, 68, 69, 71, 76, 77, 79, 81, 165
Agricultural foodstuffs, 77, 80
Albuminous, 32
Aleurone, 33, 38, 43, 44
Amphibians, 178
Animal nutrition, 159, 162
Animal physiology, 158
Animal proteomics, 159
Animal welfare, 158, 159, 167
Anthocyanins, 53
Apis mellifera, 225, 227, 229, 234, 240, 243
Aquaculture, 158, 162, 176–178, 181, 186, 188, 194, 201
Aquatic ecosystems, 175, 176

B
Basidiomycetes, 00
Biofuel, 106, 107
Bioinformatic, 2, 18, 19, 68, 227
Biotoxins, 188
Brackish, 176
Brassicaceae, 39

C
Carotenoids, 54, 56
Cattleya, 128
Celiac disease, 73
Chemical labeling, 7
Climacteric seeds, 44
Contamination, 68, 71, 75, 78, 80, 166
Cotyledon, 32, 33, 39, 41, 42, 44
Crustaceans, 177, 181, 194
Cyanobacteria, 116, 178

D
2D-DIGE, 6, 145, 160, 167, 181, 186
2-DE, 4–6, 38, 69, 72, 80, 144, 145, 147, 148, 216, 239, 241–243
De novo sequencing, 6, 70, 94, 243
Desmoglein, 212
Desmoplakin, 212

E
Egg fertilization, 240
Egg proteomics, 166
Energy crops, 106
Environment friendly, 105
Epigenomics, 90
ESI, 70, 74, 144, 219, 220
EST, 3, 18, 59, 62, 177
Ethanol, 106, 107, 111
Ethylene, 54, 60
Euphorbiaceae, 40
Ex-albuminous, 32
Expressed sequence tags, 3

F
Farm animals, 158, 159, 161, 163, 167, 168
Fishery products, 175, 176, 178, 194, 201
Fish farming, 181, 187
Fish welfare, 167, 187, 201
Food allergens, 69, 71, 72, 80
Food borne microbes, 74
Food poisoning, 74
Food proteomics, 67, 69
Food quality, 71, 76
Food safety, 71, 74, 76, 188
Fresh water, 176
Fruit development, 53–56, 58–60

Fruit proteomics, 55, 59, 62
Fruit ripening, 54

G
Ganoderma lucidum, 00
Gastrointestinal health, 159, 163
Glycosylation, 12, 16, 17, 187, 245
Green revolution, 88
Grifola frondosa, 00

H
Hallucinogenic, 00
Hemolymph, 230, 234, 241, 242
High glycine-tyrosine proteins, 212
High sulphur protein (HSP), 212
Honeybee, 225, 227, 228, 232, 233, 238–240, 242–244
Honeybee behavior, 228
Honeybee physiology, 228
Honeybee proteomics, 225, 227, 228
Honeybee salivary system, 238
HPLC-MS/MS, 70
Hypopharyngeal Gland, 236, 243

I
ICAT, 7, 182
Intermacrofibrillar, 212
iTRAQ, 6–8, 18, 133, 145, 161, 182, 186

J
Jatropha, 40, 106, 113

K
Keratin, 211, 212, 218
Keratin associated protein, 214

L
Label-free, 4, 7, 11, 12, 145, 146, 182
Label-free quantification, 186
LC-MS/MS, 11, 73, 75, 76, 81, 145, 149, 150, 160, 161, 166, 185, 186, 221, 244
Leguminosae, 41, 137
Lignocellulosic, 106, 107, 114

M
Macrofibrillar, 212
MALDI, 70, 215, 216
MALDI-TOF/TOF, 70, 132, 186
Mandibular glands, 236
Marine, 175, 176, 178
Mass spectrometry, 4, 6, 12, 58, 67, 79, 89, 94, 145, 149, 152, 185, 215, 227

Meat proteomics, 163
Medicago truncatula, 138, 150, 151
Metabolic labeling, 10
Metaproteomics, 87, 89–91, 95, 97, 163
Metatranscriptomics, 95
Methylation, 12, 15, 245
Microalgae, 116, 117, 178
Microbial community, 89, 91, 93
Microbial proteomics, 87, 89, 90
Micropropagation, 132
Milk proteomics, 165
Miscanthus, 114
Molluscs, 177, 188
Monogasteric, 163
Multidimensional protein identification technique, 6
Multiple reactions monitoring (MRM), 70

N
^{15}N labelling, 9
Nodulation, 137, 138, 145, 151
Nodule organogenesis, 137, 138, 149
Non-climacteric seeds, 44

O
Olea europaea, 53, 56
Olive drupe, 56–62
Orchidaceae, 128–130
OrchidBase, 130
Orchids, 127–130, 132, 133
Originality, 67
Ornamental plants, 127, 132
Orphan species, 3
Oryza Sativa, 3

P
Peptide mass fingerprinting, 70, 145, 212
Peribacteroid membrane, 146
Phalaenopsis, 128, 130, 133
Phosphoproteomics, 149
Phyllosphere, 91
Phytobiome, 87, 89, 95, 97, 99
Plant biomass, 106
Plant-environment interaction, 89, 90
Plant-pathogen crosstalk, 92
PMF, 70, 185
Pollination, 133, 225, 226
Polygasteric, 163
Polyphenol oxidase (PPO), 58
Post-translational modifications (PTM), 12, 15, 138, 149, 160, 181
Prion, 76

Proteogenomics, 61, 62
Proteomics, 2, 4, 12, 20, 41, 55, 57, 61, 67, 68, 71, 74, 87, 106–109, 111, 113–117, 133, 138, 144, 159, 165, 176, 184, 201, 227
PrP^{Sc}, 76
PTM, 4, 16, 17, 21, 187, 247

Q
Quantitative proteomics, 2, 5, 18, 62, 181, 186

R
Reciprocal labeling, 9, 11
Rhizobial inoculation, 149, 150
Rhizobial symbionts, 137
Rhizosphere-associated microbes, 92
Royal jelly, 226, 232, 243, 244

S
Scutellum, 38, 43, 44
Seed proteomics, 32, 38–41, 43
SILAC, 7, 8, 145, 182
Social insects, 232, 238

Sorghum, 74, 107, 111, 112
Spermatophytes, 31
Stable isotope labelling, 6, 8, 182
Sugarcane bagasse, 108, 109
Sulfenylation, 149, 150, 152
Switchgrass, 106, 114
Symbiosome, 146
Sympodial orchids, 128
Systems Biology, 88, 89

T
Tandem mass tag, 7, 186
Teguments, 33, 44
Traceability, 67, 68, 194

U
Ubiquitination, 12, 15, 16, 149, 152
Ultra-high sulphur proteins (UHSP), 212

W
Wool proteomics, 218

CPI Antony Rowe
Eastbourne, UK
February 20, 2019